I0070183

14

INTRODUCTION TO EQUATIONS

In this chapter we will begin to explore one of the (arguably) most important and relatable areas of math: mathematical equations.

In previous chapters, I have mostly dealt with mathematical expressions. As you know, a mathematical expression contains numbers and variables, and may contain mathematical operations, negative signs, and relational symbols. I have kept the math simple (up to a certain point), exploring the following key techniques, properties, and rules that govern math:

1) Properties and rules involving mathematical expressions.

2) The concept of the order of operations (PEMDAS),

3) Rounding, truncating, adding, subtracting, multiplying, and dividing **REAL** numbers.

4) How to perform mathematical operations with fractions.

5) Factors of numbers, exponents and roots, and the concept of a term.

6) COMPLEX numbers.

Now, we will focus on bringing these concepts together to work with equations.

So what is an equation anyway?

An equation is a mathematical expression that can be said to contain two sides (a left side and a right side) that are mathematically related. This idea of "mathematically related" means one of the following:

- Both sides are equal to each other (=).

- The left side is GREATER THAN the right side (>).

- The left side is SMALLER THAN the right side (<).

- The left side is GREATER THAN OR EQUAL TO the right side (≥).

- The left side is SMALLER THAN OR EQUAL TO the right side (≤).

- The left side is NOT EQUAL TO the right side (≠).

There are other relational symbols that are defined in math, but for the time being we will focus on these exclusively.

The following diagram with examples may help you fully understand the concept of a mathematical equation:

Left Side	Relational Symbol	Right Side
2(5)	=	10
10 + 5 − 2	=	13
8	>	1
3	<	100
6	≥	6

The main goal here is that you recognize each row as an equation. Of course, you don't normally see equations written as above (too much gap between the sides and the relational symbol), but rather as follows:

$$2(5) = 10$$
$$10 + 5 - 2 = 13$$
$$8 > 1$$
$$3 < 100$$
$$6 \geq 10$$

Of course, equations can have variables, and can have any defined mathematical operations. So they can be quite complex. But do not worry: most of the algebra that lies within the scope of this book is relatively simple to manage. So how do equations like those that I describe above look like? Observe:

$$2x + 3 = 9$$
$$5x - 10 = 2x + 20$$
$$2x + 4y = 10$$
$$2x^2 = 50$$
$$2x + 3y^2 = 10x^3$$
$$6 = 2x^4 + 3x^2 - x - 1$$
$$y + z^3 - 10 = 2$$
$$\sqrt[5]{4x^3 - 2x - 1} - x + |x - 3|^3 + 2 = 0$$

As you can see, equations can be very simple or very difficult to work with: it all depends on numerous factors, such as the number of variables involved, the exponents attached to those variables (an exponent of **1** is a lot easier to work with than, say, an exponent of **7** or **8**), the type of mathematical operations being used, etc...

So why do equations exist in the first place? Well, although the answer would strictly speaking lie in the realm of the philosophical, we could say that equations allow us to state ideas, problems, or relationships between phenomena **mathematically**; then, because math is a set of axioms (this means that it is a collection of formally stated assertions from which other such formally stated assertions follow by applying well-defined rules), by applying the rules and properties (theorems or, form a more modern perspective, assertions), we may manipulate the mathematical expression in order to find equivalent left-side/right-side expressions that convey useful information to whoever is using the process to begin with. For example, if I provide you the following equation to you so that you may figure out how much you would have to pay for a skiing vacation for you and your family, you could use math to manipulate the left-side/right-side (as appropriate or convenient) in order to come up with a useful datum (in this case, the cost for the trip):

$$C = 1800(t) + 300$$

where **C** is the **cost of the skiing vacation** and **t** is the **number of travelers**, not to exceed **5**.

Suppose you and **3** other members of your family are considering the trip. You want to know if you have the budget for this. Well, you can use math to obtain an answer.

The first step is to look at the equation $C = 1800(t) + 300$ and ask yourself "What variable values do I know?", and "What variables am I trying to find out?". Looking at the variables, **C** and **t**, it is easy to see that you DO NOT KNOW the cost, **C**, of this skiing trip, but that you DO KNOW the number of travelers, **t**. You know that **t** is equal to **4** (since you and **3** other members would be traveling, and therefore $t = 3 + 1 = 4$). Having realized this, you can replace the known variable values into your equation, as follows:

$$C = 1800(t) + 300$$
$$\downarrow$$
$$\boxed{\text{Replace } \mathbf{t} \text{ with } \mathbf{4}}$$
$$\downarrow$$
$$C = 1800(4) + 300$$

Now that we have the equation

$$C = 1800(4) + 300$$

we have made some progress, because even though the left side contains a variable, **C**, whose value is still unknown, we do know for a fact that, given the equality symbol that RELATES the left side with the right side, if we are able to express the right side in terms of an equivalent expression that is useful to us, we can know the value of **C**. (since as the equation states, **C** is equal to whatever the right side of the equation is equal to).

Of course, we cannot change the right side however we please... We **MUST** use the axioms (or assertions) that mathematics is based on, which means that we must use the accepted rules and properties that we have covered in the book so far (more of those to come!). The idea is that the set of all mathematical assertions are free of any contradictions, and therefore, applying any of the assertions results in equivalent (as in *equal*) expressions.

So, let us apply the assertions (or axioms, or rules or properties) to the right side, since as it stands now, even though it is expressed in terms of numbers (all variables have been replaced with the known/correct numbers), it is not very useful, for who can interpret something like "...our skiing trip will cost $1800(4) + 300$ dollars..."?

Observe the process:

$$C = 1800(4) + 300$$

...equation with all known variables replaced with their known/desired/correct values (**REAL** numbers)

$$C = 7200 + 300$$

...equivalent right side expression, after applying the product of the numbers **1800** and **4**. We know that this right side is equivalent (equal to) the right side from the previous step, since math has a set of assertions that tell us how to **MULTIPLY** any two **REAL** numbers. We are closer to having a useful expression on the right side, considering our goal of knowing the cost of the skiing vacation.

$$C = 7500$$

...equivalent right side expression, after applying the addition of the numbers **7200** and **300**. We know that this right side is equivalent (equal to) the right side from the previous step, since math has a set of assertions that tell us how to **ADD** any two **REAL** numbers. We now have a useful expression on the right side, considering our goal of knowing the cost of the skiing vacation.

And this is what we generally refer to as "solving an equation for an unknown variable". In this case, we solved for the unknown variable **C** which meant replacing the known variable **t** with the number **4** since we are considering a total of four travelers. Please observe the step-by-step progress, without comments:

$$C = 1800(t) + 300$$
$$C = 1800(4) + 300$$
$$C = 7200 + 300$$
$$C = 7500$$

Take note of the fact that **ALL** of the right sides on the equations above are equivalent to each other (from a mathematical perspective, that is). However, as far as our objective in this particular exercise is concerned (finding the cost of the skiing trip), only one of the right-side versions above provides us with the information we are looking for, without having to do any additional computations: that of the last equation, $C = 7500$. There it is, in plain English: **C**, or the cost of the skiing trip for **4** travelers, is equal to the amount of **$7,500 USD** (assuming we are using the United States Dollar as the currency of choice, which would need to be clearly specified).

And there you have it. An example of an equation that when correctly applied, and when conveniently manipulated using the rules and properties of math that we have reviewed so far, provides us with useful information (or, in every-day language, *solving* an equation for an unknown variable).

Conceptually speaking, there are several analogies we can use to help you fully master this idea of an equation. Let's begin with the following:

An equation is like a "balance" weight-scale...

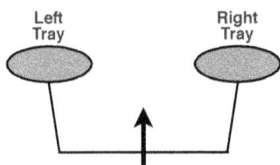

Left Tray   Right Tray

The left and right side of the equation would correspond to the left and right side of the balance weight scale; then, if the relational symbol being used is "equal to", this would correspond to the weight scale being perfectly centered. If you are unfamiliar with the workings of a **balance** weight scale, allow me to quickly explain.

A weight scale has two trays, conveniently named the "left tray" and the "right tray" (this allows us to refer to them unequivocally). Objects can then be placed on each tray, and if the objects are of the same weight, the scale will indicate this by having a pointer (the arrow in the diagram between the trays) perfectly centered, "pointing up". If, however, one side is heavier than the other, then the scale will indicate this through the arrow: it will begin to tilt (lean) towards the left if the left-side is heavier, and progressively more so (or more off-center) as that weight difference increases; similarly, it will lean towards the right if the right-side is heavier, and progressively more so (or more off-center) as that weight difference increases. The mechanism is relatively simple, and it relies fully on the force of gravity to display the equality or inequality of the respective left-side, right-side weights. Another important clue that allows you to know which side is heavier is the height of one tray with respect to the other tray: as one of the trays increases its weight with respect to the other (becomes heavier than the other), this heavier tray ends up lower with respect to the other tray (the "lighter" tray). you will see this in action shortly. The balance weight scale has been around for centuries, and when equally balanced, is one of the commonly used symbols related to justice (equality for all sides).

So what do scales have anything to do with mathematical equations? Well, they can help you make the following connection: if you start off with an equation that has a left side expressed in certain terms and a right side expressed in other certain terms, and they are related with an equal symbol, then, from that moment on you would need to make sure that whatever you do to either side of the equality symbol (to the left and right trays in the balance analogy) *MAINTAINS* the equality.

From a mathematical perspective, "doing something" to either side involves multiplying or dividing all the terms that are found on said side, or adding a term or subtracting a term to that side, or raising the entire side to a certain power, etc... Some of these "actions" would change the side beyond the mere syntax (such as replacing $3xx$ with $3x^2$) or beyond the act of simplifying (such as replacing **(5)(2)** with **10**): when this happens, we have to be careful that we *IMPACT BOTH SIDES EQUALLY*.

This last statement is key: *impact both sides equally*... whenever we change a side of an equation, if the sides are related with an equal symbol, we must change the other side as well, in the same proportion, or by applying the same factor, or by increasing or decreasing the side using equal terms, etc... And here lies the link between equations and balance weight scales. Observe the following set up:

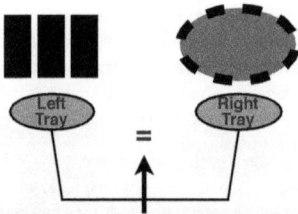

Based on the earlier explanation of how a balance weight scale works, you can determine from the set up above (assuming, of course, that nothing is hidden from this two-dimensional view)

that the left tray contains three items (presumably equal to each other, and thus having equal weight, although we cannot be absolutely certain of that from the diagram alone: think about it) and you can also determine that the right tray has only one item, clearly different from those on the other tray, and lastly, we can determine that based on the centered arrow, the item on the right tray weights EXACTLY the same as the sum of the weight of the three items that are on the left tray. What would happen to the scale's indicator arrow if I altered the set up as follows?

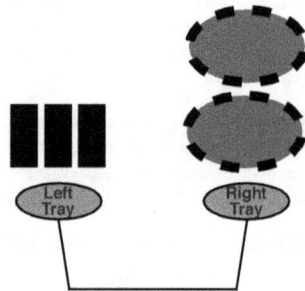

Assume that the object I added to the right tray weighs the same as the one that was initially there... Ready for the answer I'm looking for? if you responded along the lines of "...the indicator arrow will move so that it is pointing towards the right, and the right tray will end up lower than the left tray...", then you would be absolutely correct. Since the right side has more weight than the left side (twice as much, in fact, but that is not important right now), the force of gravity acting on the right tray is larger, causing it to move downwards with respect to the left tray, which in turn will cause the indicator arrow to point towards the right (the heavier tray of the two).

Observe the scale after adding the object:

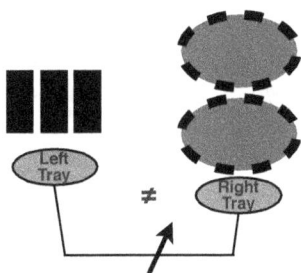

Note that I have replaced the equal symbol in the middle of the indicator with the "NOT equal" symbol, since the trays no longer weigh the same, and that the indicator arrow is pointing towards the right tray, and that the right tray is "lower" with respect to the left tray, since the right tray is heavier.

The point here is that by *adding* weight to one side (starting from a set up in which both sides weighed the same), the equality was destroyed. So, if we add something to one side, altering it, and we are not careful to impact the other side equally, the initial condition of equality can be destroyed!

Another way of destroying the equality that the initial set up possesses is to proceed as follows (observe and explain what would happen):

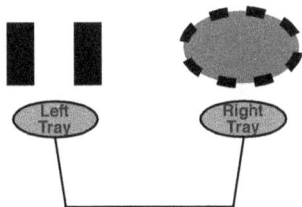

Can you predict the behavior of the arrow? What about the equality symbol? Will it be maintained or would it have to be replaced?

Once again, the arrow will end up pointing towards the right! Why? Because we removed an object from the left tray, which means we removed weight from the left tray, but left the right tray intact. And since they were initially in a state of equilibrium (equal weights), it follows that after the removal, the right tray will have more weight than the left tray. And so the equality is again destroyed. Observe the scale after this weight removal:

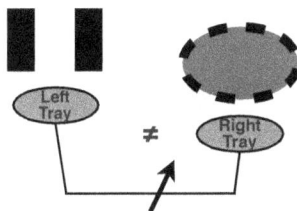

And so we can see that by *subtracting* weight from one tray but not from the other, we would again destroy the initial equilibrium state (or the initial equality condition).

So what would we need to do to make the scale's arrow point to the left? Simple: from the initial set up we have been using so far, simply *add* weight to the left tray, or *subtract* weight from the right tray. We could, of course, also *double* the weight of the left tray (by doubling the objects that are there): mathematically speaking, this means **multiplying** the existing left-tray weight times two. And if we were to do this while simultaneously leaving the right tray's weight intact, the initial equality condition would again be destroyed. Observe these modifications to the initial set up in action:

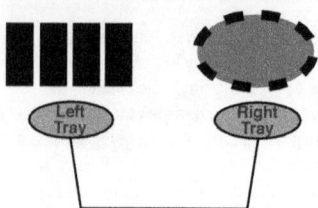

Adding weight to the left tray, leaving the right tray the same, results in the following scale state:

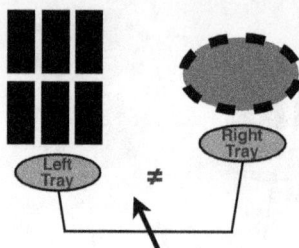

Likewise, if we were to multiply the left tray's weight times **2**, leaving the right tray's weight intact, as follows:

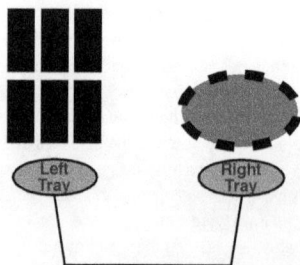

...then this would be the final scale's state:

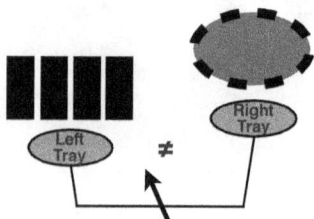

Let's now make the balance weight scale / equations connection. Consider the following initial equation:

$$2(5) + 6 \; = \; 6 + 3(2) + 4$$

If, in trying to simplify it, or solve it, or rewrite it, I were to, say, add 10 to the right side of the equation, as follows:

$$2(5) + 6 \overset{?}{=} 6 + 3(2) + 4 \boxed{+ 10}$$

Would this be acceptable? Would this *maintain* the initial equality that the original statement indicated? Or would the equality be broken?

Well, take this to the balance weight scale:

$$2(5) + 6 \qquad\qquad 6 + 3(2) + 4$$

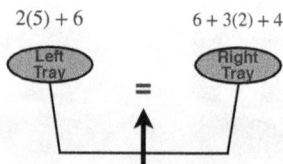

This is the initial set up; both sides are equal to each other...

But if I were to ADD **10** to the right side, leaving the left side the same, what happens to the scale? Observe:

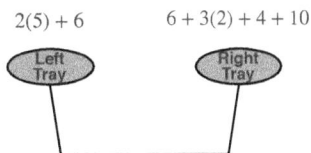

$$2(5) + 6 \qquad\qquad 6 + 3(2) + 4 + 10$$

You know the answer: the right tray will end up weighing more than the left tray, and so the equality will be destroyed!

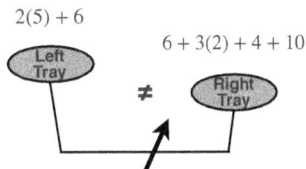

$$2(5) + 6 \qquad\qquad 6 + 3(2) + 4 + 10$$

Mathematically, that would lead to the following:

$$2(5) + 6 = 6 + 3(2) + 4$$

If I add **10** to the right side:

$$2(5) + 6 \neq 6 + 3(2) + 4 + 10$$

See how I would need to change the equal symbol with the "NOT EQUAL" symbol? The consequences of destroying the equality are far-reaching. Mainly, you end up with an expression that as far as the original goal was concerned (solving the equation for an unknown variable,

simplifying the expression, etc...), has become useless.

As a final example of how NOT to proceed when working with equations, consider two people: Tanya and Sam. They are both asked to solve the following equation for the unknown variable (which would be **x** in this case):

$$x + 4 = 23 + 1$$

Observe Tanya's step by step process:

$$x + 4 = 23 + 1$$

First, Tanya adds **23** and **1**:

$$x + 4 = 24$$

Then subtracts **4** from both sides of the equation:

$$x + 4 - 4 = 24 - 4$$

...and performs the subtraction:

$$x = 20$$

And the equation is solved. What started out being an unknown variable, Tanya has now found the value that will maintain the equality, should she replace the variable **x** with its found value of **10**. Observe how we can prove that Tanya solved this correctly:

$$x + 4 = 23 + 1$$

If we replace the **x** with **20**:

$$(20) + 4 = 23 + 1$$

...and perform the addition specified on the left side and the addition specified in the right side:

$$24 = 24$$

We can thus safely conclude that $x = 20$ is the correct solution to the original equation with one unknown variable, $x + 4 = 23 + 1$.

Now, let's observe Sam's solution process. Remember, Sam was asked to solve for the unknown variable x.

$$x + 4 = 23 + 1$$

First, Sam adds **23** and **1**:

$$x + 4 = 24$$

Then Sam subtracts **4** from the left side of the equation:

$$x + 4 - 4 = 24$$

...and performs the subtraction:

$$x = 24$$

Sam thinks the equation is solved; however, checking to see if this answer makes sense, Sam decides to replace the variable x with **24** to see if the equality is maintained: if the equation was solved correctly, then it definitely should be the case. Let's check this.

$$x + 4 = 23 + 1$$

If Sam replaces the x with **24**:

$$(24) + 4 = 23 + 1$$

...and perform the addition specified on the left side and the addition specified in the right side:

$$28 = 24$$

The inequality is broken! Why? Because Sam subtracted **4** from the left side of the equation, but left the right side the same. And just like with a balance scale, doing this destroys the scale's initial balanced-state. Adding or subtracting something other than **0** to ONLY one side is not allowed because it breaks the equality; multiplying OR dividing ONLY one side is not allowed because it also destroys the equality. There are other operations that cannot be performed only on one side of an equation because the equality would be broken, and we will explore them later. But for now, let us work with addition, subtraction, multiplication, and division.

So what can we conclude when working with equations? Think of them as balance weight scales: if you do something to one side of the equation, ask yourself: if I took this to a balance scale, where each side of the equation corresponds to a tray, would I be destroying the equality after doing whatever it is that I intend on doing?

In other words: to maintain an equality relationship, if you add a given term to one side of the equation, add the same term to the other side. If you subtract a term from one side, subtract from the other side. If you multiply an entire side by a term, multiply the other side as well by the same term, and if you divide an entire side by a term, divide the other entire side as well (remember that I will review other mathematical operations later in the book).

So let's get to it. We will begin by solving some simple equations (all of one variable: the unknown variable, of course) using a similar process to that of Tanya's.

The goal, when solving an equation of one variable (meaning that variable is equal to a number that is unknown, at least before the act of solving said equation), is to have the variable on one side, with whatever it equals to on the other. Typically, this means having the variable on one side with a **REAL** number on the other (if we are working with equations that have **REAL**-numbered solutions only).

Let's solve some equations, which will get progressively more complicated.

$$y = 4(3) + 2$$

Initial set up... multiply **4** and **3** as per PEMDAS

$$y = 12 + 2$$

Add **12** and **2** and the equation is solved:

$$y = 14$$

Simple, right? Since the left side of the equation contained the variable, already *isolated*, all we had to do was simplify the right side of the equation using PEMDAS as a guide.

$$20 \div 4 \times 5 = w$$

Initial set up... divide **20** and **4** first as per PEMDAS (remember that the MD stands for "Multiply and Divide", but it does not specify that order: the order is determined by which of these appears first from LEFT to RIGHT):

$$5 \times 5 = w$$

Multiply **5** and **5** and the equation is solved:

$$25 = w$$

I left the unknown variable on the right side on purpose for the following reason: most algebra lessons show the unknown variable on the left side with the solution on the right side. And while I appreciate the reasoning behind this (consistency, order, familiarity, etc...), it does tend to convey the wrong idea: mainly, that it is a *requirement*. It is NOT! Think about it. If I tell you that **5** = **x** is that any different than if I told you that **x** = **5**? Strictly speaking, equations in algebra are symmetric. Changing one side with the other will not alter the equality. So, we can, if we wish, keep the unknown variable on the right side, while we work with the left side to simplify it and find an answer.

In general, we say that if

$$a = b$$

then

$$b = a$$

And therefore, when solving equations, if we want to solve for an unknown variable, we may keep the variable on either of the sides.

The above principle also allows us to do the following:

$$20 \div 4 \times 5 = w$$

Initial set up... Use the above principle and exchange sides:

$$w = 20 \div 4 \times 5$$

And then solve as before...:

$$w = 25$$

But strictly speaking, we do not have to do this. As I already explained, we can solve with the variable isolated on either side.

Try to solve the following equation.

$$10x + 8 \div 4 \times 3 = 6x + 2 \times 10 \div 4$$

The problem you will face is that the unknown variable appears on **both** sides of the equation. Think about how you can collect all of the terms that contain the unknown variable on one side of the equation (hint: it requires adding or subtracting one the terms, to **both** sides of the equation so as to respect the equality).

Ready to check your answer? Observe.

$$10x + 8 \div 4 \times 3 = 6x + 2 \times 10 \div 4$$

Initial set up... Simplify as much as you can each of the sides, using PEMDAS:

$$10x + 2 \times 3 = 6x + 20 \div 4$$

Continue simplifying...

$$10x + 6 = 6x + 5$$

Since the unknown variable appears on both sides, we need to send one of them to the other side. Since the **x** term on the left is positive and its coefficient is **10**, while the **x** term on the right has a coefficient lower than **10** (it is **6** to be precise), I will subtract the term **6x** from both sides of the equation. From the balance scale analogy, this step is valid (maintains the equality) as long as I do the subtracting of the *same term* to *both sides*.

$$10x + 6 - 6x = 6x + 5 - 6x$$

The reason why I'm subtracting **6x** to both sides to begin with is simple: as you can see on the next step, this will get rid of the **6x** term that appears on the right side of the equation, effectively sending it to the left side. Now, since they are like terms, I will be able to combine them, as we saw in **Chapter 11**.

$$10x - 6x + 6 = 6x - 6x + 5$$

...rearranging the terms on each side. Strictly speaking, this step may be bypassed, but I choose to display it to help you see why we subtracted the term **6x** to both sides to begin with.

$$4x + 6 = 0 + 5$$

See how the term with the unknown variable disappeared form the right side? Well, it didn't exactly disappear into thin air: it canceled out with the subtracting **6x** that we introduced. In a way, we could say that we simply moved it, conveniently, to the left side... that's why we ended up subtracting **6x** from **10x** on the left side.

$$4x + 6 = 5$$

Simplifying the right side.

$$4x + 6 - 6 = 5 - 6$$

Subtracting 6 from both sides. Remember that the goal is to isolate the unknown variable. At this point, we need to find a way to get rid of the **6** that is adding on the left side, so that the **4x** is by itself on one side of the equation.

$$4x = -1$$

After subtracting six from both sides, the equation is almost solved. The left side contains the term with the unknown variable, while the right side contains everything else. For the last step, the question that needs to be asked is this one: how can we fully isolate the **x** on the left side? It currently has a **4** multiplying it, so we must find a mathematical operation that when applied to **both** sides will eliminate the **4**.

$$\frac{4x}{4} = \frac{-1}{4}$$

Did you realize this was coming? Dividing by **4** both sides of the equation will eliminate the **4** that is multiplying the **x** on the left side, allowing us to achieve our goal of isolating the unknown variable, **x**, on whatever side we chose to keep it.

$$x = -\frac{1}{4}$$

And the equation is solved. I could have left the negative sign on the numerator, next to the **1**, but remember that it is equally valid to have it outside of the fraction, or to shift it down to the denominator.

Can you solve the next equation?

$$8y - 20 \div 5 \times 6 - 14 = 4y + 4 \times 5 \div 4 - y$$

Hint: make sure you combine like terms on each side before trying to send terms to the opposite side.

Ready for the answer? Observe the step by step process:

$$8y - 20 \div 5 \times 6 - 14 = 4y + 4 \times 5 \div 4 - y$$

$$8y - 4 \times 6 - 14 = 4y + 20 \div 4 - y$$

$$8y - 24 - 14 = 4y + 5 - y$$

$$8y - 10 = 4y - y + 5$$

$$8y - 10 = 3y + 5$$

$$8y - 10 - 3y = 3y + 5 - 3y$$

$$8y - 3y - 10 = 3y - 3y + 5$$

$$5y - 10 = 0 + 5$$

$$5y - 10 = 5$$

$$5y - 10 + 10 = 5 + 10$$

$$5y = 15$$

$$\frac{5y}{5} = \frac{15}{5}$$

$$y = 3$$

Another equation solved! No explanation necessary, since the only difference between this and the previous equation is the step where I combined like terms that were on the same side (the **y** terms, to be specific).

For the next equation, I want you to know that I will keep the unknown variable on the right side. I do not want you to assume that it **must** be isolated on the left side. Try to solve it with that goal in mind.

$$22 - (12 \div 4 \times 2)(w) - 2 = 2w + 6 - w$$

As you solve this equation for **w** make sure that you take into account before doing something to a term if it is being subtracted or added.

Ready for the answer? Observe:

$$22 - (12 \div 4 \times 2)(w) - 2 \ = \ 2w + 6 - w$$

$$22 - (3 \times 2)(w) - 2 \ = \ 2w - w + 6$$

On the right side, I changed the order of the terms to help you see how like terms may be combined, which I do on the next step...

$$22 - (6)(w) - 2 \ = \ w + 6$$

$$22 - 6w - 2 \ = \ w + 6$$

$$22 - 2 - 6w \ = \ w + 6$$

$$20 - 6w \ = \ w + 6$$

$$20 - 6w + 6w \ = \ w + 6 + 6w$$

$$20 \ = \ w + 6w + 6$$

$$20 \ = \ 7w + 6$$

$$20 - 6 \ = \ 7w + 6 - 6$$

$$14 \ = \ 7w$$

$$\frac{14}{7} \ = \ \frac{7w}{7}$$

$$2 \ = \ w$$

Remember that I am showing steps that may not be necessary to write or show when solving a given equation. However, since we are reviewing these concepts, I explicitly provide the steps that may help you master this skill. As you practice this you will find that you may have to take less and less steps to find the solution.

Try to solve the next equation.

$$2m + 4 \ = \ \frac{m}{3}$$

The key step here is getting rid of the **3** that is dividing the unknown variable. Although there are other ways of solving it, taking that step makes for an easier solving process.

Ready to see one of the possible ways in which it may be solved?

$$2m + 4 \ = \ \frac{m}{3}$$

$$(2m + 4)(3) \ = \ (\frac{m}{3})(3)$$

Since the unknown variable on the right side is being divided by 3, I decided to multiply the entire right side by that denominator (the **3**); however, to maintain the equality, I was forced to multiply the entire left side by **3** as well.

$$6m + 12 \ = \ m$$

$$6m + 12 - m \ = \ m - m$$

$$5m + 12 \ = \ 0$$

$$5m + 12 - 12 \ = \ 0 - 12$$

$$5m \ = \ -12$$

$$\frac{5m}{5} \ = \ \frac{-12}{5}$$

$$m \ = \ \frac{-12}{5}$$

The fraction cannot be simplified, so that is the final answer to the equation.

Before moving on to another equation, let me inset a summary of the strategies covered so far:

⚓ Solving One Variable Equations

1. Equations are like balance weight scales. Make sure you keep the sides equal as you perform mathematical operations on either side of the equation.

2. You may ADD or SUBTRACT any term on either side of an equation, provided you do the same to the OTHER side:

L side + TERM = R side + TERM
or
L side - TERM = R side - TERM

3. You may multiply or divide the entire side of an equation by a term, provided you do the same to the other side:

(L side)(TERM) = (R side)(TERM)
or
$$\frac{L\ side}{TERM} = \frac{R\ side}{TERM}$$

4. When solving a one variable equation, the goal is to isolate the unknown variable on either side of the equation:

Unknown Variable = All else
or
All else = Unknown Variable

Got it? So far, I have provided examples that illustrate these principles in action. There are, of course, other scenarios that we will need to consider. Dealing with them involves mastery of all the principles and rules reviewed earlier.

Take a look at the following equation.

$$20 - n = 60$$

I will use a series of steps to solve it in order to illustrate an important situation frequently encountered when solving one variable equations. And yes, I know that this particular solution process may be replaced with another solution process. But this one specifically contains an important concept involving working with signs. Observe.

$$20 - n = 60$$

$$20 - n - 20 = 60 - 20$$

$$20 - 20 - n = 40$$

$$0 - n = 40$$

$$-n = 40$$

At this point in this specific equation-solving process, you may be wondering how to deal with a negative term containing the unknown variable we are trying to solve for in the first place. Take note of the fact that although we have *almost* solved the equation for the unknown variable, we haven't actually solved it yet, since we need to know what *n* is equal to, not *−n* which is what is currently stated. Well, there are several options. Observe:

Option 1

Multiply both sides by the number **−1**...

$$(-n)(-1) = (40)(-1)$$

This will allow us to change the sign of the unknown variable on the left side from a negative to a positive; what happens on the right side is a consequence of this sign reversal, and will allow us to find the sought-after answer...

$$n = -40$$

Equation solved. This method, which I particularly prefer over any other method, uses the principle that multiplying two negatives yields a positive (as per the Sign Table). It has the added benefit that multiplication is easy enough to carry out, assuming we are careful to multiply the *entire side* by **−1**. If we had unlike terms on the right side, we would need to multiply *all* of them by **−1**.

Option 2

Divide both sides by the number **−1**...

$$\frac{-n}{-1} = \frac{40}{-1}$$

This will allow us to change the sign of the unknown variable on the left side from a negative to a positive; what happens on the right side is a consequence of this sign reversal, and will allow us to find the sought-after answer...

$$n = \frac{40}{-1}$$

which is the same as...

$$n = \frac{-40}{1}$$

or as...

$$n = -\frac{40}{1}$$

or as...

$$n = -40$$

Equation solved. Of course, the preferred notational representation of the solution is $n = -40$ since it is fully simplified; however, if the answer had been an irreducible fraction, the second to last choice would be preferred over the others (the last choice not being an option).

I don't particularly recommend this option since it involves fractions, and it could lead to mistakes, especially when the side that does not contain the variable that we are solving for is made up of more than one term.

Option 3

Add **n** to both sides...

$$-n + n = 40 + n$$

This will allow us to change the term (variable) to the other side, making it positive in the process...

$$0 = 40 + n$$

And then simply send the unwanted terms to the other side (in this case, the **40** must be sent to the left side, which means that if we subtract **40** from both sides, this goal is accomplished)...

$$0 - 40 = 40 + n - 40$$

$$-40 = 40 - 40 + n$$

$$-40 = 0 + n$$

$$-40 = n$$

Equation solved. I find this method a bit unnecessary, since **Option 1** is easier to carry out. However, it is a perfectly valid method.

Remember, at that point, if you must have the variable on the left side (not necessary, but that is how answers are usually presented), simply flip sides as follows:

$$-40 = n$$

$$n = -40$$

The above statements are both expressing the exact same idea.

There are other options, but they are mere circumventions of those already presented, so we do not need to go over them.

In conclusion, you always have options. How you decide to move forward when solving an equation with one unknown variable will depend on your particular preference, expertise, and comfort level using the required mathematical steps.

Try to solve the next equation.

$$-3(4x + 2) + 11x + 36 = x - 20$$

Once again, there are many different paths we could take to solve this; however, they should all converge on a solution: the unknown variable on one side, the other elements on the other.

Observe the steps that may be taken in order to solve this:

$$-3(4x + 2) + 11x + 36 = x - 20$$

$$-12x - 6 + 11x + 36 = x - 20$$

$$-12x + 11x - 6 + 36 = x - 20$$

$$-x + 30 = x - 20$$

$$-x + 30 + x = x - 20 + x$$

$$-x + x + 30 = x + x - 20$$

$$0 + 30 = 2x - 20$$

$$30 = 2x - 20$$

$$30 + 20 = 2x - 20 + 20$$

$$50 = 2x$$

$$\frac{50}{2} = \frac{2x}{2}$$

$$25 = x$$

And the equation is solved. Note, once again, how I solved the equation for the unknown variable **x** keeping it on the right side of the equation. You should become very comfortable doing this, because it saves steps and is equally valid.

Try to solve these equations by yourself.

a) $2x + 34 - 12x + 6 = -2(-5x + 10)$

b) $-6y - 5 + 10 \div 5 \times 4y = 2 \times 6 \div 4 - 2y$

c) $-4b - 36 \div (4 \times 3) - b = -b + 4 \times 4$

Ready to check your answers? Observe the steps I took to solve the equations:

a) $2x + 34 - 12x + 6 = -2(-5x + 10)$

$$2x + 34 - 12x + 6 = -2(-5x + 10)$$

$$-10x + 40 = 10x - 20$$

$$-10x + 40 + 10x = 10x - 20 + 10x$$

$$-10x + 10x + 40 = 10x + 10x - 20$$

$$40 = 20x - 20$$

$$40 + 20 = 20x - 20 + 20$$

$$60 = 20x$$

$$\frac{60}{20} = \frac{20x}{20}$$

$$3 = x$$

b) $\ -6y - 5 + 10 \div 5 \times 4y = 2 \times 6 \div 4 - 2y$

$$-6y - 5 + 2 \times 4y = 12 \div 4 - 2y$$

$$-6y - 5 + 8y = 3 - 2y$$

$$-6y + 8y - 5 = 3 - 2y$$

$$2y - 5 = 3 - 2y$$

$$2y - 5 + 2y = 3 - 2y + 2y$$

$$2y + 2y - 5 = 3$$

$$4y - 5 = 3$$

$$4y - 5 + 5 = 3 + 5$$

$$4y = 8$$

$$\frac{4y}{4} = \frac{8}{4}$$

$$y = 2$$

c) $-4b - 36 \div (4 \times 3) - b = -b + 4 \times 4 + 1$

$$-4b - 36 \div (12) - b = -b + 16 + 1$$

$$-4b - (36 \div (12)) - b = -b + 17$$

$$-4b - (3) - b = -b + 17$$

$$-4b - b - 3 = -b + 17$$

$$-5b - 3 = -b + 17$$

$$-5b - 3 + 5b = -b + 17 + 5b$$

$$-5b + 5b - 3 = -b + 5b + 17$$

$$0 - 3 = 4b + 17$$

$$-3 = 4b + 17$$

$$-3 - 17 = 4b + 17 - 17$$

$$-3 - 17 = 4b$$

$$-20 = 4b$$

$$\frac{-20}{4} = \frac{4b}{4}$$

$$-5 = b$$

At this point, I would only like to add the following complexity to the concept of solving equations: incorporating fractions into the equations themselves; the principles that we have established would still apply, and the only true difference is how we deal with the terms themselves.

Observe the following example.

$$\frac{4}{5} - \frac{2x}{3} + (\frac{-1}{2} \div \frac{3}{4}) = 4x$$

As you can see, the unknown variable appears on both sides of the equation. We must therefore collect all of the terms containing x on one side, all other terms on the other. The challenge is to work with the fractions without making any mistakes. Observe one of the possible ways in which this may be solved.

$$\frac{4}{5} - \frac{2x}{3} + (\frac{-1}{2} \div \frac{3}{4}) = 4x$$

PEMDAS dictates that I simplify any grouping elements first (that's the P for parentheses, remember?). I will thus first change the way the fractions are shown dividing; using the sandwich rule is easy, and that format typically appears in algebra problems anyway. After the switch, the equation looks as follows:

$$\frac{4}{5} - \frac{2x}{3} + (\frac{\frac{-1}{2}}{\frac{3}{4}}) = 4x$$

Now, apply the sandwich rule...

$$\frac{4}{5} - \frac{2x}{3} + (\frac{(-1)(4)}{(2)(3)}) = 4x$$

$$\frac{4}{5} - \frac{2x}{3} + (\frac{-4}{6}) = 4x$$

Simplify the fraction and move the negative sign to the side, then continue to simplify...

$$\frac{4}{5} - \frac{2x}{3} + (-\frac{2}{3}) = 4x$$

$$\frac{4}{5} - \frac{2x}{3} - \frac{2}{3} = 4x$$

$$\frac{4}{5} - \frac{2}{3} - \frac{2x}{3} = 4x$$

Next, combine like terms...

$$\frac{(4)(3) - (2)(5)}{(5)(3)} - \frac{2x}{3} = 4x$$

$$\frac{12 - 10}{15} - \frac{2x}{3} = 4x$$

$$\frac{2}{15} - \frac{2x}{3} = 4x$$

At this point, since we want to have all terms that contain the unknown variable, x, on the same side, I will choose to add the term $\frac{2x}{3}$ to both sides, effectively sending it from the left to the right side of the equation...

$$\frac{2}{15} - \frac{2x}{3} + \frac{2x}{3} = 4x + \frac{2x}{3}$$

$$\frac{2}{15} = 4x + \frac{2x}{3}$$

To combine the two like terms on the right side of the equation (yes, they are like terms, since they both have the variable x raised to the same power, without any other variables whatsoever), we have the following two basic options:

Option 1

Multiply both sides of the equation by the denominator of the x term, the **3**, so that the fraction x term disappears; just make sure to distribute when appropriate:

$$\frac{2}{15} = 4x + \frac{2x}{3}$$

$$(\frac{2}{15})(3) = (4x + \frac{2x}{3})(3)$$

$$(\frac{2}{15})(3) = (4x)(3) + (\frac{2x}{3})(3)$$

$$(\frac{2}{15})(3) = 12x + (\frac{2x}{3})(3)$$

Remember how to multiply a fraction term with a non-fraction term? Simple: write the non-fraction term as a fraction by dividing it by **1** and then proceed as usual (multiply straight across: numerator times numerator for the new numerator, denominator times denominator for the new denominator)...

$$(\frac{2}{15})(\frac{3}{1}) = 12x + (\frac{2x}{3})(\frac{3}{1})$$

Continue simplifying...

$$\frac{6}{15} = 12x + \frac{6x}{3}$$

$$\frac{6}{15} = 12x + 2x$$

Strictly speaking, we could've cancelled the **3** that was multiplying the $\frac{2x}{3}$ with the **3** in this term's denominator (that was the purpose of multiplying both sides of the equation by **3** to begin with), but I am showing you explicitly why that cancellation would occur.

$$\frac{6}{15} = 12x + 2x$$

And now, we can combine the x terms without having to worry about dealing with fraction elements...

$$\frac{6}{15} = 14x$$

Divide both sides of the equation by 14 to solve for x, and after applying the sandwich rule once more, we find the answer:

$$\frac{\frac{6}{15}}{14} = \frac{14x}{14}$$

$$\frac{\frac{6}{15}}{\frac{14}{1}} = x$$

$$\frac{(6)(1)}{(15)(14)} = x$$

$$\frac{6}{210} = x$$

$$\frac{1}{35} = x$$

Equation solved. I want to emphasize the fact that I am showing you all of the steps involved to help you check your work; if you have any conceptual misunderstandings, you may have to read the sections of this book that covers them.

Let me show you the other options we have for solving this equation.

Option 2

Add (or subtract, if that had been the case) the **x** terms that appear on the right side of the equation. This involves working with fractions. Since one of the terms is not a fraction, we can divided it by **1** to have it expressed as a fraction:

$$\frac{2}{15} = 4x + \frac{2x}{3}$$

$$\frac{2}{15} = \frac{4x}{1} + \frac{2x}{3}$$

$$\frac{2}{15} = \frac{(4x)(3) + (2x)(1)}{(1)(3)}$$

$$\frac{2}{15} = \frac{12x + 2x}{3}$$

$$\frac{2}{15} = \frac{14x}{3}$$

Now, we can multiply both sides of the equation by **3** to eliminate the denominator of the **x** term...

$$(\frac{2}{15})(3) = (\frac{14x}{3})(3)$$

Unlike what I did in Option 1, I will cancel on the right side of the equation the **3** that is in the denominator with the **3** that is multiplying the fraction (I have reviewed this principle throughout the book)...

$$(\frac{2}{15})(3) = (\frac{14x}{3})(3)$$

$$(\frac{2}{15})(3) = (\frac{14x}{3})(3)$$

$$(\frac{2}{15})(3) = 14x$$

To multiply a fraction times a non-fraction (as is the case on the left side), we divide the non-fraction by **1**, and then continue from there...

$$(\frac{2}{15})(\frac{3}{1}) = 14x$$

$$\frac{6}{15} = 14x$$

From here on, the solution process is the same as before.. and so the answer is:

$$\frac{1}{35} = x$$

Equation solved. Please note that both options lead to the same simplified answer, as should be.

Checking your answers when solving equations.

After you solve an equation for an unknown variable, there is a simple way to check if your answer is correct. Simply go back to the original equation, the one that contains the unknown variable, and replace it with the answer you found using a set of parentheses. Observe the following example.

$$2x + 10 = 60$$

Try to solve this equation yourself. Ready to see my process?

$$2x + 10 = 60$$

$$2x + 10 - 10 = 60 - 10$$

$$2x = 50$$

$$\frac{2x}{2} = \frac{50}{2}$$

$$x = 25$$

So how can we know if we didn't make a mistake along the way? Just plug in **25** for **x** into the original equation, and check to see if both sides of the equation are, in fact, balanced and equal to each other. Just remember to replace the variable with a set of parentheses so that you can deal with any negative signs, exponents, or any other mathematical operations that may be having an impact on the unknown variable.

$$2x + 10 \;=\; 60$$

$$2(25) + 10 \;=\; 60$$

$$50 + 10 \;=\; 60$$

$$60 \;=\; 60$$

As you can observe, the solution was, in fact, correct, since replacing the variable with the answer we found maintains the equality.

We will continue to solve equations in other chapters of the book. But for now, this serves as an introduction into the concept of equations, and how we may use the principles and rules of algebra to solve them.

See you on the next chapter!

MORE ON EQUATIONS

On **Chapter 14** (Introduction to Equations) I briefly reviewed how to work with equations that contain one unknown variable. Now, we will look more closely at equations that have one variable and deal with equations that have two or more unknown variables as well. However, please note that in this chapter, all variables involved are raised to the power of **1**. This plays a crucial role when solving equations, as we will see in later chapters of the book, when we explore equations (and functions) with variables that have other power (exponent) values.

In general, equations with variables that are raised to the power of **1** are called "linear equations", specifically those that contain terms with only one variable each, as in the following examples:

$$y = 5x - 4$$

$$2P - 3 = 6a + 5$$

$$0 = 3w + 4x - z$$

etc...

I will begin by exploring equations that have two unknown variables. Although we will have to revisit equations with one unknown variable and those with more than two unknown variables, this will allow us to explore a concept that plays an important role in working with all kinds of equations. Observe the following prompt.

Solve the equation

$$2x + 3y = 23$$

Suppose that, based on what we did on **Chapter 14**, you said to yourself "I know... I will solve for the unknown variable **x** and take it from there"; would it be of any help? Let's give it a try.

$$2x + 3y = 23$$

$$2x + 3y - 3y = 23 - 3y$$

$$2x = 23 - 3y$$

$$\frac{2x}{2} = \frac{23 - 3y}{2}$$

$$x = \frac{23 - 3y}{2}$$

What would we do next? The equation above is telling us that **x** is equal to the **y** value multiplied times **2**, subtracting that answer from **23**, and then dividing that by **2**. But if we don't know what **y** is equal to, what good would that do? We don't have a starting point.

Maybe if we were to solve for **y** instead:

$$2x + 3y = 23$$

$$2x + 3y - 2x = 23 - 2x$$

$$3y = 23 - 2x$$

$$\frac{3y}{3} = \frac{23 - 2x}{3}$$

$$y = \frac{23 - 2x}{3}$$

This is not helpful, either. We get a similar statement, but this time one that tells us what **y** is equal to in terms of **x**.

Let's look at the original equation again:

$$2x + 3y = 23$$

Let's try something else. I can quickly come up with two pairs of numbers that would solve this equation: if $x = 10$ and $y = 1$ the equation is solved, since if we replaced the corresponding variables with the **REAL** numbers stated, we would maintain the equality (as we saw at the end of **Chapter 14**, that is the requirement if we are to claim that an equation is solved):

$$2x + 3y = 23$$

I will use a question mark above the equal symbol since we are testing the numbers as a potential solution to the equation...

$$2(10) + 3(1) \overset{?}{=} 23$$

$$20 + 3 \overset{?}{=} 23$$

$$23 \overset{\checkmark}{=} 23$$

As you can see, it works out. At this point, you may be claiming victory. As in, "That's the answer we were looking for! So what's the big deal here?" Well, the big deal is this: what if I told you that these numbers also work: $x = 4$ and $y = 5$. Don't believe me? Try them:

$$2x + 3y = 23$$

$$2(4) + 3(5) \overset{?}{=} 23$$

$$8 + 15 \overset{?}{=} 23$$

$$23 \overset{\checkmark}{=} 23$$

See how those numbers are also solutions of the original equation? Something is going on here that surpasses the concepts and principles we have been reviewing so far.

So which pair of numbers do we consider to be *the* solution to the equation? Are there more solutions that we haven't even considered? The answer to the latter is that this equation has an *infinite* pair of solutions (I will not provide any more pairs of numbers that solve it, but as a challenge, you may try to find at least one more that would work); as to the former, it is said that all of the pairs of values that solve the equation are solutions to the equation; that's right, all of them. One pair is not "better" than any other.

And so we find ourselves at a crossroads. Solving mathematical equations involves a lot more than what meets the eye. In **Chapter 14** I briefly reviewed equations, but I did not cover any special cases. And there are special cases that need to be considered. But we must proceed carefully, building on skills one at a time.

As a thought experiment, we could imagine the following scenarios when solving equations. Let's begin.

As we have already seen before, a given equation may have *one* solution and *one* solution **ONLY**. For example, the equation $2x = 10$ has only one solution; $x = 5$ is that single valid solution (simply divide both sides of the equation and you find this answer). In other words, the original equation $2x = 10$ can be assigned the number **5** in place of the x variable, and the equality will be maintained. It is, in fact, the *only* **REAL** number that we can use to replace the x variable and that maintains the equality. Any other **REAL** number that we care to use to replace the x will destroy the equality (the left side will not be equal to the right side anymore), which means that there aren't any other solutions to the original equation besides $x = 5$. I know it seems like I am repeating myself, but this is a very important concept that must be understood, and it is thus worth repeating.

Moving on, must all equations involving a variable (or variables) have a solution? Think about it... can you come up with an equation where the left side and the right side will *never* be equal to each other regardless of the **REAL** number you use to replace the unknown variable? It's actually not that difficult. Remember that equations are like weight scales, so say you have the unknown variable on **both** sides of the equation, as follows:

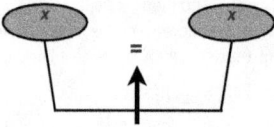

You can see how both sides are equal, and therefore balanced (in perfect *equilibrium*). But the goal is to have an equation that does not have a solution whatsoever. Think about what you could do to each side of the equation so that this goal is achieved.

Ready for one of the infinitely many possibilities that exist to accomplish this? Observe:

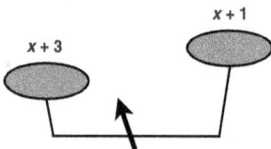

There you go! An "equation" that does not have a solution. Even though we know we should not use the equal symbol since it is obvious from the process we used to come up with this "equation" that the left and right sides are not balanced, we can imagine that someone came up with it and wants to know if it has a solution or not. Thus, we would be asked to solve the following equation:

$$x + 3 = x + 1$$

How can we know that it does not have a solution? Before involving algebra, it's actually quite simple. When we started out writing this equation, we had an **x** on the left side and an **x** on the right side. In other words, two identical elements on both sides of the equation. At this point, any number that you would care to use to replace **x** would solve the equation (hint: this is another possibility involving equations that we will explore shortly). But then, we added **3** to the left side (think about the balance weight scale analogy: we added, say, **3** pounds to the left side) and only **1** to the right side (that would be, say, **1** pound only). Now, the equality is broken! Regardless of the **REAL** number that you use to replace the **x** variable, the left side will **never** equal the right side. Observe an example of this:

$$x + 3 = x + 1$$

This is the original equation. It has ZERO solutions. If, say, $x = 1$...

$$(1) + 3 \overset{?}{=} (1) + 1$$

$$4 = 2$$

At this point you can see that regardless of the number that you use to replace the **x** variable, the left side will always be different from the right side, since on the left side you will be adding **3** to that number while on the right you would be adding only a **1**.

Someone may wonder if we can "fix" this by solving for the unknown variable, as we did in **Chapter 14**. What would we end up with? Try it by yourself (in other words, solve for **x**).

Ready to see what you should have come up with?

$$x + 3 = x + 1$$

I will first subtract **3** from both sides of the equation, since I will attempt to isolate the **x** on the left side...

$$x + 3 \underset{\cdots}{(-3)} = x + 1 \underset{\cdots}{(-3)}$$

$$x = x - 2$$

Now, I will subtract **x** from both sides based on the fact that there is a positive **x** on the right side of the equation; as we have seen before, subtracting an **x** will get rid of that term. However, remember that what we do to one side of the equation MUST be done to the other side as well, lest the equality be broken (this was, in fact, the strategy that we used to create an equation that has zero solutions, right?). I will only simplify the right side for now...

$$x \underset{\cdots}{(-x)} = x - 2 \underset{\cdots}{(-x)}$$

$$x - x = x - x - 2$$

$$x - x = 0 - 2$$

$$x - x = -2$$

Well, at this point, things may look promising. We have a **−2** on the right side, and a couple of **x** terms on the left side. However, as soon as we simplify the left side (combine like terms) we realize that we have a major problem:

$$x - x = -2$$

$$0 = -2$$

What does this expression mean? Is it possible? Does the **REAL** number **0** actually equal the **REAL** number **−2**? I know you know the answer to that question... it is, of course, a false statement. The equation is not valid. Note that we obtained this outrageous result without replacing the unknown variable with any **REAL** number whatsoever. It was through the process of solving for the unknown variable that we ended up with this false statement.

We can now safely conclude that *if* after solving for the unknown variable we end up with a statement in which the unknown variable has disappeared AND the left side is NOT equal to the right side, that the equation has ZERO solutions (in other words, that the equation does not have a solution within the **REAL** number system). If, on the other hand, the variable remains on one side and a **REAL** number on the other (as was the case in the problems we worked on in **Chapter 14**), we conclude that it has one solution, whichever was found.

But there is a third option we haven't considered (I gave you a hint of its existence earlier). What if, say, we left the equation as follows (going back to the balance weight scale analogy), and this time, we multiplied both sides by, say, the number 4...

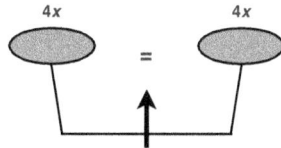

The equation we obtain would look like this:

$$4x = 4x$$

It seems a bit strange, but it is a perfectly valid equation. In fact, you can see right away that it is a true statement. The term **4x** is in fact equal to itself.

The more relevant question is this: how many answers does this equation have? One, two, ten, an infinite number of them?

Well, if we are free to use any number within, say, the **REAL** number system as we have been using, the answer is infinitely many. That's because there are an infinite number of **REAL** numbers, and any of them, when used to replace the variable **x** will maintain the equality, since essentially we would end up with a statement that says "the given number times **4** is equal to itself times **4**", which is, of course, true. Observe what happens if we test, say, the solution $x = -2$ on the equation:

$$4x = 4x$$

$$4(-2) \overset{?}{=} 4(-2)$$

$$-8 \overset{\checkmark}{=} -8$$

As expected, we end up with a true statement. No matter what number you use to replace the variable **x**, you will find that the equality will be maintained.

What happens if we were to solve the equation for the unknown variable **x**? Try it before you look at my solution process... It seems strange to try to do this, but let's see what you come up with.

Ready for my solution? Observe:

$$4x = 4x$$

As a first step, I notice the **x** term that is on the right side of the equation (and that there is an **x** term on the left side as well). I decide to collect the **x** terms on the left side of the equation, which means I need to eliminate the **4x** term that is on the right side. I can achieve this by subtracting the term **4x** that is on the right side, to both sides of the equation, of course, since I need to maintain the equality...

$$4x - 4x = 4x - 4x$$

Next, I focus on the right side exclusively. Since anything minus itself is zero, I simplify as follows...

$$4x - 4x = 0$$

Almost done. I now look at the left side, and after simplifying, we end up with the following...

$$0 = 0$$

This time, we started with an equation that contained an unknown variable, and *in the process* of trying to solve for said unknown variable, we ended up with an expression that is absolutely true, but that lacks the unknown variable we set out to solve for to begin with. As you can now imagine, when this happens we are able to conclude, unequivocally, that the equation has an infinite number of solutions. If we are told that we may use all **REAL** numbers as replacement for the unknown variable (in this case, the **x**), we know we would be able to replace the variable with any **REAL** number, and the equality would be maintained. We will explore this idea (called the *domain* and the *range*) later on in this chapter.

Essentially, whenever we set out to solve a linear equation that has *one* unknown variable, we will always encounter one of the following three possibilities:

I) The equation has only **ONE** solution. We would know this because after using the rules and properties of algebra, we would end up with a statement in the form of an equation that contains the unknown variable we set out to solve for on one side, and a **REAL** number on the other side:

Variable = REAL number
or
REAL number = Variable
For example:

$$x = 10$$
$$-145 = w$$
$$t = \frac{3}{4}$$
$$-\sqrt[4]{3} = z$$
etc...

II) The equation has **INFINITE** solutions. We would know this because after using the rules and properties of algebra, we would end up with a statement in the form of an equation that DOES NOT contain the unknown variable we set out to solve for on any side; instead, we end up with the same **REAL** number on both sides of the equation, typically the number **0**:

REAL number = Same REAL number

For example:

$$0 = 0$$
$$3 = 3$$
$$-1 = -1$$
$$\sqrt{2} = \sqrt{2}$$
etc...

It should be noted that these equations may always be reduced to the form $0 = 0$ since after the terms that contain the unknown variable cancel out on both sides of the equation and the independent terms remain (terms that are formed by number elements exclusively), these could also be added or subtracted to both sides of the equation, resulting in the fully simplified equation of $0 = 0$, signaling that it is of the INFINITE solutions type.

III) The equation has **ZERO** solutions. We would know this because after using the rules and properties of algebra, we would end up with a statement in the form of an equation that DOES NOT contain the unknown variable we set out to solve for on any side; instead, we end up with different **REAL** numbers on either sides of the equation, typically the number **0** on one of the sides, and anything else on the other:

REAL number = DIFFERENT REAL number
For example:

$$0 = 4$$
$$-7 = 3$$
$$9 = 0$$
$$0 = \sqrt{2}$$
etc...

It should be noted that these equations may always be reduced to the form $0 = $ **REAL** number or **REAL** number $= 0$ since after the terms that contain the unknown variable cancel out on both sides of the equation and the independent terms remain (terms that are formed by number elements exclusively), these could also be added or subtracted to both sides of the equation, resulting in the fully simplified equation of $0 =$ **REAL** number or **REAL** number $= 0$, signaling that it is of the **ZERO** solutions type.

Got it?

It's time for you to check if you have mastered these concepts. Try to solve the following equations being careful to check for the solution types mentioned above.

a) $4x + 8 = 2(10 - x)$

b) $2x - 3 - x - 1 = -3x - 4 + 4x$

c) $y + 2(3 + 5y) = -5y + 1 + 16y$

d) $2(x + 3) = 5x + 6$

e) $w + 1 + 2w + 2 = -w - 1 - 2w - 2$

f) $3 - 5(m + 1) = 2 - 5m$

Remember try these on your own.

Did you solve the equations? Were you able to detect what type they belong to? As you tried to solve them, I hope you were able to appreciate how the different equation types can be easily disguised as "regular equations" (in other words, as equations that have only one solution).

I will now provide you with the steps I would take to solve these equations, and what those solutions have to say about the equation types.

a) $4x + 8 = 2(10 - x)$

$$4x + 8 = 2(10 - x)$$

$$4x + 8 = 20 - 2x$$

$$4x + 8 \underline{+ 2x} = 20 - 2x \underline{+ 2x}$$

$$6x + 8 = 20$$

$$6x + 8 \underline{- 8} = 20 \underline{- 8}$$

$$6x = 12$$

$$\frac{6x}{6} = \frac{12}{6}$$

$$x = 2$$

This is, of course, a **Type I** equation. It has only **ONE** solution, $x = 2$.

b) $2x - 3 - x - 1 = -3x - 4 + 4x$

$$2x - 3 - x - 1 = -3x - 4 + 4x$$

$$2x - x - 3 - 1 = -3x + 4x - 4$$

$$x - 4 = x - 4$$

$$x - 4 \underline{- x} = x - 4 \underline{- x}$$

$$0 - 4 = 0 - 4$$

$$-4 = -4$$

At this point, I may stop and conclude that the original equation is of the infinite solutions type (the **Type II** we reviewed earlier). I could, of course, add **4** to both sides of the equation, as follows:

$$-4 = -4$$

$$-4\underset{\cdots}{(+4)} = -4\underset{\cdots}{(+4)}$$

$$0 = 0$$

But this was not actually necessary.

c) $y + 2(3 + 5y) = -5y + 1 + 16y$

$$y + 2(3 + 5y) = -5y + 1 + 16y$$

$$y + 6 + 10y = -5y + 1 + 16y$$

$$y + 10y + 6 = -5y + 16y + 1$$

$$11y + 6 = 11y + 1$$

$$11y + 6\underset{\cdots}{(-11y)} = 11y + 1\underset{\cdots}{(-11y)}$$

$$6 \overset{?}{=} 1?$$

Absolutely **FALSE**! **6**, as you well know, is NOT equal to **1**. So we conclude that the equation is a **Type III** equation, with **ZERO** solutions. I could, of course, subtract **1** from both sides, as follows:

$$6\underset{\cdots}{(-1)} = 1\underset{\cdots}{(-1)}$$

$$5 \overset{?}{=} 0?$$

Once again, this last step wasn't really necessary. I only wish to show you the way the equation would look like if **fully** simplified.

d) $2(x + 3) = 5x + 6$

$$2(x + 3) = 5x + 6$$

$$2x + 6 = 5x + 6$$

$$2x + 6\underset{\cdots}{(-2x)} = 5x + 6\underset{\cdots}{(-2x)}$$

$$2x + 6 - 2x = 5x - 2x + 6$$

$$6 = 3x + 6$$

$$6\underset{\cdots}{(-6)} = 3x + 6\underset{\cdots}{(-6)}$$

$$0 = 3x$$

At this point, since our goal is to isolate the **x** variable (the unknown variable, that is), we would need to divide by **3**. This step seems strange, since there's a **0** on the other side, but remember that dividing **0** by anything other than **0** is allowed; dividing **by** 0 is what is undefined. Therefore, we would proceed as follows:

$$\frac{0}{3} = \frac{3x}{3}$$

$$0 = x$$

Equation solved. Please note that this statement explicitly says "the answer to the equation $2(x + 3) = 5x + 6$ is $x = 0$", and is therefore a **Type I** (one solution equation). Do not confuse it with either one of the other two types when the answer is ZERO. How can you tell? Simple: the unknown variable (in this case, the **x**) shows up as part of the fully simplified answer.

e) $w + 1 + 2w + 2 = -w - 1 - 2w - 2$

$$w + 1 + 2w + 2 = -w - 1 - 2w - 2$$

$$w + 2w + 1 + 2 = -w - 2w - 1 - 2$$

$$3w + 3 = -3w - 3$$

$$3w + 3\underset{\cdots}{(+3w)} = -3w - 3\underset{\cdots}{(+3w)}$$

$$6w + 3 = 0 - 3$$

$$6w + 3 = -3$$

$$6w + 3\underbrace{(-3)} = -3\underbrace{(-3)}$$

$$6w = -6$$

$$\frac{6w}{\underbrace{6}} = \frac{-6}{\underbrace{6}}$$

$$w = -1$$

Equation solved. This is a **Type I** equation as well, a one solution equation.

f) $3 - 5(m + 1) = 2 - 5m$

$$3 - 5(m + 1) = 2 - 5m$$

$$3 - ((5)(m) + (1)(5)) = 2 - 5m$$

$$3 - (5m + 5) = 2 - 5m$$

$$3 - 5m - 5 = 2 - 5m$$

$$3 - 5\underbrace{(-5m)} = 2\underbrace{(-5m)}$$

$$-2 - 5m = 2 - 5m$$

$$-2 - 5m\underbrace{(+5m)} = 2 - 5m\underbrace{(+5m)}$$

$$?{-2} \overset{?}{=} 2\,?$$

Absolutely **FALSE!** **−2**, as you well know, is NOT equal to **2**. So we conclude that the equation is a **Type III** equation, with **ZERO** solutions. I could, of course, add **2** to both sides, as follows:

$$-2\underbrace{(+2)} = 2\underbrace{(+2)}$$

$$?0 \overset{?}{=} 4\,?$$

Once again, this last step wasn't really necessary. I only wish to show you the way the equation would look like if **fully** simplified. Since

0 is not equal to **4** we can conclude that the original equation, $3 - 5(m + 1) = 2 - 5m$, is a **Type III** equation, which means it does not have a solution.

⚓

Types of Linear Equations of One Variable

Equations may be classified as one of the following three types (the examples illustrate what the equation might look like **after** solving for the unknown variable)

Type I: Equations with **ONE** solution, as in:
$$x = 10 \quad, \quad -145 = w \quad,$$
$$t = \frac{3}{4} \quad, \text{ etc...}$$

Type II: Equations with **INFINITE** solutions, as in: $0 = 0$, $3 = 3$, $\sqrt{2} = \sqrt{2}$, etc...

Type III: Equations with **ZERO** solutions, as in: $0 = 4$, $-7 = 3$, $9 = 0$, etc...

Now that you know everything there is to know about solving equations that have one unknown variable raised to the power of **1** (remember that whenever you see a mathematical element (such as a number, a variable, a root, a grouping set of symbols, etc...) without an exponent explicitly written, it means that the exponent is **1**, not "nothing" or "blank" or "empty space", or "zero"). For example, we solved the following equation earlier:

$$4x + 8 = 2(10 - x)$$

If I ask you to specify the exponent that is attached to the **x** variable on the left side and on the right side, you should say **1** for both. This same equation may thus be written as follows (though you will never see it like this, since the power of **1** is always implied):

$$4x^1 + 8 = 2(10 - x^1)$$

We could even write the above equation as follows, and it would still be the exact same equation:

$$4^1x^1 + 8^1 = 2^1(10^1 - x^1)^1$$

I know it's taking it to the extreme, but it is, in fact, a correct application of the notational rule mentioned above.

So, if you quickly scan through all of the equations we have been solving, you will find that all of them have the unknown variable (or variables) raised to the power of **1**.

These equations are relatively easy to work with because the exponent of the unknown variable is equal to **1**. If it is not **1** (nor the trivial case of **0**), the solving process can become very challenging, very fast. We will cover some of the more common powers later on (such as **2** and **3**). For now, however, we can review a common kind of equation, called a ***literal equation***. Although these may or may not be linear equations themselves, I will, for now, explore only those that have the variables raised to the power of **1**; however, they may have multi-variable terms (multiplying or dividing each other).

Literal equations are equations that contain variables (normally more than one), which are usually letters of the alphabet, hence the "literal" part of its name (one of the definitions of literal is "related to a letter"). You may remember

equations for areas from a geometry course, or any of the physics formulas that are so common.

For example, all of the following are literal equations:

Area of a rectangle:
$$A = lw$$

Area of a triangle:
$$A = \frac{bh}{2}$$

Volume of a rectangular box:
$$V = lhw$$

Perimeter of a square with side **s**:
$$P = 4s$$

Normally, a literal equation is already solved for a commonly desired variable. Take the perimeter of a square: we normally know the side of the square, while the perimeter is the sought-after value. So, the equation $P = 4s$ has the perimeter variable **P** on the left side (just a common practice: remember that mathematically speaking, it does not matter if it is solved for the unknown variable on the left side or on the right side) with the other elements on the right. Say, for instance, that a given square has sides that measure **5 feet**; what is the perimeter of the square? You would, of course, use the perimeter formula, replacing **s** with **5**, as follows:

$$P = 4s$$

$$P = 4(5)$$

$$P = 20$$

Strictly speaking, the answer is **20 feet** since the **s** is **5 feet**, which means that **s** is **dimensional** (as is possesses a unit of time, length, speed, etc...).

But what if they gave us the perimeter instead? How would we find the length of the side, **s**? Well, you would need to solve the equation for **s** instead of leaving it solved for **P** as it currently stands. This will allow us to plug int the perimeter measurement and find the side length that yields said perimeter. In other words, you are solving literal equations.

Observe:

$$P = 4s$$

Since I want to solve for s, I will try to isolate the **s** variable; dividing both sides by **4** will achieve it:

$$\frac{P}{4} = \frac{4s}{4}$$

Simplify...

$$\frac{P}{4} = s$$

And the literal equation is now solved for **s**. Now, if they asked us to find what the side of a square must measure in order to have a perimeter of, say, **200 feet**, all we would need to do is plug in the perimeter value into the equation that we solved for **s**, as follows:

$$\frac{P}{4} = s$$

$$\frac{200 \; feet}{4} = s$$

$$50 \; feet = s$$

And we have the answer. Now, try to solve the next problem on your own.

If a triangle's area is **20 inches squared** (written as $20 \; in^2$), and its base is **10 inches** (written as $10 \; in$), what is its height? To solve this problem, remember that the area of a triangle is given by $A = \frac{bh}{2}$.

To solve this, first identify the variable (letter in this specific example) you are trying to solve for, solve the literal equation for said letter, and then and only then plug in the values given to find the height of this triangle.

Ready to check how you did? Observe my solution steps.

1) Solve the literal equation $A = \frac{bh}{2}$ for **h**, since as it stands, it is not very helpful to us (we know **A** and **b** but need to determine the value of **h**).

$$A = \frac{bh}{2}$$

First, multiply both sides by **2**:

$$(A)(2) = (\frac{bh}{2})(2) \quad \text{Cancel out}$$

$$2A = bh$$

Now, divide both sides by **b** so that we may isolate the **h** on the right side, achieving our goal of solving for that variable:

$$\frac{2A}{b} = \frac{bh}{b}$$

400

$$\frac{2A}{b} = h$$

2) Plug in the values given. Since we are told that the area is $20 \ in^2$ and that the base is $10 \ in$ we have the following:

$$A = 20 \ in^2$$

$$b = 10 \ in$$

Replacing **A** and **b** in $\dfrac{2A}{b} = \dfrac{bh}{b}$:

$$\frac{2(20 \ in^2)}{10 \ in} = h$$

$$\frac{40 \ in^2}{10 \ in} = h$$

$$4 \ in = h$$

And the problem is solved. The triangle, therefore, has a height of **4 inches**. The only part of this solution process that requires further explanation is the **dimensional analysis** that I performed on the last step. A dimensional analysis is the name given to the process of finding the units of an answer (as in time units, or distance, or speed, etc...). On the last step, I divided $40 \ in^2$ by $10 \ in$ and I said the answer of this was $4 \ in$. The **4** part I believe you can see for yourself how it came to be (**40** divided by **10** is, of course, equal to **4**); however, why did I say that the answer has **inches** as units? Was it because I happen to know that a height measurement should be in inches and not in inches squared? The answer is no. When dealing with numbers that have units (dimensions), we must perform a dimensional analysis whenever we multiply or divided them by numbers that also

have units. In this case, the dimensional analysis looks as follows:

$$\frac{in^2}{in} = \frac{(in)(i\!\!\!/n)}{i\!\!\!/n} = \frac{in}{1} = in$$

In other words, **inches squared** means **inches** times **inches**, and since it is being divided by **inches**, the denominator unit cancels with one of the numerator units, and we are left with **inches**. It is very similar to how you would simplify expressions containing variables, as you can see in the equivalent expression below:

$$\frac{x^2}{x} = \frac{(x)(x\!\!\!/)}{x\!\!\!/} = \frac{x}{1} = x$$

The only difference between the "dimensional analysis version" and the version immediately above is that I replaced the word "*inches*" with the variable or letter "*x*". In the end, the dimensional analysis tells us that in this situation our answer will be in **inches**, which makes sense considering what we set out to solve in the first place (a side measurement).

When working with literal equations (and solving for them), we face the possibility of making assumptions about the literal equation that we cannot make. This involves understanding the concept of **domain** and **range**, so this is the perfect time to talk about them. Take, for instance, the equation $P = 4s$.

As you know, $P = 4s$ is a formula that allows us to find the perimeter **P** of a square that has sides measuring **s** units. The units of **s** may be any length unit (such as inches, feet, centimeters, yards, meters, miles, kilometers, light-years, etc...), and the corresponding perimeter will be in the same units as that of the **s** units since multiplying **s** (a number with a unit dimension) times a dimensionless number (in this case the **4**) does not alter the dimension of **s**, and hence, of the result of **4s**.

What I want you to think about now is this: what **REAL** numbers do you think we may use as values for the variables **P** and, what's even more relevant, **s**? If you don't remember the exact classification of **REAL** numbers, let me copy the table we saw in **Chapter 1**:

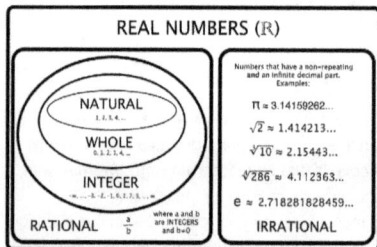

REAL NUMBERS ($\mathbb{R}$)

	Numbers that have a non-repeating and an infinite decimal part. Examples:
NATURAL 1, 2, 3, 4...	$\pi = 3.14159262...$
WHOLE 0, 1, 2, 3, 4...	$\sqrt{2} \approx 1.414213...$
INTEGER ...-4, -3, -2, -1, 0, 1, 2, 3...	$\sqrt[3]{10} \approx 2.15443...$
RATIONAL $\frac{a}{b}$ where a and b are INTEGERS and b≠0	$\sqrt[4]{286} \approx 4.112363...$
	$e \approx 2.718281828459...$
	IRRATIONAL

For example, **s** may be equal to, say, **5 inches**, or **2.33 feet**, or $\sqrt{2}$ **miles** , or **200 light years**, or even **0 any-units**. So I imagine that at this point you are realizing that **s** can be any positive **REAL** number or **0**. But can it really be **0**? You would think so, since saying that $s = 0$ (say, feet) leads to the following set up:

$$P = 4s$$

$$P = 4(0 \; feet)$$

$$P = 0 \; feet$$

Makes sense, doesn't it? if a square has a side that measures **0 feet** its perimeter will be **0 feet**. But wait. does the square exist then? Read the first part carefully: "if *a square has a side* that measures **0 feet**..."; doesn't that seem a bit odd? If the square exists, it must have a side greater than **0 feet**; if it has a side measuring **0 feet**, then the square simply doesn't exist, so what side and what square would we be talking about? Matters turn philosophical very quickly, as in "a square with a side measuring **0 any-units** does not exist; therefore, we can't talk about **the square** nor about **its side**, since neither exist...". Of course, we could simply say that **0** is a special case, that where the square disappears, and that the perimeter formula, when equal to **0**, conveys precisely that information.

But now we can explore from a mathematical standpoint which **REAL** numbers would be allowed as replacement values for **s**. And why **s** and not **P**? Because **s** would be considered the independent variable (or input variable, which is what the *domain* relates to), and **P** the dependent variable (or output variable, which is what the *range* relates to). Another clue as to why we are looking at input values for **s** is the fact that the equation is *solved for* **P**, which means that the formula is meant to calculate **P** *based on* **s**.

Now, strictly speaking, the perimeter formula would seem to allow not only positives and **0** as values for **s**, but also negative numbers. Did you see that coming? What are your thoughts? Can negative numbers be legitimately used as values assigned to a given **s**? Think about it... a *negative side measurement* for a square? Well, it does seem strange, but believe me, read enough papers on cutting-edge physics theories which attempt to explain the behavior of both the large and the small–general relativity and quantum physics respectively–and you will soon find that there may be scenarios in which we may consider that to make sense. But beyond the *meaning* of it all or the interpretation of a negative side measurement, mathematically speaking we do not encounter any forbidden set ups. Remember those? They are, up to this point in the book:

$$\frac{anything}{0} = UNDEFINED$$

$$0^0 = UNDEFINED$$

$$\sqrt[even]{negative} = NOT\ a\ REAL\ number$$

Although all of these are equally prohibited (the third case on this list when working with **REAL** numbers exclusively, as we are doing until further notice), it is the first set up that I would like to focus on for the time being.

So what is the relevance of this to the concept of solving literal equations? Well, let's go back to our perimeter equation, $P = 4s$ and solve for the **4** element this time. Seems a bit strange to do so, but there is nothing that prevents us from doing that. Observe:

$$P = 4s$$

Since I want to solve for **4**, I will try to isolate it; dividing both sides by **s** will achieve it:

$$\frac{P}{s} = \frac{4s}{s}$$

Simplify...

$$\frac{P}{s} = 4$$

And we are done. If you think about it, this simply states that whenever you have a square, dividing its perimeter by the measurement of its side will always yield **4**. This is called a *ratio*, by the way, as we saw in **Chapter 1** and in the chapter on fractions.

Now, look at the equation $\frac{P}{s} = 4$, and answer this question: must we exclude one or more **REAL** numbers when specifying the domain of this equation, based on the variable **s**? Remember the forbidden set ups as you try to answer this question.

If you answered along the lines of "...Well, you should never be allowed to replace **s** with **0**...", then you are absolutely correct. Why? Because if we were to replace the variables **s** with **0** observe what would happen:

$$\frac{P}{s} = 4$$

Since $s = 0$ ...

$$\frac{P}{0} = 4 \; ?$$

We know, of course, that this is completely false. **P** divided by zero is UNDEFINED; it will never be equal to **4**. So we must exclude **0** from the set of **REAL** numbers that may be used as input values for **s**; otherwise, we would end up with a statement that is mathematically invalid (UNDEFINED).

Well, this is what is called the *domain* of an equation or of a function. The domain tells us specifically which numbers may be used to replace specific variables (in other words, as input values) on given equations and functions, or in statements in general (I will go over functions in an upcoming chapter of the book, very relevant to the formal definition of domain, since strictly speaking, modern mathematical theory defines domain in terms of sets of ordered pairs which are how functions are built). In other words, the domain specifies the allowed *input* values. It is, of course, necessary to know which variable we are considering as the input variable (it may, of course, also be plural, as in input *variables*).

I could therefore provide the equation $\frac{P}{s} = 4$ and define its domain with respect to **s** as all the **REAL** numbers other than **0**, or, if I wish to exclude negative numbers (keep the length measurements positive) then I would define its domain with respect to **s** as all the **REAL** numbers *greater* than **0**.

We will look at the specific notation that is typically used to communicate domain when I go over functions. However, at this point in the book, you should be comfortable explaining the concept of the domain of an equation with respect to a given variable (the input variable, or independent variable).

The *range*, on the other hand, specifies the output, or what the equation or function may be equal to, which will be a direct consequence of the different **REAL** numbers that may be used as input values (in other words, the domain, remember?).

For example, in the equation $P = 4s$, if we were to define the domain–in terms of the variable **s**– as all the **REAL** numbers (allowing negative length measurements, whatever interpretation we may wish to give to them, as well allowing the

REAL number **0**), then the equation's *range* would be all **REAL** numbers as well; if, however, we defined its domain as all **REAL** numbers greater than or equal to **0**,(excluding negative numbers in an effort to avoid having to interpret a negative length measurement), then the range would be all **REAL** numbers greater than or equal to **0** as well. Observe:

Case 1.

$$P = 4s$$

Domain: All **REAL** numbers

Then it follows that its range is all **REAL** numbers

Why? Because replacing **s** with a **REAL** number that can be anything between **negative infinity** to **positive infinity**, including **0**, would result in a perimeter value, **P**, that could be anything between **negative infinity** and **positive infinity**, including **0**, as well. Observe these input/output examples.

$$P = 4s$$

If $s = -3$
$$P = 4(-3)$$
$$P = -12$$

If $s = -10.5$
$$P = 4(-10.5)$$
$$P = -42$$

If $s = -300,000$
$$P = 4(-300,000)$$
$$P = -1,200,000$$

And so the less s is, the less P will be...

See how the perimeter value, **P**, decreases as **s** decreases as well? We could keep going, replacing **s** with **REAL** numbers that are successively more to the left of **0** on the number line, tending towards negative infinity, which would cause P to decrease as well, also tending towards negative infinity, as the following diagram illustrates:

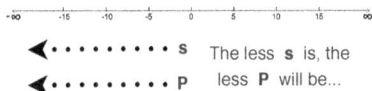

The less **s** is, the less **P** will be...

You can probably get a sense that if we may replace **s** with *any* negative **REAL** number, then the perimeter **P** may end up being equal to *any* negative number as well. Therefore, the output, or *range*, must include all negative **REAL** numbers.

Now, let's explore what happens when **s** is equal to **0**:

$$\text{If } s = 0$$
$$P = 4(0)$$
$$P = 0$$

We obtain a mathematically valid answer. Therefore, it should be clear at this point that the perimeter value, **P**, can be anything between negative infinity all the way to **0**, which means the range must span all **REAL** numbers from negative infinity to **0** as well (at this point of our analysis).

And if we start exploring positive **REAL** numbers for **s**? What kind of values do we get for the output? Well, let's explore the perimeter value (and hence the range behavior) when we use positive **s** values, increasingly greater:

$$P = 4s$$

$$\text{If } s = 1$$
$$P = 4(1)$$
$$P = 4$$

$$\text{If } s = 320$$
$$P = 4(320)$$
$$P = 1{,}280$$

$$\text{If } s = 2{,}000{,}000$$
$$P = 4(2{,}000{,}000)$$
$$P = 8{,}000{,}000$$

See how the perimeter value, **P**, increases as the value of **s** increases as well? The following diagram illustrates this tendency:

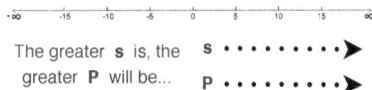

The greater **s** is, the greater **P** will be...

And so we can conclude that for the equation given, based on the variable **S**, the following is true:

$$P = 4s$$

Domain: All **REAL** numbers
Range: All **REAL** numbers

Please note that this doesn't always happen–that the domain and range are the same, that is. A careful analysis must be made in order to determine the domain of an equation, and separately, of the range of the equation. Although the domain has an impact on the range, they must be independently evaluated. Typically, it is much easier to define the domain than it is to define the range.

To finish the example we have been working with here, if we took the same equation $P = 4s$ and this time decided not to allow negative numbers as input values for **s**, the following would occur:

Case 2.

$$P = 4s$$

Domain: All **REAL** numbers greater than or equal to **0**

Then it follows that its range is all **REAL** numbers greater than or equal to **0**

It is easy to see that if **s** cannot be equal to a negative **REAL** number, then the perimeter **P** will never be negative either. Therefore, the range (output) had to be defined as above.

The obvious question at this point is: who determines the domain? Since the domain plays a key role in determining the range, it is important to ask this question so that we may determine the equation's valid input values (domain) and therefore what its possible output values will be (range).

The answer is that there isn't one simple answer. It all depends. First and foremost, however, the domain **can never include** a number that causes the equation to be undefined (as in the equation $\frac{P}{s} = 4$ in which case we said that **s** may never

be equal to **0**). Therefore, when defining the domain, one should be careful to exclude input values that would cause such states. Beyond this, it is the equation's intended use that may be a factor. If, within the context of this use, certain **REAL** numbers lack meaning, then they may be excluded from the domain for that reason and the person writing the equation (or function) would need to specify the desired domain.

Generally, we treat equations (and later on, functions), from a purely mathematical standpoint, and we generally consider all **REAL** numbers in our analysis, irrelevant of interpretation. When proceeding in this fashion, we look for numbers that lead to UNDEFINED outputs, and exclude those numbers from the domain.

As a second step, **after** defining the domain, it would be necessary to check which **REAL** numbers need to be excluded from the range as a result of having specified said domain.

To practice this, let's first focus on the domain of an equation. We will practice the concept of range when I review functions in the next chapter of the book.

Specify the **domain** of the following equations.

a) $I = \dfrac{100}{R}$; use **R** for the domain.

b) $F = 200m$; use **m** for the domain.

c) $S = \dfrac{1000}{t}$; use **t** for the domain.

d) $F = \dfrac{9C}{5} + 32$; use **C** for the domain.

Try to define the domain of these equations based on the variable indicated to their right.

Ready?

a) $I = \dfrac{100}{R}$; use **R** for the domain.

Looking at the right side of the equation, it should be easy to see that the only **REAL** number that we must exclude from the domain is **0**, since using that input value would result in $I = \dfrac{100}{0}$, an output value that is UNDEFINED.

Therefore, we would state the following:

$$I = \dfrac{100}{R}$$

Domain: all **REAL** numbers except **0**.

b) $F = 200m$; use **m** for the domain.

Looking at the right side of the equation, it should be easy to see that we do not need to exclude any **REAL** numbers from the domain.

Therefore, we would state the following:

$$F = 200m$$

Domain: all **REAL** numbers.

c) $S = \dfrac{1000}{t}$; use **t** for the domain.

It should be easy to see that the only **REAL** number that we must exclude from the domain based on the variable **t** is **0**, since using that input value would result in $S = \dfrac{1000}{0}$, an output value that is UNDEFINED.

Therefore, we would state the following:

$$S = \dfrac{1000}{t}$$

Domain: all **REAL** numbers except **0**.

d) $F = \dfrac{9C}{5} + 32$; use **C** for the domain.

Looking at the right side of this equation, we can see that since **C** in the numerator of the fraction $\dfrac{9C}{5}$, we will never encounter an UNDEFINED set up, regardless of the input value used for **C**. Therefore, we would state the following:

$$F = \dfrac{9C}{5} + 32$$

Domain: all **REAL** numbers.

At this point, it should be clearly understood why the concept of domain is relevant when working with literal equations. We may end up solving an equation for a given variable, and if we are not careful, the newly defined equation may have an UNDEFINED output value for a given input (or set of inputs). Observe the following example.

1) What is the domain of $F = ma$ with respect to m ? Solve for a; what is the domain of this newly-solved equation, also with respect to m ?

First, we can clearly see that the original equation, $F = ma$, will NOT produce any UNDEFINED outputs with respect to m since the variable is NOT in the denominator, and since it is being raised to the power of 1 (remember?). Therefore, we can state the following:

$$F = ma$$

Domain: all **REAL** numbers with respect to **m**

Next, we solve for **a**:

$$F = ma$$

Divide both sides by **m** to isolate **a**...

$$\frac{F}{m} = \frac{ma}{m}$$

Simplify...

$$\frac{F}{m} = a$$

And we are done solving the literal equation for the variable **a**.

Next, we are asked to define the domain of this newly-found equation, with respect to **m**. Notice how this time we would encounter a problem if we did not restrict the domain of the equation with respect to **m**. If we were to replace **m** with **0**, we would end up with an UNDEFINED output. Therefore, we must exclude **0** from the domain of this newly-found equation. And so we may conclude as follows:

$$\frac{F}{m} = a$$

Domain: all **REAL** numbers except **0**

To end the chapter, I would like to point out the key concepts that we reviewed.

First, we learned that linear equations (equations that have single-variable terms, with the variables themselves raised to the power of **1**)

In this chapter, we learned the following very important things about working with equations, specifically, with **_linear equations_**:

1) **Linear Equations** of one variable can have one answer, an infinite number of answers, or zero answers. It is important to know how to detect each of these possibilities; simply solve the equation for the unknown variable, and match the type of answer that you obtain to one of the three that we defined earlier in the chapter.

2) Solving **Literal Equations** involves solving equations that are typically defined in terms of more than one variable, for one of the variables that the equation is not solved for. As an example, take the formula for the area of a rectangle, usually presented as $A = lw$ (area, **A**, is equal to length, **l**, times width, **w**). If you were asked to solve the formula for **w**, you are being asked to solve a literal equation for said variable.

3) The **Domain** of an equation (or of a function) specifies the numbers that may be used as input values, with respect, of course, to the input variable. The domain prevents the equation from providing an UNDEFINED output, such as when division by zero is encountered.

4) The **Range** of an equation (or of a function) specifies the possible output values that the equation (or function) may be able to provide. The domain has a direct impact on the range.

We are now in a position to fully explore linear equations that have more than one variable. Remember the prompt that started out this chapter? Let's revisit the prompt, and see what we can say about it now, and how we can explore the concept of "systems of linear equations" and how to solve them.

The equation is the following:

$$2x + 3y = 23$$

As we saw earlier, we found a couple of pairs of values for x and y that would solve this equation. If we use $x = 10$ and $y = 1$, the equation would be correctly solved, since the left side would be equal to 23, the same as the right side. Likewise, if we use $x = 4$ and $y = 5$, the equation would also be solved, since the left side of the equation would again be equal to 23 :

$$2x + 3y = 23$$

If $x = 10$ and $y = 1$

$$2(10) + 3(1) \overset{?}{=} 23$$

$$20 + 3 \overset{?}{=} 23$$

$$23 \overset{\checkmark}{=} 23$$

$$2x + 3y = 23$$

If $x = 4$ and $y = 5$

$$2(4) + 3(5) \overset{?}{=} 23$$

$$8 + 15 \overset{?}{=} 23$$

$$23 \overset{\checkmark}{=} 23$$

So we have two pairs of values that solve the equation. And as I said at the beginning of this chapter, this equation has an infinite number of pairs of values that solve it.

Typically, however, we have what is called a **system of linear equations**. In these systems, we have more than one equation, and we need to determine if there are any **REAL** numbers (or whatever the domain may be) that may be used as inputs for the variables and that would simultaneously solve all of them.

Let's explore this idea with an example.

Solve the following system of linear equations:

$$\left.\begin{array}{l} \text{①} \quad y = 2x + 1 \\ \text{②} \quad y = 6x - 7 \end{array}\right\}$$

Typically, a system of linear equations will consist of two or more linear equations that are grouped using a particular type of bracket: **braces** or **curly brackets** (as above); it may appear to the left or to the right of the equations. Normally, the equations are numbered, so that we may refer to them clearly and unequivocally, which is why you see the circled numbers to the left of each equation above; although this is not absolutely necessary, it certainly makes the solving process easier to follow.

As you can see, we now have two linear equations, not just one as we had before. So this time, if we tried to find a pair of x and y values that solve equation **(1)**, said solution pair may or may not be a solution pair of equation **(2)**.

To illustrate this point, let's first work with equation **(1)**. Can you identify a solution pair? I will suggest we use $x = 1$ and $y = 3$ (which is, in fact, a solution pair of the first equation). Observe how the pair does, in fact, work:

$$\text{①} \quad y = 2x + 1$$

Using $x = 1$ and $y = 3$...

$$(3) \overset{?}{=} 2(1) + 1$$

$$3 \overset{?}{=} 2 + 1$$

$$3 \overset{\checkmark}{=} 3$$

At this point, you would probably think that since this solution pair works on equation **(1)** that it would also work on equation **(2)**. Let's put that theory to the test.

$$②\ \ y = 6x - 7$$

Using $x = 1$ and $y = 3$...

$$(3) \overset{?}{=} 6(1) - 7$$

$$3 \overset{?}{=} 6 - 7$$

$$?\ 3 \overset{?}{=} -1\ ?$$

Do you see how the numbers we assigned to **x** and to **y** are, in fact, a solution pair for the first equation but NOT for equation **(2)**? Likewise, we may find a solution pair that works on equation **(2)** but not on equation **(1)**; in fact, each of these equations independently posses an infinite number of solution pairs. What this means is that if we want to solve a **system of linear equations**, we are trying to determine if the system has one solution for all equations involved (and if so, which solution it is), infinite solutions (and if so, how we may determine the solution pairs), or zero solutions, just like we determined this for linear equations with one variable earlier in this chapter.

There are multiple techniques that may be used to solve systems of linear equations. However,

the most common (and those that we will focus on in subsequent chapters) are the following:

1) Substitution Method

2) Elimination Method

3) Graphing Method

A fourth method, which consists on using **matrices**, is beyond the scope of this book.

Let's begin with the substitution method.

Solving Linear Equations with the Substitution Method

Let's bring back the system we are trying to solve:

$$\left.\begin{array}{l} ①\ \ y = 2x + 1 \\ ②\ \ y = 6x - 7 \end{array}\right\}$$

When using the substitution method, the first step is to solve one of the equations for one of the unknown variables (in this case, we can choose to solve for **x** or for **y**) and then to replace the variable we solved for, on the other equation. That will typically yield a numeric answer for one of the variables in the case of a system of linear equations of two variables (the reason I say typically is because the system may have infinite answers or zero answers; we will explore how to detect this later). The second step involves finding the value for the remaining variable; this involves replacing the variable with the numeric value found in the first step in either one of the two equations, and solving for said remaining variable. A numeric value is typically found for the remaining variable at this point. The two values, collectively, represent the solution pair to the system of equations (in the case of two linear equations with two variables).

Since these equations are already solved for the variable y, we can use either one of the two equations. If we were to use, say, the first equation, we are now able to state that y is equal to $2x + 1$ and we can move on to the second step. Since we used the first equation on the first part of this step, we must now replace the y variable of the second equation with $2x + 1$, since that is what equation (1) states:

$$y = 2x + 1$$

Replacing y with $2x + 1$ into (2)

$$2x + 1 = 6x - 7$$

Our goal at this point is to solve for x which means I will collect all x terms on one side of the equation. In this case, I will subtract $2x$ from both sides of the equation, so that the x terms end up on the right side...

$$2x + 1 - 2x = 6x - 7 - 2x$$

$$2x - 2x + 1 = 6x - 2x - 7$$

$$0 + 1 = 6x - 2x - 7$$

$$1 = 4x - 7$$

Next, since there is a **subtracting** 7 on the right side, I will **add** 7 to both sides in order to isolate the x term on the right side...

$$1 + 7 = 4x - 7 + 7$$

$$8 = 4x$$

As a final part of this step, I will divide both sides by **4** to remove the **4** that is multiplying the variable x...

$$\frac{8}{4} = \frac{4x}{4}$$

$$2 = x$$

Or, if you prefer, switch left side/right side:

$$x = 2$$

And the first step of the process is complete. We now know that the solution to this system must have x equal to the **REAL** number **2**.

Moving on to the second step of the substitution method, we now need to replace the variable x with **2** in either one of the two equations that are part of the system. It does not matter which equation is used, as long as the equation contains both variables. We will explore other cases shortly.

Since equation **(1)** is easier to use than equation **(2)**, I will replace x with **2** into the first equation. Observe.

(1) $y = 2x + 1$

Replacing x with **2**

$$y = 2(2) + 1$$

$$y = 4 + 1$$

$$y = 5$$

And the system is solved! The paired values $x = 2$ and $y = 5$ represent the **only** solution to this system of two linear equations of two variables.

To prove it, we simply need to replace the variables of **both** equations with the numeric values (**REAL** numbers) found, and verify that the equations remain valid (that the equality remains unbroken). Observe.

$$\left.\begin{array}{l} ① \quad y = 2x + 1 \\ ② \quad y = 6x - 7 \end{array}\right\} \quad \begin{array}{l} x = 2 \\ y = 5 \end{array}$$

Replacing $x = 2$ and $y = 5$ into equation **(1)**, we check to see if the equality is maintained:

$$y = 2x + 1$$

$$5 \overset{?}{=} 2(2) + 1$$

$$5 \overset{\checkmark}{=} 5$$

As you can see, the equality is, in fact, maintained. Therefore, this solution pair is a solution to equation **(1)** of this system. However, we still have to check if the equality is maintained in equation **(2)** after replacing the variables with the solution pair, since a solution pair that is a solution to one of the equations is not necessarily going to be a solution pair of the other equation in the system. Let's check the other equation:

Replacing $x = 2$ and $y = 5$ into equation **(2)**, we check to see if the equality is maintained:

$$y = 6x - 7$$

$$5 \overset{?}{=} 6(2) - 7$$

$$5 \overset{?}{=} 12 - 7$$

$$5 \overset{\checkmark}{=} 5$$

The equality is maintained.

Now, we are in a position to unequivocally state that the system of two linear equations with two variables has **one** solution, and it is $x = 2$ and $y = 5$. We did not encounter any undefined set ups, nor any invalid inequalities along the way; we will see other examples later on where these do occur, so that you are prepared to deal with them.

Try to solve the following system of two linear equations with two variables by yourself.

$$\left.\begin{array}{l} ① \quad y = 5x + 3 \\ ② \quad y = 2x + 6 \end{array}\right\}$$

Remember the two steps involved in the process:

Step 1. Solve one of the equations for one of the variables. If one of the equations is already solved for one of the variables, you may use that variable (saving you this part of the step). Then, replace this variable on the **other** equation with what you found the variable is equal to from the first part above. You will normally find a numeric value for one of the variables in this second part of the step.

Step 2. Replace in either of the equations that are part of the system the variable with the numeric value (**REAL** number) found in the first step. This will normally provide the numeric value (**REAL** number) of the other unknown variable.

Using the values found, you may define the solution of the system of equations.

Where you able to solve it? What is the solution pair for this system? You should have found the following answer:

$$\left.\begin{array}{l} ① \; y = 5x + 3 \\ ② \; y = 2x + 6 \end{array}\right\} \quad \begin{array}{l} x = 1 \\ y = 8 \end{array}$$

Observe my solution process.

Step 1. I will use equation **(1)** as my starting point. I now know that y is equal to $5x + 3$. Now, I replace y in equation **(2)** with $5x + 3$ and I solve for x as follows:

$$② \quad y = 2x + 6$$

Replacing y with $5x + 3$

$$5x + 3 = 2x + 6$$

$$5x + 3 - 2x = 2x + 6 - 2x$$

$$5x - 2x + 3 = 2x - 2x + 6$$

$$3x + 3 = 0 + 6$$

$$3x + 3 = 6$$

$$3x + 3 - 3 = 6 - 3$$

$$3x = 3$$

$$\frac{3x}{3} = \frac{3}{3}$$

$$x = 1$$

Step **1)** is now complete. I know that $x = 1$ is part of the solution to this system of equations.

Step 2. I will now replace $x = 1$ into either one of the two equations that are part of this system. I will choose equation **(2)** since it is the easiest of the two to work with. As part of this step, I will solve for the remaining unknown variable, in this case, the variable y. Observe.

$$② \quad y = 2x + 6$$

Replacing x with ①

$$y = 2(1) + 1$$

$$y = 2 + 1$$

$$y = 3$$

And the system is now fully solved. The pair of values $x = 1$ and $y = 3$ forms the solution of this system of two linear equations with two variables.

I will not prove it, but you may want to check that the solution is, in fact, correct. Simply replace the variables with the numeric values found above into each of the two equations, and verify that the equality is maintained on **both** equations. If the solution does not work on both, then it cannot be a solution to the **system of equations**.

Let's try to solve another system of linear equations. This time, however, the equations will not be solved for one of the variables, so you must work on that on the first step of the process.

Try this by yourself before looking at my solution.

$$\left.\begin{array}{l} ① \; 2y - 1 = 2x + 3 \\ ② \; 3y + 10 = 12x - 11 \end{array}\right\}$$

Of course, our solution process may differ, but I encourage you to try to solve it by yourself regardless of how I may choose to solve it.

If you would prefer to solve the system using the steps I am going to use so that you may compare your work directly, then I can tell you that in Step 1, I will solve equation **(1)** for **y**, replace this into equation **(2)** to find the numeric value of variable **x**; then, on the second step, I will replace the **x** value found in Step 1 into equation **(1)**. However, note that there will be two versions of equation **(1)**: the original version as specified in the system of equations and the one that is obtained when solved for **y** in Step 1 of the solving process. Since we need to determine the numeric value of **y** it is much easier to use the latter version than to use the former and have to solve for **y** all over again.

Ready to check your answer?

The solution is:

$$x = 3 \text{ and } y = 5$$

Observe my solution process.

Step 1. I will solve equation **(1)** for **y**, and then replace **y** in equation **(2)**.

$$①\ 2y - 1 = 2x + 3$$

$$2y - 1 + 1 = 2x + 3 + 1$$

$$2y = 2x + 4$$

$$\frac{2y}{2} = \frac{2x + 4}{2}$$

$$y = \frac{2x}{2} + \frac{4}{2}$$

$$y = x + 2$$

Now, I will replace **y** in equation **(2)** with $x + 2$ and solve for the other variable (**x** in this case)...

$$②\ 3y + 10 = 12x - 11$$

$$\boxed{y = x + 2}$$

$$3(x + 2) + 10 = 12x - 11$$

$$3x + 6 + 10 = 12x - 11$$

$$3x + 16 - 3x = 12x - 11 - 3x$$

$$3x - 3x + 16 = 12x - 3x - 11$$

$$0 + 16 = 9x - 11$$

$$16 + 11 = 9x - 11 + 11$$

$$27 = 9x$$

$$\frac{27}{9} = \frac{9x}{9}$$

$$3 = x$$

The first step is complete. We know that $x = 3$ is part of the solution of this system of two linear equations with two variables. All we need now is the value for **y**.

Step 2. I will now replace $x = 3$ into either one of the two equations that are part of this system. I will choose equation **(1)** but not the original version of equation **(1)** since I would need to solve for **y**. A version of this same equation has already been solved for **y** in Step 1, so I will use that one instead. Observe.

$$y = x + 2$$

Replacing **x** with **3**

$$y = (3) + 2$$

$$y = 3 + 2$$

$$y = 5$$

System solved. The solution is $x = 3$ and $y = 5$ and it is the **only** solution pair of this system of linear equations.

Let's try another system of linear equations. This time, I will involve fractions as part of the solution process. It is important to keep your fraction skills at an optimal level.

$$\left.\begin{array}{l} ①\quad 2y + 5 = -3x + 1 \\ ②\quad -4y - 2 = 2x - 6 \end{array}\right\}$$

I suggest you attempt to solve it by yourself.

If you want to follow my solution process, I will be solving equation **(1)** for **y** and then replacing **y** from equation **(2)** in order to find **x**.

Ready to check your work?

The correct answer to this system of equations is $x = -3$ and $y = \dfrac{5}{2}$. If you did not obtain this answer, you may want to revisit your solving process and try to identify your mistake (or mistakes) before looking at my solution process.

Ready? Let me show you the steps I used to solve the system.

Step 1. I will solve equation **(1)** for **y**, and then replace the **y** from equation **(2)**.

$$①\quad 2y + 5 = -3x + 1$$

$$2y + 5 - 5 = -3x + 1 - 5$$

$$2y = -3x - 4$$

$$\frac{2y}{2} = \frac{-3x - 4}{2}$$

$$y = \frac{-3x}{2} - \frac{4}{2}$$

$$y = \frac{-3x}{2} - 2$$

Now, I will replace **y** in equation **(2)** with $\dfrac{-3x}{2} - 2$ and solve for the other variable (**x** in this case)...

$$②\quad -4y - 2 = 2x - 6$$

$$\boxed{\frac{-3x}{2} - 2}$$

$$-4(\frac{-3x}{2} - 2) - 2 = 2x - 6$$

Remember that the **−4** is multiplying everything inside the set of parentheses; therefore, we need to multiply **−4** with $\dfrac{-3x}{2}$ and with **−2**, as follows (in other words, distribute):

$$\frac{(-4)(-3x)}{(1)(2)} - (-4)(2) - 2 = 2x - 6$$

The **−4** only multiplies the numerator of $\dfrac{-3x}{2}$

because when you multiply a non-fraction (in this case the **-4**) times a fraction, we can convert the non-fraction into a fraction ($\frac{-4}{1}$) and then multiply the fractions straight across (numerator with numerator, denominator with denominator), as we reviewed in **Chapter 8**. Now, we can continue simplifying, combining like terms, etc...

$$\frac{(-4)(-3x)}{(1)(2)} - (-4)(2) - 2 = 2x - 6$$

$$\frac{12x}{2} - (-8) - 2 = 2x - 6$$

$$6x + 8 - 2 = 2x - 6$$

$$6x + 6 = 2x - 6$$

$$6x + 6 - 2x = 2x - 6 - 2x$$

$$6x - 2x + 6 = 2x - 2x - 6$$

$$4x + 6 = 0 - 6$$

$$4x + 6 = -6$$

$$4x + 6 - 6 = -6 - 6$$

$$4x = -12$$

$$\frac{4x}{4} = \frac{-12}{4}$$

$$x = -3$$

The first step is complete. We know that $x = -3$ is part of the solution of this system of two linear equations with two variables. All we need now is the value for **y**.

Step 2. I will now replace $x = -3$ into either one of the two equations that are part of this system. I will choose equation **(1)** but not the original version of equation **(1)** since I would need to solve it for **y** all over again. A version of this same equation has already been solved for **y** in Step 1, so I will use that one instead. Observe.

$$y = \frac{-3x}{2} - 2$$

Replacing **x** with **-3**

$$y = \frac{-3(-3)}{2} - 2$$

$$y = \frac{9}{2} - 2$$

$$y = \frac{9}{2} - \frac{2}{1}$$

$$y = \frac{9}{2} - \frac{4}{2}$$

$$y = \frac{5}{2}$$

System solved. The solution is $x = -3$ and $y = \frac{5}{2}$ and it is the **only** solution pair of this system of linear equations.

The complexity of solving this specific system lies in the correct application of the mathematical rules and properties that we have reviewed throughout the book: distributing, working with fractions, solving for a variable, etc... The steps involved in solving a system of linear equations using the substitution method is, itself, relatively simple.

I would now like to show you what happens when a system of linear equations has infinite solutions or ZERO solutions.

Let's work with the following system:

$$\left.\begin{array}{l} \text{(1)} \quad y = 4x + 6 \\ \text{(2)} \quad 2y + 2 = 8x + 14 \end{array}\right\}$$

I suggest we work on this problem together. Since equation **(1)** is already solved for y I can replace y in equation (2) with $4x + 6$, as follows:

$$2y + 2 = 8x + 14$$

$$\boxed{y = 4x + 6}$$

$$2(4x + 6) + 2 = 8x + 14$$

$$8x + 12 + 2 = 8x + 14$$

$$8x + 14 = 8x + 14$$

At this point, this set up should ring some alarm bells; if not outright alarm bells, you should at least find this familiar (do you know why?). I will continue simplifying so you can see what I mean:

$$8x + 14 = 8x + 14$$

I will attempt to collect the **x** terms on the left side; to do that, I will subtract $8x$ to both sides of the equation in order to eliminate the $8x$ from the right side of the equation...

$$8x + 14 - 8x = 8x + 14 - 8x$$

$$8x - 8x + 14 = 8x - 8x + 14$$

$$0 + 14 = 0 + 14$$

$$14 = 14$$

At this point, the variable that we were trying to solve for has disappeared. We have already encountered this situation when working with linear equations of one variable, so you should know what we can conclude: since the statement is true, (**14** is equal to **14**), we can say that this system has an infinite number of solutions. This happens because both equations are **the same**.

If at this point you are wondering if I am reading the system correctly, allow me to prove it to you:

Equation **(1)** is: $y = 4x + 6$

Equation **(2)** is: $2y + 2 = 8x + 14$

To prove it, I will solve equation **(2)** for **y** so that we may compare them directly.

$$2y + 2 - 2 = 8x + 14 - 2$$

$$2y = 8x + 12$$

$$\frac{2y}{2} = \frac{8x + 12}{2}$$

$$y = 4x + 6$$

See how equation **(2)** is actually the same as equation **(1)**? And since they are the same (equation **(2)** was simply **presented differently** in the system of equations, "hiding" the fact that it was the same equation as **(1)**...), whatever solution pair works for equation **(1)** will also, automatically, work for equation **(2)**.

Furthermore, since we already said that a given linear equation always has an infinite amount of solutions (pairs, if it is a two variable linear equation; triplets, if three; quadruplets, if four, etc...), then it follows that a system that consists of linear equations that are exactly the same will also have an infinite amount of solutions.

And what are those solutions? Well, those solution pairs that work for $y = 4x + 6$ or for $2y + 2 = 8x + 14$ or for any of the intermediate versions that we found as we solved the latter for **y** (as in $2y = 8x + 12$, etc...).

And how do we find the solution pairs? Simple. Since equation **(1)** and **(2)** are the same (the latter in any of its versions), we can use equation **(1)**, and solve it for one of the variables. In this system, this is already the case (the equation is solved for **y**). Next, we can create a table that includes as many **input values** as we care to include (in this case, they would correspond to the variable **x** since that would be considered the independent variable), replace **x** with the number in turn, solve to find **y**, and explicitly write the solution pair. For example:

Input Number x	Evaluate y=4x+6	Solution pair
−2	y=4(-2)+6=-2	x= −2 , y= −2
−1	y=4(-1)+6=2	x= −1 , y= 2
0	y=4(0)+6=6	x= 0 , y= 6
1	y=4(1)+6=10	x= 1 , y= 10
2	y=4(2)+6=14	x= 2 , y= 14

The above table is specifying part of what is called **the solution set** of the system of equations, since it lists individual solution pairs (one pair is $x = -1$ and $y = 2$, for example; another solution pair would be $x = 0$ and $y = 6$, etc...).

And so we are finally in a position to know what solves the system of linear equations under discussion:

$$\left. \begin{array}{l} \text{①} \quad y = 4x + 6 \\ \text{②} \quad 2y + 2 = 8x + 14 \end{array} \right\}$$

The answer, of course, is that it has an **INFINITE** number of solutions, and that the solution pairs are those that solve either one of the equations (since they are in essence the same, it is irrelevant which one is used to specify this, although equation **(1)** would be the logical choice since it is simplified and already solved for one of the unknown variables). It may be expressed as follows:

This system has an infinite number of solutions. The solution pairs are all the solution pairs of
$$y = 4x + 6$$

It is possible that we will have to check if a given pair of values is, in fact, part of the solution set of a system of equations. For example, on this particular system of equations we are currently working with, can you determine if $x = 5$ and $y = 16$ form a solution pair of said system?

As you may recall, earlier in this chapter I stated that there is a technique that allows us to check if a solution is correct. It involves replacing both variables with the values defined in the solution pair, and checking to see if the equality is maintained.

Using equation **(1)**, and replacing the variables with the values specified, we can now attempt to answer the question. Observe:

$$y = 4x + 6$$

Replacing $x = 5$ and $y = 16$

$$(16) \overset{?}{=} 4(5) + 6$$

$$16 \overset{?}{=} 20 + 6$$

$$?\,16 = 26\,?$$

Since the equality is broken we can conclude that the **x** and the **y** values given do not form a solution pair to this system of equations. Had the equality remained intact, we would have concluded otherwise.

At this point, you may be wondering how we can identify the no-solution case (systems of linear equations that have ZERO solutions). Let's explore this next system to discover what that looks like (although if you remember how we were able to determine this for linear equations with one variable, you should have a very good idea of what we are about to encounter).

$$\left. \begin{array}{l} ①\quad y = -3x + 1 \\ ②\quad y + 5x = 2x + 4 \end{array} \right\}$$

Try to solve this system by yourself. Keep your eyes open for a (what should be) familiar set up.

Did you try it? What answer did you obtain? Was the answer reasonable or valid?

Observe my solution process.

Step 1. I would need to solve one of the equations for one of the unknown variables. However, since equation **(1)** is already solved for **y**, I can simply replace the **y** from equation **(2)** with $-3x + 1$:

$$② \quad y + 5x = 2x + 4$$

Equation **(1)** states that $y = \boxed{-3x + 1}$

$$(-3x + 1) + 5x = 2x + 4$$

$$-3x + 1 + 5x = 2x + 4$$

$$-3x + 5x + 1 = 2x + 4$$

$$2x + 1 = 2x + 4$$

Do you notice anything strange at this point? Think about what this equation is **saying**... Is it possible to plug in an REAL number that would maintain the equality? Let's continue...

$$2x + 1 = 2x + 4$$

$$2x + 1 - 2x = 2x + 4 - 2x$$

$$2x - 2x + 1 = 2x - 2x + 4$$

$$0 + 1 = 0 + 4$$

$$?\,1 \overset{?}{\neq} 4\,?$$

Now do you see what's wrong about this equation? The number **1** is, of course, not equal to the number **4**. We could keep going, but the equality is broken and that is not going to change:

$$1 = 4$$

$$1 - 1 = 4 - 1$$

$$?\ 0 \overset{?}{\neq} 3\ ?$$

As you can see, this equation is NOT true. And since the variable we were hoping to solve for (in this case, *x*) is gone, we know we are dealing with the no-solution case.

We did not need to continue with the solving process once we encountered either one of the following set ups (since they break the equality):

$$2x + 1 = 2x + 4$$

or

$$1 = 4$$

Regardless of how far you have to go in the solving process in order to identify that the equality is broken, once you do identify that the equality is broken, you may stop and unequivocally conclude that you are dealing with a system of equations that DOES NOT have a solution, or the ZERO-solution case.

You can see that solving a linear equation of one variable and solving a system of equations of two or more variables entails a similar process; furthermore, the set ups that you may encounter along the way communicates the same thing in both cases: if you have a variable with a numeric value, the one solution case; otherwise, you have either the infinite solution case or the zero solution case, as the following table states.

⚓

Types of Systems of Linear Equations of Two or More Variables

Systems of linear equations may be classified as one of the following three types (the examples illustrate the kind of set ups that you may encounter during the solution process, based on the substitution method discussed in this section of the chapter):

Type I: Equations with **ONE** solution, as in: $x = 10$ and $y = 3$, or $-1 = w$ and $3 = y$ and $z = 2$, etc...

Type II: Equations with **INFINITE** solutions; $0 = 0$, $3 = 3$, $\sqrt{2} = \sqrt{2}$, etc... In all these set ups, the equality is not broken, but the variable that you were solving for did disappear.

Type III: Equations with **ZERO** solutions, as in (when partially or fully simplified): $0 = 4$, $-7 = 3$, $9 = 0$, etc... The equality is broken, and the variable that you were solving for has disappeared.

Before moving on, you need to be ready to solve systems of equations that although at first they may seem strange to work with, after reviewing them you should be able to see why they work out the way they do. Try to solve this system:

$$\left. \begin{array}{l} ① \quad 4x + 2y = 3x + 2 \\ ② \qquad x = 3 \end{array} \right\}$$

The first time you see a system like this, you may wonder if the system is missing something. After all, equation **(2)** looks strange, doesn't it?

There is nothing wrong with this system of linear equations. In fact, this system is particularly easy to solve because if you think about it, one of the variables already has a numeric value assigned to it, which is one of the main goals of solving a system of equations to begin with. Hence, when you focus on equation **(2)**, it may be interpreted as saying that x is equal to **2**, which will be part of the solution pair for this system.

This ultimately means that Step 1 of the solution process is done. All we really have to do is execute Step 2 of the solving process. Observe:

Step 2. I will now replace $x = 3$ into equation **(1)** and solve it for the remaining unknown variable, **y**. Observe.

$$4x + 2y = 3x + 5$$

Replacing **x** with **3**

$$4(3) + 2y = 3(3) + 5$$

$$12 + 2y = 9 + 5$$

$$12 + 2y = 14$$

$$12 + 2y - 12 = 14 - 12$$

$$12 - 12 + 2y = 14 - 12$$

$$0 + 2y = 2$$

$$2y = 2$$

$$\frac{2y}{2} = \frac{2}{2}$$

$$y = 1$$

The system is now solved, and given the type of answer found, we know this system has only one solution. The solution pair of this particular system is $x = 3$ and $y = 1$ which can be easily checked by replacing the variables **x** and **y** on equation **(1)**; if the equality is maintained, it corroborates the conclusion. Observe:

$$4x + 2y = 3x + 5$$

Using $x = 3$ and $y = 1$

$$4(3) + 2(1) = 3(3) + 5$$

$$12 + 2 = 9 + 5$$

$$14 = 14$$

As you can see, the solution pair is, in fact, a solution of the given system of two linear equations.

Notice that equation **(2)** does not have the variable **y**. This simply means that equation **(2)** is explicitly specifying the value assigned to the variable **x**.

Try to solve the following system of equations. I will use variables other than **x** and **y** because I do not want you to assume that all systems **must** be in terms of those letters; systems may be specified in terms of any combination of variables.

$$\left. \begin{array}{ll} ① & m = 2 \\ ② & 8m - 2n = 5m + 10 \end{array} \right\}$$

Ready to see my solution process? Observe.

Step 1. There is nothing to be done in this step, since this system contains an equation that is already solved for one of the two unknown variables: equation **(1)** specifically states that the variable *m* is (and must be) equal to **2**.

Step 2. I will now replace $m = 2$ into equation **(2)** and solve it for the remaining unknown variable, *n*. Observe.

$$8m - 2n = 5m + 10$$

Replacing *m* with **2**

$$8(2) - 2n = 5(2) + 10$$

$$16 - 2n = 10 + 10$$

$$16 - 2n = 20$$

$$16 - 2n - 16 = 20 - 16$$

$$16 - 16 - 2n = 4$$

$$0 - 2n = 4$$

I would like to remind you at this point that when you simplify the left side, read as "ZERO minus two-n", the answer is "***negative*** two-n" because subtracting any term from ZERO will yield a negative term (as we reviewed in the first chapters of the book)

$$-2n = 4$$

$$\frac{-2n}{-2} = \frac{4}{-2}$$

$$n = -2$$

The system is now solved, and given the type of answer found, we know this system has only one solution. The solution pair of this particular system is $m = 2$ and $n = -2$ which can be easily checked As we reviewed on the previous problem.

Of course, systems of linear equations that contain an equation already solved for one of the unknown variables may still have an infinite number of solutions or ZERO solutions; you would be able to detect this in Step 2 of the solution process. After replacing the variable (or variables in the case of systems with more than two variables) and during the process of solving for the remaining variable, you would find one of the set ups we already discussed: in both cases, the variable would disappear; in the infinite solution case, the equality would remain intact and valid (as in $-7 = -7$), while in the ZERO solution case, the equality would be broken (as in $4 = 1$).

Try to solve the following system of linear equations by yourself.

$$\left. \begin{array}{l} ①\quad 2x + y - 3 = 2x + 2 \\ ②\qquad\quad y = 5 \end{array} \right\}$$

Keep your eyes open so that as you attempt to solve this system of equations you may identify if the system has one solution, infinite solutions, or ZERO solutions.

Ready to check your answer? Observe.

Step 1. There is nothing to be done in this step, since this system contains an equation that is already solved for one of the two unknown variables: equation **(2)** specifically states that the variable *y* is (and must be) equal to **5**.

Step 2. I will now replace $y = 5$ into equation **(1)** and solve it for the remaining unknown variable, *x*. Observe.

$$2x + y - 3 = 2x + 2$$

Replacing **y** with **5**

$$2x + (5) - 3 = 2x + 2$$

$$2x + 5 - 3 = 2x + 2$$

$$2x + 2 = 2x + 2$$

Do you notice anything familiar about this equation? Can you specify which type of system we are dealing with at this point? One solution, infinite solutions, or ZERO solutions?

$$2x + 2 - 2x = 2x + 2 - 2x$$

$$2x - 2x + 2 = 2x - 2x + 2$$

$$0 + 2 = 0 + 2$$

$$2 = 2$$

It should be obvious at this point that we are dealing with a system of linear equations that has an infinite number of solutions, since the variable we were trying to solve for (in this case, *x*) has disappeared, and the equality is maintained.

The issue at this point is identifying potential solution pairs of this system. The problem, however, is that equation **(2)** only contains the variable *y*, while equation **(1)** is not solved for the other variable, the *x*; therefore, specifying solution pairs would seem impossible. We could

try to solve equation **(1)** for *x* to see if that allows us to create valid input output pairs of values, as we did earlier. Solving equation **(1)** for *y* would be fruitless, since equation **(2)** already states that *y* is equal to **5**, and because we already found that we are dealing with an infinite solution system of equations, equation **(1)** must state the same regarding the value of variable *y*. So, once again, let's solve for *x*:

$$2x + y - 3 = 2x + 2$$

Let's add **3** to both sides to eliminate the **3** on the left side of the equation...

$$2x + y - 3 + 3 = 2x + 2 + 3$$

Now subtract **y** to both sides to eliminate the **y** on the left side of the equation...

$$2x + y - y = 2x + 5 - y$$

$$2x = 2x + 5 - y$$

Now, our goal is to collect all *x* terms to one side of the equation; I will choose to collect them on the left side. Because the right side contains the term   I will subtract that same term from **both sides** so that it disappears from the right side. From this step onwards, I will simplify the right side only, leaving the left side pending. Observe:

$$2x = 2x + 5 - y$$

$$2x - 2x = 2x + 5 - y - 2x$$

Next, I will rearrange the terms on the left side, so that the *x* terms are next to each other (not strictly-speaking necessary, but helps readers trying to master this simplification process grasp the technique)...

$$2x - 2x = 2x - 2x + 5 - y$$

$$2x - 2x = 0 + 5 - y$$

$$2x - 2x = 5 - y$$

I will now perform a minor cosmetic change on the right side: the variable terms are usually listed first (always in formal math notation); since the **y** term is subtracting, we can rewrite it as $5 + (-y)$ and then rearrange, which leads to this next step below...

$$2x - 2x = -y + 5$$

Smooth sailing, right? Well, this is where we encounter a new set up... Let me finish solving for **x** so you can see what I mean.

$$2x - 2x = -y + 5$$

$$0 = -y + 5$$

$$0 + y = -y + 5 + y$$

$$y = -y + y + 5$$

$$y = 0 + 5$$

$$y = 5$$

So what just happened here? We ended up with the same equation as equation **(2)**! So how do we create the solution pairs if we cannot solve the equation for **x** without having it disappear?

Simple: ANY **x** value will work on this particular system. As long as **y** is equal to **5**, the system will consist of equalities that are maintained. In other words, the following solution pairs are part of the infinite solution set of this system of two linear equations:

x	y = 5	Solution pair
:	:	:
−4	y=5	x=−4 and y=5
−3	y=5	x=−3 and y=5
−2	y=5	x=−2 and y=5
−1	y=5	x=−1 and y=5
0	y=5	x=0 and y=5
1	y=5	x=1 and y=5
2	y=5	x=2 and y=5
3	y=5	x=3 and y=5
4	y=5	x=4 and y=5
:	:	:

Of course, the values we are using for **x** do not need to be **INTEGERS**. They may consist of fractions, decimals and irrationals. It is implied, unless otherwise stated, that all **REAL** numbers may be used, and since we can assign **x** any value and the equality will be maintained, we can thus assign any **REAL** number to **x**.

To prove that we may assign any **REAL** number to **x**, simply replace any of the solution pairs included in the table above, and replace them in equation **(1)**.

Let's first use $x = -2$ and $y = 5$; if the equation is maintained, we know the solution pair is, in fact, part of the solution set for this system of linear equations...

$$2x + y - 3 = 2x + 2$$

$$2(-2) + (5) - 3 \overset{?}{=} 2(-2) + 2$$

$$-4 + 5 - 3 \overset{?}{=} -4 + 2$$

$$1 - 3 \overset{?}{=} -2$$

$$-2 \overset{\checkmark}{=} -2$$

This pair of values is, in fact, a solution pair of this system of equations.

Let's try another pair: $x = 4$ and $y = 5$

$$2x + y - 3 = 2x + 2$$

$$2(4) + (5) - 3 \overset{?}{=} 2(4) + 2$$

$$8 + 5 - 3 \overset{?}{=} 8 + 2$$

$$8 + 2 \overset{?}{=} 8 + 2$$

$$10 \overset{\checkmark}{=} 10$$

Once again, the equality is maintained. It does not matter which solution pair you choose: as long as **y** is equal to **5**, **x** may be any **REAL** number, and the pair will be a solution pair of this system.

Let's try another system of linear equations. Try to solve the following:

① $x = 4$

② $y + 2 = x + y + 1$ }

Try to identify the system's type during the solution process, since that is always key when

solving a system of linear equations. Remember what to look for.

Ready to see my steps? Observe.

Step 1. There is nothing to be done in this step, since this system contains an equation that is already solved for one of the two unknown variables: equation **(1)** specifically states that the variable **x** is (and must be) equal to **4**.

Step 2. I will now replace $x = 4$ into equation **(2)** and solve it for the remaining unknown variable, **y**. Observe.

$$y + 2 = x + y + 1$$

Replacing **x** with **4** and solving for **y**

$$y + 2 = (4) + y + 1$$

$$y + 2 = 5 + y$$

$$y + 2 - 2 = 5 + y - 2$$

$$y = 5 - 2 + y$$

$$y = 3 + y$$

Since the goal is to solve for **y**, I will subtract the term **y** from both sides (since the term **y** that is currently on the right side would have to be "sent" to the other side). This will allow me to collect all **y** terms on the left side...

$$y - y = 3 + y - y$$

$$y - y = 3$$

And now, I believe, you can see the problem... When we focus on the left side and attempt to simplify it, the **y** variable will disappear from the equation, and we will end up with a broken equality... In other words, the equality will no longer be true:

$$0 = 3$$

You can now see that this system has ZERO solutions... There isn't a pair of **x** and **y** values that will simultaneously solve equation **(1)** and equation **(2)** when used to replace said variables.

There is one last possible system configuration that you may encounter when solving a system of two linear equations that we have not explored. It is, however, rather trivial. Observe this next example.

$$\left.\begin{array}{l} ① \quad x = 1 \\ ② \quad y = 3 \end{array}\right\}$$

It may seem strange to call each of these an "equation" since each of the statements above essentially assigns a value to **x** and to **y**, the two so-called **unknown** variables of the system (yes, I realize that in this case they **are** explicitly known, right from the start, no manipulation/ simplification necessary). But they are perfectly valid equations from a mathematical standpoint. Furthermore, since they are part of a "system", we can immediately determine the solution pair of the system: $x = 1$ and $y = 3$ is the solution pair.

We have now covered every possible configuration for a system of two linear equations. Try to solve the following exercises (remember to look back at the examples we worked on), but don't look at my solution process until you find yourself stuck and cannot fathom a way out.

a)

$$\left.\begin{array}{l} ① \quad y - 7 = -2x + 4 \\ ② \quad 2x + 5 = 2y + 7 \end{array}\right\}$$

b)

$$\left.\begin{array}{l} ① \quad y - 2 = -3x \\ ② \quad 6x - 12 = -2y - 8 \end{array}\right\}$$

c)

$$\left.\begin{array}{l} ① \quad 2z - 12 = 10w \\ ② \quad -5w + 4 = -z + 2 \end{array}\right\}$$

d)

$$\left.\begin{array}{l} ① \quad y - 8 = 4x - 3 \\ ② \quad \quad x = -2 \end{array}\right\}$$

e)

$$\left.\begin{array}{l} ① \quad 2x - y = 8 - y \\ ② \quad y + x = 6 + x \end{array}\right\}$$

f)

$$\left.\begin{array}{l} ① \quad 3x + 4y = -9 + 4y \\ ② \quad y + 2x = y - 12 \end{array}\right\}$$

It is important to identify the type of system that you are working with; the possibilities are:

Type I. One Solution

Type II. Infinite Solutions

Type II. ZERO Solutions

It may not be obvious from the start, so look for the set ups we reviewed earlier.

426

Ready to check your answers? Please note that I will not show every part of the solution process. It is expected that at this point, you are able to figure out what needs to be done when solving for an unknown variable, for instance, or when simplifying an equation. You may always look back at the previous examples to figure out what had to be done.

a)

$$\left.\begin{array}{ll} ① & y - 7 = -2x + 4 \\ ② & 2x + 5 = 2y + 7 \end{array}\right\}$$

This system has ONE solution (it is a **Type I** system). As I already stated earlier, there are multiple ways in which you may correctly solve this system, so our paths may be different; however, if your solution process is correct, you should be able to match my answer.

Step 1. After solving equation **(1)** for **y**, I obtain the following:

$$y = -2x + 11$$

I can now replace the **y** from equation **(2)** with $-2x + 11$.

Step 2. Replacing **y** with $-2x + 11$ in equation **(2)** results in the following:

$$2x + 5 = 2(-2x + 11) + 7$$

$$2x + 5 = -4x + 22 + 7$$

$$6x = 24$$

$$x = 4$$

Now, I can use the equation obtained in Step 1 that is already solved for **y** and replacing **x** with the **4** obtained above, I can find the solution pair of the system:

$$y = -2(4) + 11$$

$$y = 3$$

System solved. The solution pair is:

$$x = 4 \text{ and } y = 3$$

b)

$$\left.\begin{array}{ll} ① & y - 2 = -3x \\ ② & 6x - 12 = -2y - 8 \end{array}\right\}$$

This system has infinite solutions (it is a **Type II** system).

Step 1. After solving equation **(1)** for **y**, I obtain the following:

$$y = -3x + 2$$

I can now replace the **y** from equation **(2)** with $-3x + 2$.

Step 2. Replacing **y** with $-3x + 2$ in equation **(2)** results in the following:

$$6x - 12 = -2(-3x + 2) - 8$$

$$6x - 12 = 6x - 4 - 8$$

$$6x - 12 - 6x = 6x - 12 - 6x$$

$$-12 = -12$$

At this point, since the variable **x** I was trying to solve for has been eliminated and the equality is valid, I know that the system has an infinite number of solution pairs. To create a sample list of solution pairs, I can use either one of the equations, use any **REAL** number as an input value, and compute the output value: each pair is a solution pair of the system. Using equation **(1)** (the version that has already been solved for *y*) and assigning the input values to *x*, we can figure out *y* and specify the solution pair:

Input Number x	Evaluate y = –3x+2	Solution pair
–2	y = –3(-2)+2 = 8	x=–2 , y= 8
–1	y = –3(-1)+2 = 5	x=–1 , y= 5
0	y = –3(0)+2 = 2	x= 0 , y= 2
1	y = –3(1)+2 = –1	x= 1 , y= –1
2	y = –3(2)+2 = –4	x= 2 , y= –4

As I stated earlier, the input numbers used may be any **REAL** number whatsoever; I am using **INTEGERS** in the sample table above because they are easier to work with, but I could have used values for *x* such as **–.456**, $\frac{17}{8}$, $\sqrt{2}$, etc...

I would like to point out a very important aspect of working with two (or more) variable equations. Which of the unknown variables that are part of the system should be called the **INPUT** variable– and hence, which variable should be called the **OUTPUT** variable?

The answer is simple: you are free to assign them as you wish. Carefully analyze the following summary box.

⚓

In a two variable equation, which variable is the INPUT variable?

You may consider the **INPUT** variable to be any of the variables that are part of an equation (or of a system of equations). However, for that system or for that equation, once you define the INPUT variable, that must stayed fixed for as long as you work with that equation or that problem in a given context, to work out a given problem, or to answer a specific question regarding the equation or the system. Furthermore, the remaining variable is automatically considered the **OUTPUT** variable.

When listing solution pairs, always specify the INPUT variable *first*, followed by the OUTPUT variable's value.

Usually, when the variables *x* and *y* are used:

x is the INPUT variable
y is the OUTPUT variable

As the above states, whenever an equation (or a system) uses *x* and *y*, it is practically a given and expected that *x* will be the INPUT variable and that *y* will be the OUTPUT variable (unless otherwise stated).

Let's move on to the next problem on the list.

c)

$$\left.\begin{array}{l} ① \quad 2z - 12 = 10w \\ ② \quad -5w + 4 = -z + 2 \end{array}\right\}$$

This system has ZERO solutions (it is a **Type III** system).

Step 1. After solving equation **(1)** for **z**, I obtain the following:

$$z = 5w + 6$$

I can now replace the **z** from equation **(2)** with $5w + 6$.

Step 2. Replacing **z** with $5w + 6$ in equation **(2)** results in the following:

$$-5w + 4 = -(5w + 6) + 2$$

$$-5w + 4 = -5w - 6 + 2$$

$$-5w + 4 = -5w - 4$$

$$-5w + 4 + 5w = -5w - 4 + 5w$$

$$? \; 4 \neq -4 \; ?$$

At this point, since the variable **w** I was trying to solve for has been eliminated and the equality is **not** valid, I know that the system has ZERO solution pairs. In other words, regardless of the values that you assign to **w** and **z** you will never find a pair that will simultaneously solve both equations; you may find solution pairs for equation **(1)**, but none of those solution pairs would solve equation **(2)**, and vice-versa: those that are solution pairs of equation **(2)** will not be solution pairs of equation **(1)**.

d)

① $\quad y - 8 = 4x - 3$

② $\quad\quad x = -2$

This system has ONE solution (it is a **Type I** system).

Step 1. Equation **(2)** is already solved for **x**, so I can move on to Step 2. Unless the system is a ZERO solution type, solution pair(s) must state that **x** is equal to **-2** based on equation **(2)**.

Step 2. Replacing **x** with **-2** in equation **(1)** results in the following:

$$y - 8 = 4(-2) - 3$$

$$y - 8 = -11$$

$$y = -3$$

System solved. The solution pair is:

$$x = -2 \text{ and } y = -3$$

e)

① $\quad 2x - y = 8 - y$

② $\quad y + x = 6 + x$

This system has ONE solution (it is a **Type I** system).

Step 1. After solving equation **(1)** for **x**, I obtain the following:

$$x = 4$$

Please note that even if you tried to solve for **y**, you would still end up with this answer. The reason is that both sides of the equation contain the same **y** term, so they will cancel out during the process of collecting them on either side and combining them.

I can now replace the **x** from equation **(2)** with **4**.

Step 2. Replacing **x** with 4 in equation **(2)** results in the following (please note that I need to replace all of the **x** variables that appear on the equation, regardless of which side they may appear on):

$$y + (2) = 6 + (2)$$

$$y + 2 = 8$$

$$y + 2 - 2 = 8 - 2$$

$$y = 6$$

Because equation **(1)** gave us the value of **x** directly, we have solved the system. The solution pair is:

$$x = 4 \text{ and } y = 6$$

f)

$$\left.\begin{array}{l} \textcircled{1} \quad 3x + 4y = -9 + 4y \\ \textcircled{2} \quad y + 2x = y - 12 \end{array}\right\}$$

This system has ZERO solutions (it is a **Type III** system).

Step 1. Looking at these equations, I would try to solve equation **(2)** for **y**, because it seems to be the easier course of action. Observe, however, what happens when attempting this:

$$y + 2x - 2x = y - 12 - 2x$$

$$y = y - 12 - 2x$$

$$y - y = y - 12 - 2x - y$$

$$0 = -12 - 2x$$

At this point, you may be wondering what happened. We were trying to solve for **y** and yet the **y** variable disappeared (it was canceled during the simplification process). Well, we could simply move forward, and solve for the remaining variable instead (in this case, **x**). Let's do that.

$$0 + 2x = -12 - 2x + 2x$$

$$2x = -12$$

$$\frac{2x}{2} = \frac{-12}{2}$$

$$x = -6$$

Now, I will replace **x** with **−6** to find the value of **y** using the other equation (equation **(1)** in this case); it is important to use the other equation, since when solving for **y** in equation **(2)** the variable was cancelled out (it would happen again if we replaced **x** with **−6**).

Step 2. Replacing **x** with **−6** in equation **(1)** results in the following:

$$3(-6) + 4y = -9 + 4y$$

$$-18 + 4y = -9 + 4y$$

$$-18 + 4y - 4y = -9 + 4y - 4y$$

$$-18 = -9$$

Houston, we have a problem. As you can see, the **y** variable was cancelled on both sides and the equality is broken, the unequivocal signs of a **Type III** system of linear equations. Therefore, this system has ZERO solutions.

We are almost done with this chapter. However, I want you to see how we can extend the principles and rules that we applied to systems of linear equations with two variables to systems of linear equations with three or more variables.

When dealing with systems that have three or more variables, the number of equations that define the system must be equal to the number of variables that define the system. In other words, a three variable linear system must have three equations, a four variable system four equations, etc... Having less equations would prevent us from finding a solution (and even a system Type), while having more would require radically altering how we interpret and approach such a system.

We are, of course, still operating under the **substitution** technique. So, let's explore how to extend what we learned to a three variable linear system (after that, you would be able to extend it to a four, five, six, etc... variable linear system).

Let's begin with the following linear system.

$$\left. \begin{array}{l} ① \quad 2x + 3y = 3z - 2 \\ ② \quad x + 5y = 2z + 7 \\ ③ \quad 2x + 2y = 2z \end{array} \right\}$$

As you can observe, this is a linear system of three variables; furthermore, since it consists of three equations, we know we will be able to determine if it is a **Type I**, **Type II**, or **Type III**.

So how do we extend the two-variable, two-equation substitution technique that we reviewed earlier to a three-variable, three equation system? Can you foresee how to do it?

You will need to use three steps to solve the system instead of two. As before, labeling each equation is important. The following steps work for systems in which all three equations have all three variables (as in the one we are working with); when one (or more) of the equations have less variables, the process becomes easier because some of the steps would become shorter, just like we saw on some of the two variable systems. The steps are as follows:

Step 1. Solve one of the equations for one of the unknown variables. This will typically provide you with the variable that you solved for on one side, and the other two unknown variables on the other side.

Step 2. Replace the variable that you solved for in Step 1 in either one of the two remaining equations (at this point, it leaves you with a third equation that is said to be unused). As with systems of two variables, you cannot use the same equation that you used in Step 1. Solve for one of the two remaining variables. Typically, this will provide you with the variable that you solved for on one side, and the other unknown variable on the other side (usually accompanied by a coefficient and an independent term adding or subtracting).

Step 3. This step is crucial, and it can be confusing. You need to replace all instances of the variable that you solved for in Step 1 into the equation that has been unused up to this point, and then follow this by replacing all instances of the variable that you solved for on the Step 2 into the equation as well. This results in a single equation of one variable, which you can now solve for said unknown variable.

Step 4. This step is relatively easy. Since in Step 3 you found the value of one of the three unknown variables, you can now scan the equations that you obtained in Steps 1 and 2, and you will notice that the equation that you obtained in Step 2 can be solved for the as-yet unknown variable that you solved for in that step, by replacing the variable with its value found in Step 3.

Step 5. Now that you have solved two of the three variables, you can use any of the three equations (whichever is easiest) and replace the now-known variables with their values, and solve for the third and last variable. *Always check your answer by replacing the variables with their values in all three equations.*

Observe how I apply these steps to the system we are currently working with (you should try to solve it by yourself before continuing):

$$\left. \begin{array}{l} \text{①} \quad 2x + 3y = 3z - 2 \\ \text{②} \quad x + 5y = 2z + 7 \\ \text{③} \quad 2x + 2y = 2z \end{array} \right\}$$

Step 1. The instructions read "Solve one of the equations for one of the unknown variables. This will typically provide you with the variable that you solved for on one side, and the other two unknown variables on the other side."

I will solve equation **(3)** for **z**, since it is the easiest equation to solve for one of the unknown variables and it will not have a fraction part.

$$2x + 2y = 2z$$

$$\frac{2x + 2y}{2} = \frac{2z}{2}$$

(a) $x + y = z$

I have labeled the resulting equation as **(a)** so that we can refer to it easily.

Step 2. In this step, the instructions read as follows: "Replace the variable that you solved for in Step 1 in either one of the two remaining equations (at this point, it leaves you with a third equation that is said to be unused). As with systems of two variables, you cannot use the same equation that you used in Step 1. Solve for one of the two remaining variables. Typically, this will provide you with the variable that you solved for on one side, and the other unknown variable on the other side (usually accompanied by a coefficient and an independent term adding or subtracting)."

I will replace $z = x + y$ into equation **(2)** and then I will solve it for **x**.

$$x + 5y = 2z + 7$$

Replacing **z** with $x + y$...

$$x + 5y = 2(x + y) + 7$$

And solving for **x**...

$$x + 5y - 2y = 2x + 2y + 7 - 2y$$

$$x + 3y = 2x + 7$$

$$x + 3y - 7 - x = 2x + 7 - 7 - x$$

(b) $3y - 7 = x$

I can now move on to Step 3.

Step 3. In this step the instructions state the following: "This step is crucial, and it can be

432

confusing. You need to replace all instances of the variable that you solved for in Step 1 into the equation that has been unused up to this point, and then follow this by replacing all instances of the variable that you solved for on the Step 2 into the equation as well. This results in a single equation of one variable, which you can now solve for said unknown variable."

I will start by replacing **(a)** into **(1)**, and then and only then **(b)** into the resulting equation (making sure to replace any instances of the variable x, in this case)...

$$2x + 3y = 3z - 2$$

Replacing z with $x + y$...

$$2x + 3y = 3(x + y) - 2$$

And now, replacing x with $3y - 7$...

$$2(3y - 7) + 3y = 3((3y - 7) + y) - 2$$

Now, I can solve for y. Note that what's left is one equation with one variable...

$$2(3y - 7) + 3y = 3((3y - 7) + y) - 2$$

$$6y - 14 + 3y = 9y - 21 + 3y - 2$$

$$9y - 14 = 12y - 23$$

$$9y - 14 + 23 - 9y = 12y - 23 + 23 - 9y$$

$$9 = 3y$$

$$\frac{9}{3} = \frac{3y}{3}$$

$$3 = y$$

We have found the first part of the solution of this system; one down, two to go. Now, we may move on to Step 4, to find the value of the other two variables.

Step 4. This step reads as follows: "This step is relatively easy. Since in Step 3 you found the value of one of the three unknown variables, you can now scan the equations that you obtained in Steps 1 and 2, and you will notice that the equation that you obtained in Step 2 can be solved for the as-yet unknown variable that you solved for in that step, by replacing the variable with its value found in Step 3."

I will replace the variable y with **3** in equation **(b)** (the equation we found in Step 2), to find the value of the variable x.

$$3y - 7 = x$$

Replacing y with **3**...

$$3(3) - 7 = x$$

$$9 - 7 = x$$

$$2 = x$$

And we now know the values of two of the three variables that will form the solution triplet of the system. Let's move on to the next step.

Step 5. The final step reads: "Now that you have solved two of the three variables, you can use any of the three equations (whichever is easiest) and replace the now-known variables with their values, and solve for the third and last variable."

Since we now know that the solution of the system must consist of $y = 3$ and $x = 2$ we may replace x and y with their respective values into any of the equations that we were either given as part of the initial system, or that we ended up with after solving for a particular variable. Since i need to establish the value of **z**, the remaining unknown variable, I will use equation **(a)** since it is already solved for **z**. Observe.

$$x + y = z$$

Replacing $x = 2$ and $y = 3$...

$$(2) + (3) = z$$

$$5 = z$$

And we now know the solution of this linear system of equations of three variables. The answer may be expressed as follows:

$$
\left.
\begin{array}{ll}
① & 2x + 3y = 3z - 2 \\
② & x + 5y = 2z + 7 \\
③ & 2x + 2y = 2z
\end{array}
\right\}
\quad
\begin{array}{l}
x = 2 \\
y = 3 \\
z = 5
\end{array}
$$

Or, simply state that the solution of this system of linear equations of three variables is $x = 2$, $y = 3$, and $z = 5$. *I leave it to you to check that the answer is correct: simply replace each of the variables with the values obtained, in each equation, and make sure that you obtain an equality that is true.*

There are many variants to a linear system of equations of three variables (similar to those that I pointed out to you and reviewed for linear systems of equations of two variables). I will not explore each possibility, because essentially, they represent an iteration of the possibilities for linear systems of equations of two variables.

Just remember that if, say, an equation is already solved for a given variable as given in the initial set up, it just means that your workload has been reduced in the appropriate step because you would not have to solve it yourself for that given variable. In addition to this, if one or more of the equations of the system contain a single variable, then you may solve it and directly find that variable's value. It, once again, reduces your work load.

Do not be confused if two or all three variables have the same value. That is a perfectly valid solution triplet for a given linear system of equations with three variables.

Finally, what is essential is that you look for and recognize the following situations that you may encounter as you solve the system:

1) If the variable that you are solving for is cancelled, and you end up with an equation without variables, check to see if the equality remains true or if it has been broken. If it remains true, you may continue solving the system using the steps already defined, and simply assume that you may be looking at a **Type II** system (infinite solutions). However, you must check that all three equations are used and checked before being able to conclude that the system has, in fact, an infinite number of solutions.

2) If the variable that you are solving for is cancelled, and you end up with an equation without variables, check to see if the equality remains true or if it has been broken. If the equality has been broken, you automatically know that the system is a **Type III** system (ZERO solutions). There is no need to check the other equation.

Now, try to solve the following systems of linear equations with three variables. I am not going to mislead you: solving these can get tricky, and it is easy to get lost in the process of applying the steps we reviewed. Just make sure to label the equations you come up with and to follow all the steps, from beginning to end. If you get stuck solving one of the systems, move on to the next. Try all three, and then compare my solving process so that you can learn from the experience..

a)

① $4x + 8y = 2z$

② $2x + 4y + 2 = z$

③ $2x + 3y = z$

b)

① $2x + y = z + 2$

② $x + 4y = 2z$

③ $2x + 2z = 4y + 6$

c)

① $x - 2y = -z + 3$

② $3x + 3z = 6y + 9$

③ $4x - 8y = 12 - 4z$

How did you do? Were you able to identify the type of systems that each of these belong to? Compare your work.

a)

① $4x + 8y = 2z$

② $2x + 4y + 2 = z$

③ $2x + 3y = z$

Step 1. Because equation **(3)** is already solved for **z**, all I need to do is identify it as equation **(a)**. Please note that I could have used equation **(2)**, but I chose **(3)** instead, for no particular reason.

ⓐ $2x + 3y = z$

Let's go on to Step 2.

Step 2. I will replace $z = 2x + 3y$ into equation **(2)** and then I will solve it for **x**.

$$2x + 4y + 2 = z$$

Replacing **z** with $2x + 3y$...

$$2x + 4y + 2 = 2x + 3y$$

And solving for **x**...

$$2x + 4y + 2 - 3y = 2x + 3y - 3y$$

$$2x + y + 2 = 2x$$

$$2x + y + 2 - 2x = 2x - 2x$$

$$y + 2 = 0$$

Please note that the **x** variable has cancelled out. We have encountered similar situations before, so we know that we may be working with a system that may allows us to use any **x** value we wish to use. However, we cannot conclude anything until all equations are considered. For now, we solve for **y** and obtain the answer:

$$y = -2$$

I can now move on to Step 3.

Step 3. I will start by replacing **(a)** into the yet-to-be-used equation **(1)**, and then and only then replacing $y = -2$ into the resulting equation (making sure to replace any instances of the variable **y**, in this case). Let's see what happens:

$$4x + 8y = 2z$$

Replacing **z** with $2x + 3y$...

$$4x + 8y = 2(2x + 3y)$$

And now, replacing **y** with **−2**...

$$4x + 8(-2) = 2(2x + 3(-2))$$

$$4x - 16 = 2(2x - 6)$$

$$4x - 16 = 4x - 12$$

$$4x - 16 - 4x = 4x - 12 - 4x$$

$$-16 = -12$$

And this is where I stop. The variable disappeared (it was cancelled during the simplifying/solving process), and the equality has been broken. Therefore, I know unequivocally, that this is a **Type III** system (ZERO solutions).

We do not have to check the solution because we did not obtain a value for all three variables.

Let's move on to the next problem.

b)

$$\begin{array}{l} \text{①} \quad 2x + y = z + 2 \\ \text{②} \quad x + 4y = 2z \\ \text{③} \quad 2x + 2z = 4y + 6 \end{array} \Bigg\}$$

Step 1. I will solve equation **(2)** for **x**, since it is the easiest equation to solve for one of the unknown variables, one that will not have a fraction part.

$$x + 4y = 2z$$

$$x + 4y - 4y = 2z - 4y$$

$$\text{ⓐ} \quad x = 2z - 4y$$

I have labeled the resulting equation as **(a)** so that we can refer to it easily.

Step 2. I will replace $x = 2z - 4y$ into equation **(3)** and then I will solve it for **z**.

$$2x + 2z = 4y + 6$$

Replacing **x** with $2z - 4y$...

$$2(2z - 4y) + 2z = 4y + 6$$

And solving for **z**...

$$4z - 8y + 2z = 4y + 6$$

$$6z - 8y = 4y + 6$$

$$6z - 8y + 8y = 4y + 6 + 8y$$

$$6z = 12y + 6$$

$$\frac{6z}{6} = \frac{12y + 6}{6}$$

(b) $z = 2y + 1$

I can now move on to Step 3.

Step 3. I will start by replacing **(a)** into **(1)**, and then and only then **(b)** into the resulting equation (at that point, making sure to replace any instances of the variable **z**, in this case, with $2y + 1$)...

$$2x + y = z + 2$$

Using **(a)** we replace **x** with $2z - 4y$...

$$2(2z - 4y) + y = z + 2$$

Using **(b)**, replacing **z** with $2y + 1$...

$$2(2(2y + 1) - 4y) + y = (2y + 1) + 2$$

Now, I can solve for **y**. Note that what's left is one equation with one variable...

$$2(4y + 2 - 4y) + y = 2y + 1 + 2$$

$$8y + 4 - 8y + y = 2y + 3$$

$$4 + y = 2y + 3$$

$$4 + y - y - 3 = 2y + 3 - y - 3$$

$$1 = y$$

This is the first part of the solution of this system; one down, two to go. Now, we may move on to Step 4, to find the value of the other two variables.

Step 4. I will replace the variable **y** with **1** in equation **(b)** (the equation we found in Step 2), to find the value of the variable **z**.

$z = 2y + 1$

Replacing **y** with **1**...

$$z = 2(1) + 1$$

$$z = 2 + 1$$

$$z = 3$$

And we now know the values of two of the three variables that will form the solution triplet of the system. Let's move on to the next step.

Step 5. Since we now know that the solution of the system must consist of $y = 1$ and $z = 3$ we may replace **y** and **z** with their respective values into any of the equations that we were either given as part of the initial system, or that we ended up with after solving for a particular variable. Since I need to establish the value of **x**, the remaining unknown variable, I will use equation **(a)** since it is already solved for **x**. Observe.

$$x = 2z - 4y$$

Replacing $y = 1$ and $z = 3$...

$$x = 2(3) - 4(1)$$

$$x = 6 - 4$$

$$x = 2$$

And we now know the solution of this linear system of equations of three variables.

Remember to always check that the solution is, in fact, a solution of the entire system, by replacing the solution values into all three equations and simplifying each one to verify that the equalities remain. Only then would you be able to conclude that the system is, in fact, solved by the set of values found.

Let's move on to the last problem you had to solve.

c)

$$\left.\begin{array}{ll} \text{①} & x - 2y = -z + 3 \\[2mm] \text{②} & 3x + 3z = 6y + 9 \\[2mm] \text{③} & 4x - 8y = 12 - 4z \end{array}\right\}$$

Step 1. I will solve equation **(1)** for **x**, since it is the easiest equation to solve for one of the unknown variables, one that will not have a fraction part.

$$x - 2y = -z + 3$$

$$\text{ⓐ} \quad x = -z + 3 + 2y$$

I have labeled the resulting equation as **(a)** so that we can refer to it easily.

Step 2. I will replace $x = -z + 3 + 2y$ into equation **(2)** and then I will solve it for **z**.

$$3x + 3z = 6y + 9$$

Replacing **x** with $-z + 3 + 2y$...

$$3(-z + 3 + 2y) + 3z = 6y + 9$$

And solving for **z**...

$$-3z + 9 + 6y + 3z = 6y + 9$$

$$9 + 6y - 6y = 6y + 9 - 6y$$

$$9 = 9$$

At this point, because the variables have cancelled out and the equality is true, we may assume that we are dealing with a **Type II** system of equations (infinite solutions). However, we have not used equation **(3)**, and since the system is a three equation system, we *must* check if using equations **(1)** and **(3)** or **(2)** and **(3)** also gives us the same result (a variable-less true equation).

Step 2 revisited. I will replace $x = -z + 3 + 2y$ into equation **(3)** and then I will solve it for **z**.

$$4x - 8y = 12 - 4z$$

Replacing **x** with $-z + 3 + 2y$...

$$4(-z + 3 + 2y) - 8y = 12 - 4z$$

$$-4z + 12 + 8y - 8y = 12 - 4z$$

$$-4z + 12 = 12 - 4z$$

At this point, you should be able to see that we do not need to keep solving... the left side is exactly the same as the right side; adding $4z$ to both sides will cancel the variables, and we would be left with an equation that reads $12 = 12$ which is true.

Now, and only after having checked that the third equation also yielded a similar result, were we able to conclude, unequivocally, that the system is a **Type III** system. It has an infinite number of solutions. And how would we be able to create solution sets of this system? We saw how to do this for two variable systems, bit not for three, so I will review how to do this.

The principle is relatively simple. Simply take one of the equations, whichever is easiest to work with (not, of course, a requirement, but certainly a welcome endeavor), ideally one that is already solved for a variable. You can then replace any values you want for two of the three variables (exclude the variable that the equation is solved for, should that be the case) and then find the third variable's value. Since there is an infinite number of combinations of **REAL** numbers that you may use for the two variables, you can readily see how the system would yield an infinite number of solutions.

To illustrate this, I will use equation **(a)**, since it is already solved for one of the variables. It is, after all, simply a modified version of equation **(1)**.

$$x = -z + 3 + 2y$$

I can use any **REAL** number imaginable, both for **z** and for **y**. We can take a systematic approach (if we were to limit this exercise to, say, replacing the two variables with positive **INTEGERS** (remember how they are called? **NATURAL** numbers, right?). We could then assign **1** to variable **y** and then starting with **1** for variable **z**, systematically increase its value by **1**, and compute the resulting value for **x**. Then, we could change the value of **y** by adding **1**, and reset the value of **z** to **1** and systematically increase it by **1** once again, computing x for every iteration. And on an on and on, **ad nauseam** (remember its meaning as explained in **Chapter 1**?). This process would look something like this:

Setting $y = 1$, and then starting with $z = 1$ we may compute **x**:

$$x = -z + 3 + 2y$$

$$x = -(1) + 3 + 2(1)$$

$$x = 4$$

So, our first solution triplet would be:

$$x = 4 \ , \ y = 1 \ , \ z = 1$$

Then, adding one to **z** and evaluating **x** again:

$$x = -z + 3 + 2y$$

$$x = -(2) + 3 + 2(1)$$

$$x = 3$$

Thus, our second solution triplet would be:

$$x = 3 \ , \ y = 1 \ , \ z = 2$$

Iterating this process, we would end up with the following first **6** triplets (solutions to this system of equations):

$$x = 4 \ , \ y = 1 \ , \ z = 1$$

$$x = 3 \ , \ y = 1 \ , \ z = 2$$

$$x = 2 \ , \ y = 1 \ , \ z = 3$$

$$x = 1 \ , \ y = 1 \ , \ z = 4$$

$$x = 0 \ , \ y = 1 \ , \ z = 5$$

$$x = -1 \ , \ y = 1 \ , \ z = 6$$

$$\vdots \qquad \vdots \qquad \vdots$$

Observe that the fifth and sixth triplets (and all subsequent triplets) are no longer made up of **NATURAL** numbers, since the variable **x** is equal to **0** and **−1** respectively. We have no control over this because it is the equation that is providing the value of **x** based on the inputs being used. Not that there is anything wrong with this; unless otherwise stated (and unless the real-world application being used specifically calls for a specific type of number, such as **NATURALS**, **WHOLE**, **INTEGERS**, etc...), we are always defining the domain to be all **REAL** numbers, at least as far as linear equations are concerned.

Shifting gears, we may change the value of **y** to **2** and begin the process again (setting **z** equal to **1** to compute **x** to find the triplet, and then adding **1** to **z** to compute the new triplet, etc...). This would give us the following solutions:

Using $x = -z + 3 + 2y$ once again,

$$x = 6 \ , \ y = 2 \ , \ z = 1$$

$$x = 5 \ , \ y = 2 \ , \ z = 2$$

$$x = 4 \ , \ y = 2 \ , \ z = 3$$

$$x = 3 \ , \ y = 2 \ , \ z = 4$$

$$x = 2 \ , \ y = 2 \ , \ z = 5$$

$$x = 1 \ , \ y = 2 \ , \ z = 6$$

$$\vdots \qquad \vdots \qquad \vdots$$

We do not have to exclude decimals (which means fractions) nor IRRATIONALS (numbers such as $\sqrt{2}$ or $\sqrt[3]{4}$ or π, but unless the context of the problem requires us to use such inputs, it would be over doing it a bit, since the computations can become quite annoying to perform. Observe:

Setting $y = \sqrt{2}$, and then starting with $z = \sqrt[3]{4}$, for example, we would need to do the following in order to compute **x**:

$$x = -z + 3 + 2y$$

$$x = -(\pi) + 3 + 2(\sqrt[3]{4})$$

$$x = -\pi + 3 + 2\sqrt[3]{4}$$

And now what? How do we figure out the value of **x**? As you may recall, **IRRATIONAL** numbers (π and $\sqrt[3]{4}$ in this case) possess an infinite decimal part, that is also *non-repeating*. We cannot express **IRRATIONALS** as **RATIONAL** numbers (fractions, that is), and so we would have to either approximate their values (as in π is approximately equal to **3.141592** and do the same for $\sqrt[3]{4}$), or leave the value of **x** expressed as above (unhelpful if you actually need a number that you can work with in the real world: if the variable **x** represented the length–in feet–of a beam of wood to support a roof, we would be forced to find an approximation). We could, therefore, express it as follows:

$$x = -\pi + 3 + 2\sqrt[3]{4} \ , \ y = 1 \ , \ z = 1$$

Or, using a precision of **5** decimals (truncating), expressing **x** as...

$$x = 2.49742 \ , \ y = 1 \ , \ z = 1$$

It must be clear that this is an approximate value of **x**. We have lost the precise value the minute we used a calculator (or any other computational device) to figure out a decimal form of the irrational elements that are part of the value of **x** in the equation above. This is where knowing what the number will be used for becomes crucial, as I already discussed in **Chapter 6**; different applications will require different levels of precision. If we didn't need more than one decimal, we could use the following triplet instead (truncating, for example):

$$x = 2.4 \ , \ y = 1 \ , \ z = 1$$

If we chose to use the rounding method, we would obtain the following triplet:

$$x = 2.5 \ , \ y = 1 \ , \ z = 1$$

So which of the triplets is correct? Only the triplet that kept the elements π and $\sqrt[3]{4}$, of course. The other two are mere **approximations**. Good enough, perhaps, for a given application, but in the idealized world of math, not **actual solutions** to the system of equations under discussion.

At this point, you should have a very clear understanding of how to solve systems of linear equations, regardless of the number of variables that it may possess. Just remember that if it is to be solvable, it must consist of as many equations as there are variables. Furthermore, remember to use all equations during the solution process to guarantee that the system is in fact the Type you may have identified it to be before using all the equations, as in the last example we recently worked on (problem **(c)**, **Type II**–infinite solutions). Last but not least, remember to check the solution set that you obtain (if the system is a **Type I**) by replacing all variable values defined by said solution set in all of the equations that are part of the system, and then verifying that the equations still hold true (in other words, that the equality is not broken). If in that process one of the equalities is broken, you would be able to conclude **ipso facto** (Latin for "by the very fact or act") that the system is a **Type III** (in other words, of ZERO solutions).

Before I end this review of solving systems of linear equations using the substitution method, I would like to provide you with an example of how you could approach a system of, say, **5** variables (and therefore, of **5** equations). Simply start with equation **(1)**, solve for one of the variables, let's call it **v**; then replace what this variable is equal to in equation **(2)**, and solve for a second variable, let's call it **w**; then replace what the first variable, in this case **v**, was equal to, and then what **w** is equal to, in equation **(3)**, and solve for a third variable, let's call it **x**; then replace what **v** and then what **w** are equal to, along with **x** in equation **(4)**, and solve for a fourth variable, let's call it **y**; finally, replace (in succession) what **v**, then **w**, then **x**, and then **y** is equal to in equation **(5)**, and solve for the fifth and final variable, let's call it **z**. At this point, if the system is a **Type I** system (one solution), you will have found the value of the variable **z**. You then backtrack, replacing **z** into the second-to-last equation that you found, to obtain the value of **y**, and you keep backtracking until you find the solution quintuple (five variable solution).

Another chapter concludes. We may now move on to the elimination method. See you there!

16

ELIMINATION METHOD

As I briefly mentioned in the previous chapter, systems of equations of two or more variables may be solved using the so called *elimination method*. This method is sometimes more useful to use than the substitution method, although the substitution method is usually more straight-forward to apply.

Before I dive right into the elimination method, however, it's a good idea to review a few principles that apply to equations in general. We have indirectly worked with them before, but it is important that you see them in action explicitly, one at a time.

Say we have the following equation:

$$y = 5x - 4$$

We could, if we wanted to, "collect" all of the terms on the left side (we could collect them on the right side as well, but usually, the left side is used). In this case, we would need to subtract $5x$ and add 4 to both sides of the equation in order to achieve this.

$$y = 5x - 4$$

$$y - 5x + 4 = 5x - 4 - 5x + 4$$

$$y - 5x + 4 = 0$$

At this point, we have achieved our goal of collecting all terms on the left side. The right side, however, is not left empty. The **0** that remains is the result of $5x - 4 - 5x + 4$ which, as you know, becomes $5x - 5x - 4 + 4$. This, in turn, is equal to $0 + 0$, which is **0**.

In math terms, we have "**set the equation to zero**" or simply "**set to zero**", since we have made the equation be equal to zero by collecting all terms on one side.

Another important principle involves manipulating an equation by multiplying or dividing the entire both sides of the equation. Of course, if our equation is set to zero, the zero will remain unchanged (since multiplying **0** with anything is **0**, and dividing **0** with anything other than **0** is equal to **0** as well). Observe how I create different (and yet equivalent) versions of a given initial equation.

$$y = 5x - 4$$

Multiplying by **3** both sides:

$$(3)(y) = (5x - 4)(3)$$

$$3y = 15x - 12$$

In other words, we know that the equations $y = 5x - 4$ and $3y = 15x - 12$ are equivalent to each other. It is the same principle that allows us to say that $\sqrt{4}$ and $\frac{8}{4}$ both represent the same number, the number **2**. Why would we want to find equivalent versions of an equation? That will be explained shortly.

Another example:

$$y - 5x + 4 = 0$$

Multiplying by **3** both sides:

$$(3)(y - 5x + 4) = (0)(3)$$

$$3y - 15x + 12 = 0$$

And so we can conclude that $y - 5x + 4 = 0$ and $3y - 15x + 12 = 0$ are equivalent. Do you recognize the starting equation above? It is itself an equivalent version of the original equation from the previous example (set equal to zero). To prove it, I could take the equation $3y = 15x - 12$ and set it to zero so that you may compare them:

$$3y = 15x - 12$$

$$3y - 15x + 12 = 15x - 12 - 15x + 12$$

$$3y - 15x + 12 = 0$$

Can you see it now? I simply wanted to reinforce the idea of equivalency between equations.

Let me show you some additional examples of finding equivalent equations.

Starting equation: $-8y + 20x + 7 = 0$

Multiplying by -1 both sides:

$(-1)(-8y + 20x + 7) = (0)(-1)$

Equivalent equation: $8y - 20x - 7 = 0$

Starting equation: $y - 7x + 2 = 0$

Multiplying by 4 both sides:

$(4)(y - 7x + 2) = (0)(4)$

Equivalent equation: $4y - 28x + 8 = 0$

Starting equation: $-8y + 20x + 7 = 0$

Multiplying by -1 both sides:

$(-1)(-8y + 20x + 7) = (0)(-1)$

Equivalent equation: $8y - 20x - 7 = 0$

Starting equation: $-w + y - z = 0$

Multiplying by -10 both sides:

$(-10)(-w + y - z) = (0)(-10)$

Equivalent equation: $10w - 10y + 10z = 0$

By now you should be able to see how we can use multiplication to find equivalent version of a given equation. We could, of course, multiply the equations using *any* **REAL** number. For example:

Starting equation: $8y + 12x - 4z = 0$

Multiplying by $\dfrac{1}{2}$ both sides:

$(\dfrac{1}{2})(8y + 12x - 4z) = (0)(\dfrac{1}{2})$

$\dfrac{8y + 12x - 4}{2} = 0$

Equivalent equation: $4y + 6x - 2 = 0$

Starting equation: $-x + 2w - z + 4 = 0$

Multiplying by π both sides:

$(\pi)(-x + 2w - z + 4) = (0)(\pi)$

Equivalent equation: $-\pi x + 2\pi w - \pi z + 4\pi) = 0$

Remember that we may also divide both sides by any **REAL** number other than ZERO and find equivalent versions of the same equation. Observe.

Starting equation: $-6x + 21y + 9 = 0$

Dividing by -3 both sides:

$\dfrac{-6x + 21y + 9}{-3} = \dfrac{0}{-3}$

$\dfrac{-6x}{-3} + \dfrac{21y}{-3} + \dfrac{9}{-3} = \dfrac{0}{-3}$

$2x + (-7y) + (-3) = 0$

Equivalent equation: $2x - 7y - 3 = 0$

444

I want to point out that a given equation has potentially an infinite number of variations, since we could, if we wanted to, multiply or divide both sides of the equation by any **REAL** number, and since there are an infinite number of **REAL** numbers, that conclusion follows. Observe.

Starting equation: $-4x + 8y - 12 = 0$

First equivalent version: dividing by **−2** both sides:

$$\frac{-4x + 8y - 12}{-2} = \frac{0}{-2}$$

$$2x + (-4y) - (-6) = 0$$

Equivalent equation: $2x - 4y + 6 = 0$

Second equivalent version: dividing by **−4** both sides:

$$\frac{-4x + 8y - 12}{-4} = \frac{0}{-4}$$

$$x + (-2y) - (-3) = 0$$

Equivalent equation: $x - 2y + 3 = 0$

Third equivalent version: dividing by **−4** both sides:

$$\frac{-4x + 8y - 12}{-4} = \frac{0}{-4}$$

$$x + (-2y) - (-3) = 0$$

Equivalent equation: $x - 2y + 3 = 0$

Based on the principles mentioned earlier, all four equations above are equivalent, which means they represent the same mathematical relationship. In other words, the following equations are equal to each other:

$$-4x + 8y - 12 = 0$$

$$2x - 4y + 6 = 0$$

$$x - 2y + 3 = 0$$

$$x - 2y + 3 = 0$$

At this point, you should be able to set an equation to ZERO and to find equivalent versions of the same equation, which allows us to review the elimination method.

Solving a system of linear equations of two or more variables using the elimination method.

The elimination method rests on the idea of adding or subtracting equations from each other in order to eliminate (cancel) unknown variables, to the point where we may determine the value of a chosen unknown variable. We have not done anything like this before, so allow me to illustrate.

Suppose you have the following two equations:

① $2x + 4y - 8 = 0$

② $4x + 8y + 2 = 0$

We may, if we chose to do so, add the equations. This is what the process would look like:

$$2x + 5y + 7 = 0$$

$$\mathbf{+} \quad (4x + 3y + 2 = 0)$$

$$6x + 8y + 9 = 0$$

Can you see how I added the two equations? It basically involves adding like terms on each of the two sides of the equations. Since **2x** plus **4x** is equal to **6x**, **5y** plus **3y** is equal to **8y**, and **7** plus **2** is **9**, and finally, since **0** plus **0** is **0**, the resulting equation is $6x + 8y + 9 = 0$.

It is important that you only combine like terms that are on the same side of the equations. Let me illustrate the do's and don'ts of this technique with the following set up.

(1) $\quad 3x = 4y + 5$

(2) $\quad 2y + 4 = 6x + 1$

If we had to add these equations, but left them exactly the way they are expressed, we would face the following situation:

$$3x = 4y + 5$$
$$\mathbf{+} \quad (2y + 4 = 6x + 1)$$

If you look at the equations closely, you will notice that the left side of equation **(1)** consists of only an **x** term, while the left side of equation **(2)** has two terms, but not an **x** term whatsoever. Therefore, the left side of these equations do not have any like terms that may be combined. This does not mean we cannot add the equations, it simply means that after adding the equations we will end up with three terms on the left side, all "**non-alike**", and which cannot be combined. On the other hand, the right side of equation **(1)** has a **y** term and an independent term (the "**5**"), while equation **(2)** has an **x** term and also an independent term. Therefore, we may combine the independent terms, but after adding these

equations we will end up with an **x** term and a **y** term on the right side. Since they are not like terms, they may not be combined. Observe the process.

$$3x = 4y + 5$$
$$\mathbf{+} \quad (2y + 4 = 6x + 1)$$

$$3x + 2y + 4 = 6x + 4y + 6$$

As you can see, we were only able to combine the like terms **5** and **1**, which when added, gave us the **6** that you see on the resulting equation. We could, of course, take the resulting equation and add and subtract the terms on either side in order to solve for a given variable, just like we saw in the previous chapter. But do note that we had to add each of the sides independently.

As I already mentioned earlier, the elimination method rests not only on the idea of adding equations but on subtracting them as well. Observe the subtraction process using the following example.

$$2x + 5y + 7 = 0$$
$$\mathbf{-} \quad (4x + 3y + 2 = 0)$$

$$-2x + 2y + 5 = 0$$

Can you see how I subtracted the two equations? You must be careful to **distribute** the subtraction operation as you perform the subtraction, just as I reviewed in earlier chapters of the book. What we are doing is this:

$$2x + 5y + 7 \; - \; (4x + 3y + 2)$$

and

$$0 \; - \; (0)$$

Therefore, the resulting left side is derived from the following addition/subtraction of terms:

$$2x + 5y + 7 - 4x - 3y - 2$$

Why? Because remember that when you subtract a collection of terms, you have to subtract each term as per to the Sign Table. At this point, you may proceed and combine like terms. That leads to the same resulting expression obtained earlier: $-2x + 2y + 5 = 0$.

Observe another example.

$$3x - 6y - 8 = 0$$

$$- \; (-3x + 5y - 9 = 0\,)$$

━━━━━━━━━━━━━━━━━

$$6x - 11y + 1 = 0$$

It is a good idea to use a set of parentheses to remind you that you have to subtract each term that is part of a given side, as follows:

$$3x - 6y - 8 = 0$$

$$- \; (-3x + 5y - 9 = 0\,)$$

━━━━━━━━━━━━━━━━━

$$6x - 11y + 1 = 0$$

Regardless of the technique that works best for you, just remember these principles as you add or subtract equations. Before we see the elimination in action, observe one last example in which the equations have terms both on the left side and the right side (other than a single **0**).

$$5x - 2y + 7 = 4x + y$$

$$- \; (-2x + y - 4 = 3y - 4\,)$$

━━━━━━━━━━━━━━━━━━━━

$$7x - 3y + 11 = \; 4x - 2y + 4$$

As you can see, the left side of the resulting equation was determined by performing the following operation:

$$5x - 2y + 7 \; - \; (-2x + y - 4)$$

$$5x - 2y + 7 + 2x - y + 4$$

$$7x - 3y + 11$$

While the right side was obtained by doing the following operation:

$$4x + y - (3y - 4)$$

$$4x + y - 3y + 4$$

$$4x - 2y + 4$$

Which is how the following resulting equation was obtained:

$$7x - 3y + 11 \; = \; 4x - 2y + 4$$

Finally, we are ready to solve a linear system of equations using the elimination method. Take the following system of two equations.

① $2x + y = 8$

② $x + y = 6$

The system consists of two unknown variables. In order to solve it, we want to cancel or eliminate one of the unknown variables, so that we are left with an equation that contains the other unknown variable, and a value (independent term), that would allow us to determine part of the solution of the system.

With this in mind, I will subtract equation **(2)** from equation **(1)** because it will eliminate the term that contains the **y** in the resulting equation. Observe.

$$2x + y = 8$$

$$- \quad (x + y = 6)$$

$$\text{------------------}$$

$$x = 2$$

We have found the value of one on the two unknown variables that solve both equations simultaneously. To find the other variable's value, we simply have to replace the **2** for **x** into one of the other two equations, and solve for the remaining unknown variable, **y**. I will use equation **(2)** for this. Observe.

$$x + y = 6$$

Replacing **2** for **x** and solving for **y**:

$$(2) + y = 6$$

$$2 + y - 2 = 6 - 2$$

$$y = 4$$

At this point, we are almost ready to conclude that the system is **Type I** (remember this from the previous chapter?) and that the solution would appear to be $x = 2$ and $y = 4$. In order to unequivocally state this, however, we must replace the variables with the values found into **both** equations and verify that the equality is, in

fact, maintained. It is the only way to determine the system type when using the elimination method. So let's test the solution found.

$$2x + y = 8$$

Using $x = 2$ and $y = 4$:

$$2(2) + (4) \overset{?}{=} 8$$

$$4 + 4 \overset{?}{=} 8$$

$$8 \overset{\checkmark}{=} 8$$

The solution did work out on the first equation. But what about on the second equation?

$$x + y = 6$$

Using $x = 2$ and $y = 4$:

$$(2) + (4) \overset{?}{=} 6$$

$$6 \overset{\checkmark}{=} 6$$

The solution worked on the second equation as well. At this point, we are ready to state the the the system of equations is **Type I**, and that the solution is $x = 2$ and $y = 4$.

The advantages of the elimination method with respect to the substitution method is that for some systems, the process of finding the solution is relatively easier, since if adding or subtracting the equations we are able to eliminate some of the unknown variables and end up with an equation with a single unknown variable, we can determine the variable's values very easily. However, we do need to check the solution that we obtain on all equations, so there is that to consider.

Try to use the elimination method to solve the following system of linear equations.

① $4x + 3y = 10$

② $4x + 5y = 14$

You may use the previous problem as a guide.

Ready to check your work? Let me solve it to show you one of the paths you could have taken to find the solution of this system of equations.

Taking into account the fact that both equations have the term **4x**, if we subtract equation **(2)** from equation **(1)** we will eliminate the **x** term from the resulting equation, allowing us to solve for **y** (we could subtract **(1)** from **(2)** instead, and it would still work). Observe.

$$4x + 3y = 10$$

$$- \ (\ 4x + 5y = 14 \ )$$

$$-2y = -4$$

At this point, all we need to do is solve for **y**...

$$-2y = -4$$

$$\frac{-2y}{-2} = \frac{-4}{-2}$$

$$y = 2$$

Now, we try to determine the value of **x** by replacing **y** with **2** on either one of the two equations of the system. I will use equation **(1)**.

$$4x + 3y = 10$$

Replacing **y** with **2** and solving for **x**:

$$4x + 3y = 10$$

$$4x + 3(2) = 10$$

$$4x + 6 = 10$$

$$4x + 6 - 6 = 10 - 6$$

$$4x = 4$$

$$\frac{4x}{4} = \frac{4}{4}$$

$$x = 1$$

We seem to have determined that the system is, once again, **Type I**, and that the solution of the system is $x = 1$ and $y = 2$. Remember, however, that you need to check the solution by replacing the variables in both equations and checking that the equality is maintained. I will test the solution on equation **(1)** and then on equation **(2)**.

① $4x + 3y = 10$

Using $x = 1$ and $y = 2$:

$$4(1) + 3(2) \overset{?}{=} 10$$

$$4 + 6 \overset{?}{=} 10$$

$$10 \overset{\checkmark}{=} 10$$

The solution did work out on the first equation. But what about on the second equation?

$$4x + 5y = 14$$

Using $x = 1$ and $y = 2$:

$$4(1) + 5(2) \overset{?}{=} 14$$

$$4 + 10 \overset{?}{=} 14$$

$$14 \overset{\checkmark}{=} 14$$

The solution worked on the second equation as well. At this point, we can safely state that the solution of this system is is $x = 1$ and $y = 2$, and that it is **Type I**.

Let's try another system.

① $2x - 4y = 2$

② $-4x + 2y = -10$

I suggest you try to solve it by yourself and then compare your result (and your conclusion) to mine (I will give you a hint: you need to find an equivalent version of either one of the equations, so that you can add or subtract one from the other, canceling one of the variables in the process).

Ready?

First, I will multiply equation **(1)** by the number **2** in order to obtain an equivalent equation in which the **x** term is **4x**; this will allow me to cancel the **x** by **adding** this new version of equation (1) which I will call equation (1') (one **prime**) with equation **(2)**. Remember that both sides must be multiplied by the same number in order to maintain the equality.

①   **(2)**$(2x - 4y = 2)$

①' $4x - 8y = 4$

Now, if I add equations **(1')** and **(2)**, I will achieve the goal of eliminating one of the unknown

variables. If you are concerned that I changed equation **(1)**, remember that equations **(1)** and **(1')** represent **the exact same** mathematical relationship, and are thus equal to each other.

$$4x - 8y = 4$$

$$\mathbf{+} \ (-4x + 2y = -10)$$

$$-6y = -6$$

At this point, all we need to do is solve for **y**...

$$-6y = -6$$

$$\frac{-6y}{-6} = \frac{-6}{-6}$$

$$y = 1$$

Now, we try to determine the value of **x** by replacing **y** with **1** on either one of the two equations of the system. I will use equation **(1)**; please note that I could have also used equation **(1')**, since it is, in essence, the same as equation **(1)**.

$$2x - 4y = 2$$

Replacing **y** with **1** and solving for **x**:

$$2x - 4(1) = 2$$

$$2x - 4 = 2$$

$$2x = 6$$

$$\frac{2x}{2} = \frac{6}{2}$$

$$x = 3$$

We seem to have determined that the system is, once again, **Type I**, and that the solution of the system is $x = 3$ and $y = 1$. Remember, however, that you need to check the solution by replacing the variables in both equations and checking that the equality is maintained. I will test the solution on equation **(1)** and then on equation **(2)**.

$$2x - 4y = 2$$

Using $x = 3$ and $y = 1$:

$$2(3) - 4(1) \overset{?}{=} 2$$

$$6 - 4 \overset{?}{=} 2$$

$$2 \overset{\checkmark}{=} 2$$

The solution did work out on the first equation. But what about on the second equation?

$$-4x + 2y = -10$$

Using $x = 3$ and $y = 1$:

$$-4(3) + 2(1) = -10$$

$$-12 + 2 = -10$$

$$-10 = -10$$

The solution worked on the second equation as well. At this point, we can safely state that the solution of this system is is $x = 3$ and $y = 1$, and that it is **Type I**.

Remember that a given system of linear equations can be either a **Type I** (one solution), **Type II** (infinite solutions) or **Type III** (zero solutions). So how do you identify a Type II and a Type III? Observe the next example. I suggest you try to solve it on your own, and see if you can

determine the type based on the review we did on the previous chapter. The only hint I will give you is that sometimes, in order to use the elimination method, you must multiply and/or divide **both** equations, not just one of them.

$$① \quad -4x + 2y = 8$$

$$② \quad -6x + 3y = 18$$

Ready to see the solution process I would use to solve this?

First, notice that it is not possible to eliminate one of the variables by adding or subtracting one of the equations from the other one the way they are currently expressed, because the equations do not have the same variable on the same side of the equation with the same (negative or positive) coefficient. If I focus on the, say, x term, I can make both equations have the same x coefficient by multiplying equation **(1)** by **3** and equation **(2)** by **2**. Observe.

$$① \quad \mathbf{(3)}(-4x + 2y = 8)$$

$$② \quad \mathbf{(2)}(-6x + 3y = 18)$$

Which leads to the following alternate versions of equations **(1)** and **(2)**:

$$①' \quad -12x + 6y = 24$$

$$②' \quad -12x + 6y = 36$$

Now, I can subtract equation **(2')** from equation **(1')** so that the x terms end up having opposite signs, effectively canceling out the x term (eliminating it).

Just remember that using a set of parentheses will help distribute the subtraction sign throughout the expression, on **both** sides of the equation:

$$-12x + 6y = 24$$

$$\mathbf{-}\ (-12x + 6y = 36)$$

- - - - - - - - - - - - - - - - -

$$0x + 0y = -12$$

$$0 = -12$$

At this point, we encounter a problem. In the process of eliminating one of the unknown variables (in this case, the x variable), both variables that make up the system of equations were canceled out. Furthermore, we ended ip with an expression that reads "zero is equal to negative twelve", which is, of course, not true. Therefore, based on the same principles that I reviewed previously, we may conclude that this system is **Type III**: it has ZERO solutions. There is nothing to check because we did not find a value for any of the variables.

What about this next system of linear equations? Give it a try. Before I let you go, though, I want to give you a hint: the elimination method is easier to apply if all of the equations that form the system are in the same format (for example, all the variables on the same side of the equation and the independent term on the other side, or all terms on one side and thus equal to zero, etc...). So, if the equations are not in the same format, just tweak them so they are by sending the appropriate terms to the side you want them to be in, so that both equations are in the same format.

① $-1 = -3x + y$

② $6x - 2y = 2$

Ready to check your solution process? Here are the steps I would follow to solve this system.

First, I will change the format of equation **(1)**. As I mentioned earlier, the elimination method works best when all equations that form the system have corresponding terms on the same side of the equation. Using equation **(2)** as a reference, I want to have the x and y terms of equation **(1)** on the left side of the equation, and the independent term on the right side of the equation. I can achieve this in one of two ways: I can add **3x**, subtract y and add **1** to both sides of the equation, or I can simply flip the sides and write the equation as $-3x + y = -1$ (this works because if we say that **a = b** we can say that **b = a**, remember?). I will, of course, use the latter method. Therefore, our system of equations now looks as follows:

① $-3x + y = -1$

② $6x - 2y = 2$

Moving on to the next step, I want to check if any of the like terms from both equations have the same coefficient (positive or negative); since this is not the case, I need to multiply or divide one or both of the equations in order to make it happen. In this case, focusing on the x terms of both equations, multiplying equation **(1)** by the number **2** will do the trick, since doing this will change the $-3x$ into $-6x$, which means its coefficient will be the same (though opposite-signed) as the x term of equation **(2)**. We don't really care how the other terms in equation **(1)** will be affected by multiplying said equation by **2** although we must remember to multiply all of the terms.

① $(2)(-3x + y = -1)$

Leads to the following alternate version of equation **(1)**:

①' $-6x + 2y = -2$

Now, I can add equations **(1')** and **(2)** to eliminate the **x** term.

$$-6x + 2y = -2$$

$$\boldsymbol{+} \quad (6x - 2y = 2)$$

$$0x + 0y = 0$$

$$0 = 0$$

Unlike the previous system of equations, this time, even though all of the variables that make up the system were eliminated (we were only trying to eliminate one of them, the **x** variable, but as an unintended consequence of our efforts, the other variable, the **y**, was also eliminated) we end up with a statement that reads "zero is equal to zero", which is true. We can therefore conclude that the system is a **Type II** system (in other words, it has an infinite number of solutions). We can come up with solution pairs just as we reviewed in the previous chapter.

Solving linear systems of equations of three or more variables (and thus of the corresponding three or more equations) using the elimination method is not as straight-forward as using the substitution method that we reviewed in the previous chapter. Therefore, I will not review the elimination method for linear systems of three or more variables.

And that brings another chapter to a close. See you on the next one!

17

INTRODUCTION TO FUNCTIONS

Functions play a major role in math. They are actually quite simple to work with, and they provide a powerful link between the idea of having an entity that receives input values (typically **REAL** numbers) and provides specific output values (again, typically **REAL** numbers as well).

I quickly mentioned them in **Chapter 1**, but in this chapter, we will go over them in detail.

Functions should be seen as a mathematical "entity" (or construct), formally a *relation*, that specifically relates a set of inputs with a set of permissible outputs, with a very important restriction: each input may only be related to exactly one output. Functions do not necessarily have to fully or partially involve numbers. Although this may surprise you, a vending machine may be seen as being a function (the input is the amount of money that you insert into the appropriate slot, while the output is the item selected). Observe the diagram.

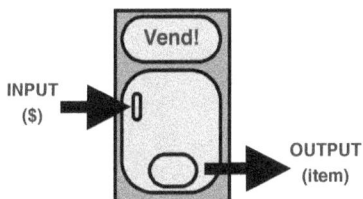

INPUT
($)

OUTPUT
(item)

Of course, in order to satisfy the condition that for a given input, only *one* output may be assigned, the vending machine would need to have a different price for all the products being sold. Thus, we would see the following table for our hypothetical (and *healthy!*) vending machine:

Input $	Output (item)
$0.50	Celery sticks
$1.00	Plain yogurt
$1.50	Fresh salad
$2.00	Fruit salad
$3.00	Raw walnuts

We could represent the same relationship by *mapping* the inputs with their corresponding outputs. Observe.

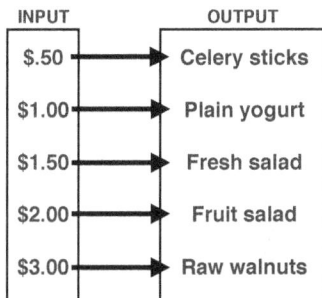

INPUT	OUTPUT
$.50	Celery sticks
$1.00	Plain yogurt
$1.50	Fresh salad
$2.00	Fruit salad
$3.00	Raw walnuts

Let's explore another example (this time, fully of the non-number kind)... Take a look at this input/output relationship:

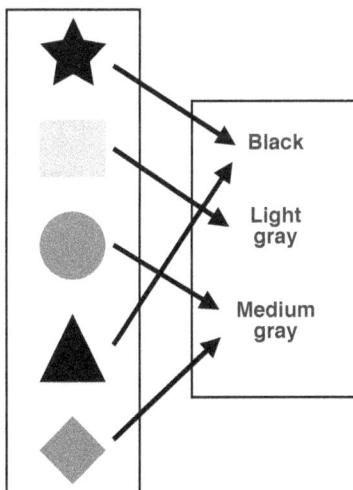

Black

Light gray

Medium gray

The diagram represents a relationship (which is what a function is defined to be), and it satisfies the condition that for a given input, only one output should be assigned. If you think the rule is being broken because two different inputs have the same output, then you are not reading the rule clearly. It only prohibits having for *one input*, two or more outputs. For example, star (this is the input) is related with black exclusively (the output). The fact that another figure, the triangle, also happens to have black as its output does not imply that the rule for functions has been broken.

Thus, this relationship is, in fact, a function. The simple idea behind this function is to consider as the input the color of the object, regardless of its shape, and the output is its color (in this case, a gray scale). Normally, you would see the following table (as opposed to the diagram presented initially):

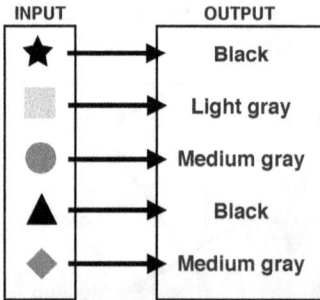

INPUT	OUTPUT
★	Black
■	Light gray
●	Medium gray
▲	Black
◆	Medium gray

Analyzing the table above (which represents the same function that the diagram represents), you can see that for a given input (take your pick) *one and only one* output is related. It is thus a function.

Take a look at the following diagram.

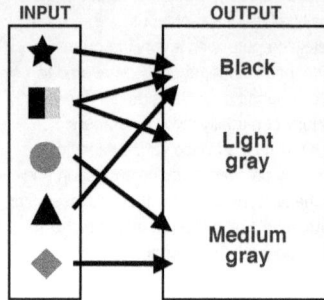

Can you see if there is any reason why the relationship above should *not* be called a function? Look closely at the input/output pairings...

Well, it turns out that the above relationship cannot be called a function. The reason is simple: there is one input (the two-shaded square) that is related to two (different, of course) outputs; and that breaks the rule for functions. Think of it this way: using the relationship above, if I asked you for the output value given an input of "black-shaded star", you would unequivocally say "black"; however, if I asked you for the output value (note that it is singular) given an input of "two-shaded square", what would you answer? You would waver between "black" and "light gray", and you cannot unequivocally state that there is a single answer to that question. And because of that, the relationship *cannot* be deemed to be a function.

To sum up, remember that a function must have, for any of its given (defined) input values, one and only one output. There may, however, be different inputs that have the same output value related (for this does not break the rule, think about it).

Test your mastery of this concept by looking at the following relationships and identifying each one of them as a function or as a non-function.

a)

Input	Output
1	red
2	blue
3	red
4	blue
5	blue

b)

Input	Output
a	1
b	2
c	3 or 5
d	4
e	9

c)

Input	Output
@	+
%	-
&	>
$	/
#	<

d)

Input	Output
-2	4
-1	1
0	0
1	1
2	4

So what did you conclude? Can you correctly state if each of the above relationships may be called a function?

The first relationship a) is, in fact, a function. Why? Because look at any one of the five inputs that it contains: they each have a single output assigned to them. So, even though the same outputs appear more than once, that, as I explained earlier, is not an issue.

The next relationship, example **b)**, cannot be called a function because one of its inputs, **c**, has two possible outputs: either **3** or **5**. As I explained earlier, functions are not allowed to have such a relationship between the input value and its output value.

Finally, examples **c)** and **d)** qualify as functions because they adhere to the rule that for a given input, only one output may be assigned. Again, having the same output for two or more given inputs does not prevent a relationship from being considered a function.

Let's explore the more common type of function that is used in algebra, involving mathematical expressions with variables and mathematical operations.

First, we need to know how function notation works (in other words, the symbols and structure used in expressing functions).

I'll start by showing you an example of how a function is written mathematically:

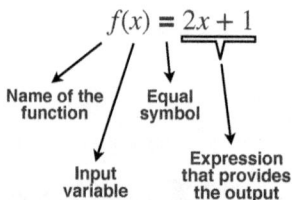

$$f(x) = 2x + 1$$

Name of the function

Equal symbol

Input variable

Expression that provides the output

It is important to understand that the set of parentheses on the left side of the equal symbol surround the input variable, and that it should be considered a "label" and not a mathematical expression (in other words, it should **NOT** be

interpreted as stating "**f** times **(x)**". Instead, the left side reads as follows:

$$f(x)$$
"**f** of **x**" ; in other words, "**ef of ex**"

In this example, the function's name is "**f**", and it is defined in terms of the variable "**x**". To the right of the equal symbol, we find the actual function itself; it is the mathematical expression that provides a way to compute an output, based on a given input. In this example, the expression involves multiplying the variable's input value (represented by **x**) times **2** and then adding **1** to the product.

Remember how functions are defined as a relational set, in which for a given input there is one and only one output? Well, function notation allows us to compute any output for a given input, as long as the input is part of the domain of the function.

In this example, we could create as many input/ output paris as we needed or cared to make. To do this, simply consider an input value, and replace that input value throughout the left and right side of the equal symbol wherever the input variable appears, using a set of parentheses when appropriate. In the example that we are working with, we would replace **x** with the input value, and then compute the output using the rules and principles that we have already reviewed (order of operations, etc...). For example:

$$f(-1) = 2(-1) + 1 = -2 + 1 = -1$$

$$f(0) = 2(0) + 1 = 0 + 1 = 1$$

$$f(.5) = 2(.5) + 1 = 1 + 1 = 2$$

$$f(\pi) = 2(\pi) + 1 = 2\pi + 1$$

$$f(\frac{11}{3}) = 2(\frac{11}{3}) + 1 = \frac{22}{3} + 1 = \frac{25}{3}$$

$$\vdots$$

Which means that this function, called "**f**", yields the following input/output pairs (**x** , **f(x)** pairs):

$$f(x) = 2x + 1$$

Input x	Output f(x)
-1	-1
0	1
0.5	2
π	2 π + 1
$\frac{11}{3}$	$\frac{25}{3}$

Of course, we could keep going... this particular function's domain could be defined as being all **REAL** numbers, which means that we could replace the input variable **x** with any **REAL** number that spans from negative infinity to positive infinity. The above table is only a sampling of the input/output pairs that the function defines (**relates**). I used both **RATIONAL** and **IRRATIONAL** numbers as input values to remind you that they may all be used as input values in a given function (unless the domain specifically states otherwise).

To help you fully understand how function notation works, let me show you some additional examples of functions. The list will include functions that are named differently, and defined

using various mathematical expressions. The input variable used will change as well.

Observe.

$$f(x) = 2x + 1$$
$$g(x) = x - 4$$
$$h(x) = x^3 - 2x^2 + x - \pi$$
$$i(y) = 3y - 2$$
$$f(m) = m$$
$$j(x) = 2$$

As you can see from the list above, the name of the function appears to the left of the equal symbol ("**f**", "**g**", "**h**", "**i**", "**f**", and "**j**") along with the input variable being used ("**x**", "**x**", "**x**", "**y**", "**m**", and "**x**", respectively), while on the right side of the equal symbol we find the function itself, which allows us to figure out the output based on a given input value that would replace the specified input variable.

Notice that the last function is defined in terms of the variable **x** , and yet the right side does not contain any variables; this is a perfectly valid arrangement. In the case of this function, $j(x) = 2$, regardless of the input value used, the output will always be equal to **2**. For example, using the following input values, the input/output table would look as follows:

$$j(x) = 2$$

Input x	Output j(x)
-3	2
-2	2
-1	2
0	2
1	2

At this point, you should be able to look at a function that is expressed using mathematical function notation and identify its name, the input variable that is used to define it, and the mathematical expression that provides the means of figuring out the output value given a specific input value that one may wish or need to use.

It is worth mentioning that the act of finding an output value given a specific input value can be referred to as "evaluate the function when the variable is equal to ---" or as "plug in the number --- into the function and evaluate" or simply as "find (**function name**) of (**input value used**)". This last version is used the most often, so it is worth clarifying it. Let's say that you are asked to evaluate the following function when the input value is equal to **4** :

$$f(x) = 3x + 10$$

I could simply ask you to "find **ef** of **four**", which is how "**f(4)**" is read out-loud. Which leads to the following computation:

$$f(4) = 3(4) + 10$$
$$f(4) = 12 + 10$$
$$f(4) = 22$$

Based on the above result, "**ef** of **four** is equal to **twenty-two**" which is how "$f(4) = 22$" is read out-loud.

Please note that from now on, I will use–for the most part– **f**, **g**, and **h** as names of functions, since those are the letters that are most commonly used in most math courses and text books.

Try to solve the following problems on your own.

a) $f(x) = 5x - 3$; find $f(1)$

b) $g(y) = 2y + 4$; find $g(-2)$

c) $h(a) = -a^2 + 3$; find $h(3)$

d) $f(w) = -w^3 - 1$; find $f(-1)$

e) $g(m) = m^2 + 4m - 2$; find $g(2)$

Ready to check your answers?

a) $f(x) = 5x - 3$; find $f(1)$

$$f(1) = 5(1) - 3$$
$$f(1) = 5 - 3$$
$$f(1) = 2$$

b) $g(y) = 2y + 4$; find $g(-2)$

$$g(-2) = 2(-2) + 4$$
$$g(-2) = -4 + 4$$
$$g(-2) = 0$$

c) $h(a) = -a^2 + 3$; find $h(3)$

$$h(3) = -(3)^2 + 3$$
$$h(3) = -(9) + 3$$
$$h(3) = -6$$

d) $f(w) = -w^3 - 1$; find $f(-1)$

$$f(-1) = -(-1)^3 - 1$$
$$f(-1) = -(-1) - 1$$

$$f(-1) = 1 - 1$$

$$f(-1) = 0$$

e) $g(m) = m^2 + 4m - 2$; find $g(2)$

$$g(2) = (2)^2 + 4(2) - 2$$

$$g(2) = 4 + 8 - 2$$

$$g(2) = 10$$

So how did you do? If you did not solve all of these problems correctly, make sure that you identify your mistakes and reread this section in order to make sure that you have mastered these concepts.

So far, we have seen several ways in which the input/output values for a given function are expressed. We may use tables, diagrams, or simply express the related values using function notation (as in "$g(2) = 10$"). But there is another option, one that is commonly used in math and algebra in general. It involves the concept of an ordered pair of values (in the case of functions with one input variable). In general, an ordered pair is defined as follows:

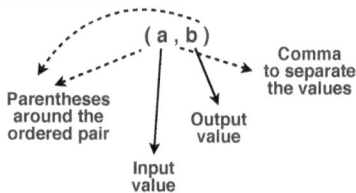

It is important to understand two key principles involving this notation: first, in an ordered pair, a set of parentheses **must** be used around the pair of values, and second, the **input** value goes first,

while the **output** value must be placed after the comma.

Revisiting the set of problems that we solved earlier, we could express their answers as follows:

a) (1 , 2)

b) (–2 , 0)

c) (3 , –6)

d) (–1 , 0)

e) (2 , 10)

Therefore, we may say that for functions in general, ordered pair notation implies the following:

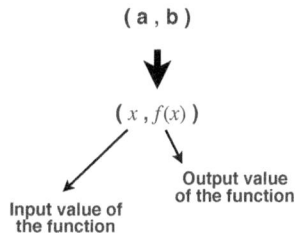

If the expression $(x , f(x))$ confuses you, remember that for a given function (I am using $f(x)$ to represent any function in general), the part "$f(x)$" is one of the ways in which we express the function's output value, as in the example we encountered earlier $f(1) = 2$. In said example, the ordered pair would be $(1 , f(1))$, which is to say $(1 , 2)$, since in said example $f(1) = 2$. Seems a bit redundant, but it is a powerful mathematical notational tool.

Observe these additional examples that should help you understand the notation $(x, f(x))$.

a) $f(x) = x^2 - x$; find $(3, f(3))$

$$f(3) = (3)^2 - (3)$$

$$f(3) = 9 - 3$$

$$f(3) = 6$$

...therefore, this function contains the following ordered pair as part of its solution set:

$$(3, f(3))$$

↓

$$(3, 6)$$

...which means that in the case of this function, and **input** of **3** results in an output of **6**

b) $g(x) = -3x - 10$; find $(2, g(2))$

$$g(2) = -3(2) - 10$$

$$g(2) = -6 - 10$$

$$g(2) = -16$$

...therefore, this function contains the following ordered pair as part of its solution set:

$$(2, g(2))$$

↓

$$(2, -16)$$

...which means that in the case of this function, and **input** of **2** results in an output of **–16**

c) $h(x) = -3$; find $(0, f(0))$

$$f(0) = -3$$

...therefore, this function contains the following ordered pair as part of its solution set:

$$(0, f(0))$$

↓

$$(0, -3)$$

...which means that in the case of this function, and **input** of **0** results in an output of **–3**

By now you should be very comfortable using function notation (to express a function itself, to evaluate a function, and to express the input/output values that have been computed).

As I specified earlier, functions are usually named *f*, *g*, and *h*; furthermore, the input variable is usually *x*. This does not mean that other letters/variables cannot be used, but rather that for the sake of simplicity and consistency, only said letters and variables are typically tapped.

The domain and range of a function

In **Chapter 14** I mentioned to you the importance of considering the domain and the range of literal equations (and equations in general). The same principle applies to functions as well.

From a purely mathematical standpoint, we typically look at functions and consider the **REAL** number system as the source of the function's domain (the allowed input values): we then exclude any number or numbers that would yield an undefined output. Please note, however, that there are other numbers systems that may be considered, such as the **IMAGINARY** numbers that we reviewed in **Chapter 12** which, combined with the **REAL** number system, is referred to as the **COMPLEX** number system.

From a real-world application standpoint, functions may have a different domain than the domain derived from a purely mathematical standpoint (with the former being smaller). For example, if the function $f(x) = 4x$ is used to determine the perimeter of a perfect square where x represents the length of its sides, we would specify the domain of this function as all **REAL** numbers greater than or equal to 0 (allowing 0 as an input and interpreting it as the **absence** of a square) since allowing negative numbers wouldn't make any **real-world** sense (strictly speaking and without bringing in additional properties such as orientation, what would a measurement of, say, **–10 feet** mean?); note that from a purely mathematical standpoint, the domain of the function $f(x) = 4x$ is all **REAL** numbers, since using any **REAL** number as an input would never yield an undefined output (in other words, we would never end up with something divided by zero, or zero raised to the power of zero, or the even root of a negative).

Generally speaking, and unless otherwise stated, use the purely mathematical perspective when specifying the domain of a function.

The range, as you may remember, depends entirely on the domain (the allowed input values) and the function itself. It is usually a lot more difficult to specify the range. I will not review the range at this point in the book; instead, I will cover this after we explore the concept of graphing functions.

Take a look at the following functions, and try to figure out their domain (remember, from a purely mathematical standpoint).

a) $f(x) = x^2$

b) $g(x) = \dfrac{2}{x}$

c) $h(x) = \sqrt{2x}$

d) $f(x) = \dfrac{2x}{5}$

Ready?

When specifying the domain of a function, simply try to identify if there are any numbers that if they were used as input values in the function, they would yield an undefined output. Remember that under the **REAL** number system, the following setups yield undefined answers:

Dividing by 0:

$$\dfrac{anything}{0} = \textbf{UNDEFINED}$$

Raising 0 to the power of 0

$$0^0 = \textbf{UNDEFINED}$$

The even root of a negative

$$\sqrt[even]{negative} = \textbf{UNDEFINED}$$

The first function, $f(x) = x^2$, will never yield an undefined output for any **REAL** number that is used as an input. Therefore, its domain is: all **REAL** numbers.

The second function, $g(x) = \dfrac{2}{x}$, can potentially yield an UNDEFINED output. Can you see what

REAL number needs to be excluded from the domain? If you said **0** , that is correct; any other REAL number will not present a problem. Therefore, the domain of this function is defined as "all REAL numbers, **x ≠ 0**", or using the more advanced mathematical notation

$$\{\, g(x) \mid \boldsymbol{x} \in \mathbb{R} \,;\, \boldsymbol{x \neq 0} \,\}$$

which may be read as

"the domain of "gee" of "ex" is all "exes" such that "ex" belongs to the REAL numbers, with the exception that ex cannot be zero"

This notation uses the following symbols that merit an explanation: the vertical bar, " | ", which separates the function's name from the domain itself; "∈", a symbol used in set notation, which stands for "belongs to"; and finally, the symbol $\mathbb{R}$ which represents the REAL number system; the brackets surrounding the entire statement are used to enclose it.

Moving on to the next function, $h(x) = \sqrt{2x}$, can you determine if there are any numbers that we would need to exclude from the domain? Remember that we are currently only considering the REAL number system.

Where you able to determine that if you allowed the variable **x** to be replaced with a negative REAL number, the function would have an UNDEFINED output? Observe some examples:

$$h(-1) = \sqrt{2(-1)}$$
$$h(-1) = \sqrt{-2}$$
$$h(-1) = \textbf{UNDEFINED}$$

$$h(-2) = \sqrt{2(-2)}$$
$$h(-2) = \sqrt{-4}$$
$$h(-2) = \textbf{UNDEFINED}$$

⋮

As you can see, if we are only considering the REAL number system, this function needs to exclude all negative numbers from its domain, otherwise, it would result in an UNDEFINED output. Try to express the domain using the more advanced mathematical notation that I presented to you on the previous function...

Did you try it? If so, check to see if you were able to figure it out:

$$\{\, h(x) \mid \boldsymbol{x} \in \mathbb{R} \,;\, \boldsymbol{x \geq 0} \,\}$$

which may be read as

"the domain of "h" of "ex" is all "exes" such that "ex" belongs to the REAL numbers, such that ex must be greater than or equal to zero"

What about the last function on the list?

$$f(x) = \frac{2x}{5}$$

What is its domain? This time, can you see that we may replace **x** with any REAL number whatsoever? Don't be thrown off by the denominator: since the numerator is being divided by 5 , it really doesn't matter what the numerator computes to for a given **x** input. Observe:

$$f(10) = \frac{2(10)}{5}$$

$$f(10) = \frac{20}{5}$$

$$f(10) = 4$$

$$f(1) = \frac{2(1)}{5}$$

$$f(1) = \frac{2}{5}$$

$$f(1) = .4$$

$$f(0) = \frac{2(0)}{5}$$

$$f(1) = \frac{0}{5}$$

$$f(1) = 0$$

In the case of $f(0)$, remember that dividing **0** by **anything other** than **0** is perfectly acceptable: it is always equal to **0**. What is not defined is dividing **by 0**, including **0** divided by **0**.

$$f(-1) = \frac{2(-1)}{5}$$

$$f(1) = \frac{-2}{5}$$

$$f(1) = -.4$$

$$\vdots$$

So how is this function's domain expressed? Well, since it includes all **REAL** numbers, simply as:

$$\{ f(x) \mid \textbf{\textit{x}} \in \mathbb{R} \}$$

At this point you should be able to read and write domain notation, and you should be able to evaluate functions, expressing the input/output pairs in the various methods that we reviewed in this chapter. To check your ability to do all these tasks correctly, try to solve the following problems.

Express the domain of the function using the advanced mathematical notation used above, evaluate the function based on the input specified, and express the input/output as an order pair.

a) $f(x) = x^2 + 1$; $f(3)$

b) $g(x) = \sqrt{4x}$; find $g(0)$

c) $h(x) = \dfrac{3 + x^2}{2x}$; find $h(1)$

d) $f(x) = \dfrac{x}{x + 2}$; find $f(0)$

e) $g(x) = \sqrt{2x + 20}$; find $g(-2)$

When you try to solve problems d) and e), make sure you think about the setup that you need to avoid, and what input would lead to said setup... If you fond those two problems confusing, evaluate them using their respective specified inputs, and maybe that will help you to "*see the light*" (figure out what to do).

Ready to check your work?

a) $f(x) = x^2 + 1$; $f(3)$

Domain: $\{ f(x) \mid \textbf{\textit{x}} \in \mathbb{R} \}$

$$f(3) = (3)^2 + 1$$

$$f(3) = 9 + 1$$

$f(3) = 10$

$$\downarrow$$

$$(\,3\,,10\,)$$

b) $g(x) = \sqrt{4x}$; find $g(0)$

Domain: $\{\,g(x) \mid \textbf{\textit{x}} \in \mathbb{R}\;;\textbf{\textit{x}} \geq \textbf{0}\,\}$

$$g(0) = \sqrt{4(0)}$$

$$g(0) = \sqrt{0}$$

$$g(0) = 0$$

$$\downarrow$$

$$(\,0\,,0\,)$$

c) $h(x) = \dfrac{3 + x^2}{2x}$; find $h(1)$

Domain: $\{\,h(x) \mid \textbf{\textit{x}} \in \mathbb{R}\;;\textbf{\textit{x}} \neq \textbf{0}\,\}$

$$h(1) = \frac{3 + (1)^2}{2(1)}$$

$$h(1) = \frac{3 + 1}{2}$$

$$h(1) = \frac{4}{2}$$

$$h(1) = 2$$

$$\downarrow$$

$$(\,1\,,2\,)$$

d) $f(x) = \dfrac{x}{x + 2}$; find $f(0)$

Domain: $\{\,f(x) \mid \textbf{\textit{x}} \in \mathbb{R}\;;\textbf{\textit{x}} \neq \textbf{-2}\,\}$

In this case, we need to exclude all **REAL** numbers that would make the denominator $x + 2$ be equal to **0**; this sets up the following expression:

$$x + 2 \neq 0$$

Remember, we can't have the denominator be equal to **0**, which is why the expression is written the way it is. Subtracting **2** to both sides of the expression solves it...

$$x + 2 - 2 \neq 0 - 2$$

$$x \neq -2$$

Therefore, the domain excludes this number. If you cannot see why, observe what happens if we were to allow it as part of the domain and tried to evaluate the function using **−2** :

$$f(-2) = \frac{(-2)}{(-2) + 2}$$

$$f(-2) = \frac{-2}{-2 + 2}$$

$$f(-2) = \frac{-2}{0}$$

$$f(-2) = \text{UNDEFINED}$$

As you can see, the function cannot contain **−2** as part of its domain. Moving on to the other part of the problem (evaluating $f(0)$):

$$f(0) = \frac{(0)}{(0) + 2}$$

$$f(0) = \frac{0}{2}$$

$$f(0) = 0$$

↓

(0 , 0)

e) $g(x) = \sqrt{2x + 20}$; find $g(-2)$

Domain: { $g(x)$ | $x \in \mathbb{R}$; $x \geq -10$ }

In this case, we need to exclude all **REAL** numbers that would make the inside of the root symbol take on a negative value. This sets up the following equation that can be used to find said number:

$$2x + 20 \geq 0$$

...subtracting 20 to both sides of the expression...

$$2x + 20 - 20 \geq 0 - 20$$

$$2x \geq -20$$

...dividing both sides by 2 ...

$$\frac{2x}{2} \geq \frac{-20}{2}$$

$$x \geq -10$$

Therefore, the domain may only *include* **REAL** numbers that are greater or equal to **−10** . Any input less than **−10** will result in a root of a negative number, which is UNDEFINED. Observe what happens if we were to use **−11** :

$$g(-11) = \sqrt{2(-11) + 20}$$

$$g(-11) = \sqrt{-22 + 20}$$

$$g(-11) = \sqrt{-2}$$

$$g(-11) = \text{UNDEFINED}$$

As you can see, the function cannot contain any **REAL** numbers less than **−10** as part of its domain, because you would end up with an UNDEFINED output (in this case, with even roots of negative numbers). Moving on to the other part of the problem (evaluating $g(-2)$):

$$g(-2) = \sqrt{2(-2) + 20}$$

$$g(-2) = \sqrt{-4 + 20}$$

$$g(-2) = \sqrt{16}$$

$$g(-2) = 4$$

↓

(−2 , 4)

As a final example, consider the following function and try to define its domain.

$$f(x) = \frac{-2}{4x + 5}$$

It is similar to several of the examples solved previously, so you may use them as a guide to help you tackle this one.

Ready to see how to correctly figure out the domain of the function?

First, you should take note of the fact that this function has the variable *x* on the function's denominator. This means that there exists a particular **REAL** number that if allowed to replace the x variable, the denominator would end up evaluating to zero, and as we have previously determined, that would lead to an UNDEFINED output. The question is, how do we find it so we may exclude it from the domain?

Well, similar to how we determined this in previous examples, we can find this particular **REAL** number by setting up the following expression:

$$4x + 5 \neq 0$$

Again, we cannot allow the denominator to be equal to zero, so we set up that precise possibility, find which input value would cause it, and then simply exclude it from the domain.

Now, solving the above expression for x...

$$4x + 5 \neq 0$$

$$4x + 5 - 5 \neq 0 - 5$$

$$4x \neq -5$$

$$\frac{4x}{4} \neq \frac{-5}{4}$$

$$x \neq \frac{-5}{4}$$

And we have found the **REAL** number that must be excluded from this function's domain. Thus, we may define the following:

$$f(x) = \frac{-2}{4x + 5}$$

$$\{ f(x) \mid x \in \mathbb{R} ; x \neq \frac{-5}{4} \}$$

which may be read as

"the domain of "ef" of "ex" is all "exes" such that "ex" belongs to the **REAL** numbers, with the exception that ex cannot be negative five fourths"

Functions defined by using set notation

There is another format that can be used to define a given function, one that uses set notation. It is very similar to the first examples that I presented to you at the beginning of this chapter.

Let's revisit the following function (abbreviating the names of the items):

Input $	Output (item)
$0.50	C (Celery)
$1.00	Y (Yogurt)
$1.50	S (Salad)
$2.00	F (Fruit)
$3.00	W (Walnuts)

As you may recall, this table represents the "healthy" vending machine that can be thought of as a function. The input is the amount of money that you have to put into the machine, and the output is the corresponding item that you may thus be able to purchase.

We could represent the exact same function using the concept of a set. Mathematically speaking, and keeping it simple for the moment, a set may be seen as a collection of distinct objects. Now, using the concept of paired values (ordered sets, as in (input , output) which I reviewed earlier) and that of a set, we can define the function as follows (let's call it $f(x)$):

$f(x) = \{(.5,C), (1,Y), (1.5,S), (2, F), (3,W)\}$

Can you see how this set notation works? It uses the concept of ordered pairs, grouping them between a set of curly brackets. Since this function consists of only 5 possible input/output pairings, we see that same number of ordered pairs between the brackets.

It is important to remember that the input value is always listed first within the ordered pair followed by the the output value, separated by a comma:

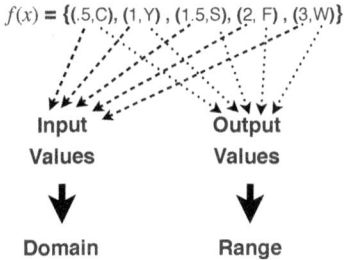

$f(x) = \{(.5,C), (1,Y), (1.5,S), (2, F), (3,W)\}$

Input Values   **Output Values**

↓   ↓

Domain   **Range**

As you can see from the diagram above, the input values define the function's **domain**, while the output values define its **range**. As you may recall, I mentioned the fact that defining the range of a function is typically more challenging than defining the domain and this is true when the function is defined using mathematical operations; however, when using the set notation above, it is very easy to define the both.

In the example above, the domain is:

.5, 1, 1.5, 2, and 3

While the range is:

C, Y, S, F, W

Let's explore the following relations. The goal is to identify if they may be called functions (remember the rule?) and to specify their domain and range. Try to answer those questions yourself before reading on.

a) $f(x) = \{ (0 , 1),(1 , 2),(2 , 5),(3 , 10) \}$

b) $g(x) = \{ (-1 , 5),(0 , 5),(1 , 5),(2 , 5) \}$

c) $h(x) = \{ (1 , 2),(1 , 3),(1 , 4),(1 , 5) \}$

Ready?

The first relationship, $f(x)$, is certainly a function. Each input is mapped to one and only one output. The domain is **0, 1, 2, and 3** while the range is **1, 2, 5, and 10**.

The second relationship, $g(x)$, is also a function. Each input is mapped to one and only one output (even though they all map to the same output value of **5**: remember, that is acceptable for a function). The domain is **–1, 0, 1, and 2** while the range is **5**.

Finally, did you notice something important about the third relationship that would not allow us to call it a function? The same input maps to several different outputs. And that, as we have already seen, is **not** allowed for functions. Therefore, the relationship $h(x)$ is **not** a function. Its domain is **1**, while its range is **2, 3, 4, and 5**.

At this point, I am going to end the chapter, since it is only meant to be an introduction to functions. Over the next several chapters, we will be exploring this in more detail, including how to graph a function and how to find the inverse of a function.

See you soon!

18

LINEAR EQUATIONS

In the previous chapter I briefly introduced to you the concept of a function. I am now going to talk about a particular type of function, called "linear functions". They are also known as "equations of lines" or simply "linear equations", if presented as an equation and not in function form.

I will begin with the equation form of these linear constructs, and then transition to the function form; you will see how simple it is to do this transition, since both represent the exact same mathematical relationship. The only thing that changes is the way in which the relationship is presented. It's similar to, say, how we can represent the **REAL** number **2** as "**2**" (**INTEGER** form), or as $\frac{2}{1}$ (fraction or **RATIONAL** form); in essence, they both represent the same idea, that of the number two. Well, these mathematical "constructs" called lines can be presented in equation form or in function form. Either way, both point to the same construct.

So what are these "lines" anyway? Well, the problem with this question is that typically, humans asking it are thinking in colloquial terms (in other words, from their every-day, common-sense, perspective). Fortunately, the *mathematical lines* that we need to explore here closely adhere to this "common sense, every-day meaning" that most think about upon hearing that word. There are, however, a few things that we need to clearly specify and define in order to avoid any confusions and misunderstandings regarding lines (from a mathematical point of view, that is).

The simplest way to define a line, adhering to the scope of this book (which is the definition typically used in *analytic geometry*), is to say that a *line* is the set of points whose coordinates satisfy a given "*linear equation*". Seems redundant? It's not... you see, we need to make a distinction between a "line" and a "linear equation". Let's begin by defining a linear equation.

We say that an equation is linear if it is of the following form:

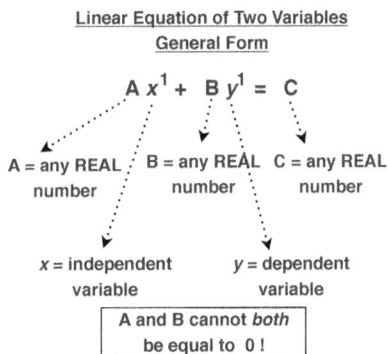

Linear Equation of Two Variables
General Form

$$A\,x^1 + B\,y^1 = C$$

A = any REAL number   B = any REAL number   C = any REAL number

x = independent variable   y = dependent variable

A and B cannot *both* be equal to 0 !

Don't be overwhelmed by the diagram... a linear equation of two-variables is simply made up of two variables raised to the power of **1** (or, with an exponent of **1**)—we generally use **x** and **y** to denote said variables, but you could use any symbol you wanted, if the symbols are defined as being variables—where the variables are adding each other, being equal to any **REAL** number whatsoever. Furthermore, the variables themselves may have any **REAL** number multiplying them (these are called **constants**, as opposed to variables, because they are not meant to change once a specific linear equation is defined, whereas variables are not *value-fixed*). These constants are called coefficients of the variables, remember? For example:

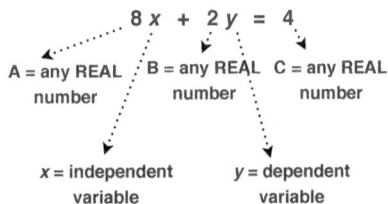

$$8\,x + 2\,y = 4$$

A = any REAL number   B = any REAL number   C = any REAL number

x = independent variable   y = dependent variable

Observe that $8x + 2y = 4$ is in fact linear since it adheres to (satisfies) the definition; in this case, the assigned values of **A**, **B**, and **C** are **8**, **2**, and **5**, respectively; the variables happen to be **x** and **y**, though this need not be the case, and they are raised to the power of **1** (that is to say, they have an exponent equal to **1**).

Of course, since you now know how to manipulate equations, you should be able to see that any of the following "versions" of this linear equation **represent** the **exact same** linear equation; they simply have the terms "collected" in different sides of the equation:

Original version

$$8x + 2y = 4$$

⬇

Modified versions (all **equivalent**)

(1) $2y = -8x + 4$

(2) $8x = -2y + 4$

(3) $8x + 2y - 4 = 0$

(4) $2y + 8x = 4$

All of the above equations represent the exact same mathematical relationship; they **seem** different because the terms have been moved around in subsequent versions of the same equation (for example, alternate version (1) is obtained by taking the original version and subtracting **8x** from both sides of the equation; alternate version (2) is obtained by taking the original version and subtracting **2y** from both sides of the equation, etc...). This is the type of

equation manipulation that we reviewed in recent chapters of the book, remember?

The defining characteristic about a linear equation is that the variables **MUST** have an exponent of **1** and that the coefficients of both the **x** and the **y** variable (or whatever variables are being used) cannot **both** be **0** for a given equation. This basically means that at least one of the two variables must be present in the equation.

Applying the rules "**to the letter**" leads to some interesting linear equations. For example:

(a) $y = x$

(b) $y = 0$

(c) $x = -5$

(d) $0 = x$

(e) $y = 3^2$

(f) $y = (2.5)^4 x + 3$

(g) $y = -1$

(h) $x = 1$

Before you complain, let me warn you that yes, **all** of these **are** linear equations, believe it or not! The first example, **(a)**, certainly qualifies: the variables have an exponent of one, and the constant **C** has a value of **0**.

Example **(b)** seems odd since one of the variables is missing; however, this just means that the missing variable's **coefficient** is equal to ZERO! Observe:

$$y = 0$$

is the same as

$$0x + y = 0$$

Remember, the coefficients **A**, **B**, and **C** may be equal to **any** REAL number, including ZERO, although **A** and **B** cannot both be equal to **0** simultaneously. Can you now see why this equation is, in fact, a linear equation?

At this point, before exploring the remaining examples, I wish to clarify something important. It is absolutely necessary to have proper context in order to know what type of mathematical construct you are dealing with. Take, for example, the equation we discussed above, $y = 0$. If you were walking down the street, enjoying the morning sun and the crisp spring weather, and you suddenly bumped into it (never mind the psychological implications of said event), what would you call it? A linear equation based on what we reviewed moments ago? But maybe it's the answer of an equation that is expressed in terms of *y*, or part of the solution of a **system of equations**, like we reviewed in **Chapters 14**, **15** and **16**; or maybe it's an "instruction", telling you to assign the value of **0** to the variable *y* in a given expression, or–well, you get the idea, I hope: there are numerous (and endless?) possibilities. So without proper context (as in, knowing where this is being used, or for what, or where it came from, or why it was created, etc...), it would be impossible to know what to call it.

Therefore, if you are working with linear equations, and in that context you are asked to work with $y = 0$, you should immediately recognize it as a linear equation and not as some of the other equally viable mathematical

constructs that it may represent as well. And so based on the context (it is a linear equation), it is immediately obvious that it is missing a variable. Which variable would be obvious based on the context of the problems being explored (if *x*'s and *y*'s are being used, then you would know that the missing variable is *x* and that it has a coefficient of **0** , which is why it is not shown, since **0** times *x* is equal to **0**); if, however, *y*'s and *z*'s are being used, then you would know that the missing variable is *z*, with a coefficient of **0** , etc...). In other words, "know the context, know it all".

Moving on, example **(c)**, $x = -5$, is also a linear equation, and based on the context (and the fact that I'm using *x*'s and *y*'s only), I am sure you can now say what the missing variable is and the value of its coefficient... That's right, the missing variable would be *y* and its coefficient is **0** : $x + 0y = 0$.

Example **(d)**, $0 = x$, is also a linear equation. The term $0y$ is not explicitly written since it is equal to **0** ; furthermore, the constant *c* is equal to **0** . There is absolutely nothing wrong with this. Remember that the only restriction regarding the value of **0** is that it cannot be simultaneously assigned (on the same equation) to both of the equation's variables. But one of the variable's coefficients **and** *c* may both be equal to **0** , simultaneously, on a particular equation.

Example **(e)**, $y = 3^2$, can be confusing because when the exponent (a **2** in this case) is noted, it is assumed that it is breaking the rule that linear equations **must** have variables with exponents of **1** . But read the rule carefully: *only* the *variables* are the elements that have the restriction, not the constants! Therefore, it is a linear equation. In fact, it can be simplified, of course, as $y = 9$.

After reading this, I am sure that if you revisit example **(f)**, you will be able to identify it as a linear equation: $y = (2.5)^4 x + 3$. The exponent, **4**, is attached to the coefficient of the **x** term; therefore, the equation does adhere to the rules regarding linear equations. If we computed $(2.5)^4$ we would obtain 39.0625, so the same equation may be written as $y = 39.0625x + 3$, which is clearly a linear equation.

Finally, **(g)** and **(h)** both can be classified as linear equations for the same reasons that we have classified other similar equations as linear in this exercise.

It is always helpful to see some non-examples as well. Can you determine why the following *cannot* be called linear equations?

(a) $\qquad y = x^2$

(b) $\qquad \sqrt{y} = x + 3$

(c) $\qquad x = \dfrac{5}{y}$

(d) $\qquad (2x + 3y)^2 = 4$

(e) $\qquad xy = 5$

Before I explain it, can you state why these equations are *not* linear? Which rule are they breaking? Ready? Let's explore it together.

Equation **(a)** is not linear because one of the variables has an exponent different than **1**. It is as simple as that.

Equation **(b)** is not linear because one of the variables has an exponent that is also *not* **1**; if you can't see it, let me rewrite the equation, switching from radical notation to fraction-power notation (remember **Chapter 10?**):

$$\sqrt{y} = x + 3$$

$$\downarrow$$

$$y^{\frac{1}{2}} = x + 3$$

Can you see now why it is not linear? If a *simplified* equation has a variable inside a root symbol, then it does not qualify as linear.

Equation **(c)** is not linear for the same reason that **(a)** and **(b)** are *not* linear. Look closely at the variable **y**: it is part of a denominator. This means that we could re-write the equation as follows:

$$x = \dfrac{5}{y}$$

$$\downarrow$$

$$x = 5y^{-1}$$

Therefore, since one of the variables has an exponent that is not equal to **1**, it means that it is *not* linear.

Equation **(d)** has both variables with an exponent of **1** (at least *locally*); so why couldn't we call it linear? Well, because they are contained inside a set of parentheses, and the parentheses has an exponent different than **1**. If we expand the left side of the equation (as per the technique that we reviewed earlier in the book), we would end up with the following (equivalent) version of the equation:

$$(2x + 3y)^2 = 4$$

$$\downarrow$$

$$(2x + 3y)(2x + 3y) = 4$$

$$(2x)(2x) + (2x)(3y) + (3y)(2x) + (3y)(3y) = 4$$

$$4x^2 + 6xy + 6xy + 9y^2 = 4$$

$$4x^2 + 12xy + 9y^2 = 4$$

Do you now see why this equation is not linear?

Finally, equation **(e)**, $xy = 5$, is perhaps the most difficult to understand why it is **not** linear. But if you remember what I said earlier about what it takes to be a linear equation, I explicitly stated that the variables **must** have an exponent of **1** and must be **adding** (or **subtracting**) each other. **Not multiplying each other** (as in this case), **not dividing** each other, not raised to the other's power (as in x^y or y^x). To prove this, simply solve for one of the variables, and watch what happens:

$$xy = 5$$

$$\downarrow$$

$$\frac{xy}{x} = \frac{5}{x}$$

$$y = 5x^{-1}$$

As you can observe, one of the two variables has an exponent different than **1** ; therefore, it is **not** linear.

The concept of what makes a two-variable equation linear can be expressed as follows: a linear equation represents a relationship between two variables in which when one of the variable changes, the other variable must also change, but always at a same "**rate of change**". I will explain this concept in detail shortly.

Let's explore, for example, the following linear equation:

$$y = 2x + 1$$

It is customary to call **x** the input variable and **y** the output variable, as we have seen in earlier chapters, and, as we saw recently as well, we can represent specific pairs of input and output values in coordinate form, using the format **(input , output)** which corresponds to **(x , y)** when these two variables are used. The order in which the values of the variables are presented is crucial, since all must agree on the order; otherwise, a mismatch would occur, leading to incorrect solution pairs.

For this equation, if we assign **x = 1** and solve for **y** , we obtain the following:

$$y = 2(1) + 1$$

$$y = 2 + 1$$

$$y = 3$$

$$(1 , 3)$$

Now, let's figure out the value of **y** if **x = 2** :

$$y = 2(2) + 1$$

$$y = 4 + 1$$

$$y = 5$$

$$(2,5)$$

Let's think about what happened here. When we assigned a value of **1** to **x**, the output was **3**; when we assigned a value of **2** to **x**, the output was **5**; can you see the "**linear pattern**" here? The "**rate of change**"? Can you guess what the outcome will be if we assign a value of **3** to **x**? If you answered **7**, then you are on the right track to mastering the principle of "linearity". Observe:

$$y = 2(3) + 1$$

$$y = 6 + 1$$

$$y = 7$$

$$(3,7)$$

The pattern is thus obvious: every increase of **1** unit of the input variable produces a **change** in the output variable of **2** units; therefore, if I ask you for the expected output if the input is **4** (for this particular linear relationship), I hope you would be able to answer correctly: **9**...

We can see the "linear" relationship in action very clearly by graphing the linear relationship (in other words, by graphing the linear equation). So how do we graph it? Well, by plotting the points (solution pairs) that correspond to the given linear equation. In this case, we have found that $y = 2x + 1$ produces the following three solution pairs (or coordinates):

$$(1,3)$$
$$(2,5)$$
$$(3,7)$$

Watch what happens if we plot these points, using the format **(x , y)** as we already specified earlier:

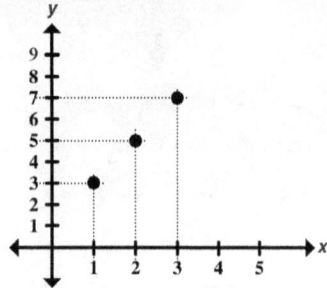

Can you visualize what would happen if we connected these points? If you answered "...we would end up with a **line**...", then you are absolutely correct. Take a look:

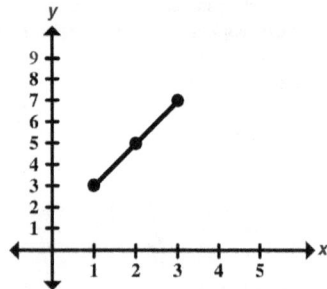

And now, because the equation is "linear", we could, if we wanted to, extend the line at both of the endpoints, indefinitely, as far as we wanted or needed to:

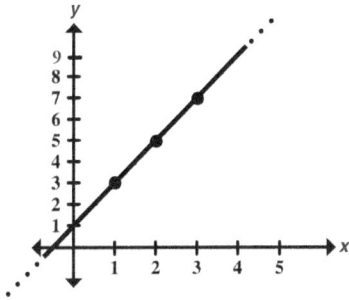

Can you now graphically see the *linear* relationship of the equation we are working with? Its graph corresponds to a *line* (colloquially speaking a "straight line", although mathematically, we can't really say that unless we were to provide a formal definition of "straight").

The interesting thing about extending the line (and connecting the three points to begin with) is that this allows us to consider the following:

A graph of an equation encompasses the solution pairs of said equation. It is formed by its solution pairs (coordinates), and therefore, all points of the graph of the equation must be part of the solution set of that equation...

What this means is that if, for example, I wanted to know the output of this equation if the input value x is equal to 4, I can use a two-step process using the graph of the equation to determine the output. **Step 1** requires locating the input value being considered on the horizontal axis (since this is, unless otherwise stated, the input axis, in this case, the x-**axis**) and finding the point that is part of the line of the

equation that has that specific input value. Let me illustrate the process:

Step 1: Locate the desired input value (in this case, $x = 4$), and locate the point on the line that corresponds to said input value:

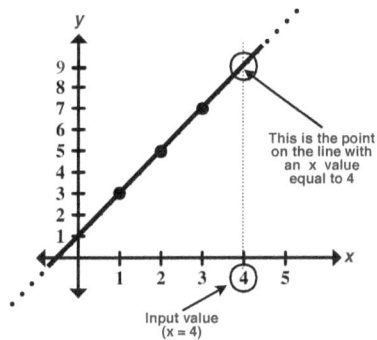

This is the point on the line with an x value equal to 4

Input value (x = 4)

Step 2: Having located the point on the line that corresponds to the input value, the next step is to identify the "height" of that point on the line (remember that this point on the line corresponds to the solution pair **(input , output)** of the equation, for the specific input value being considered), by using the **y-axis** as the frame of reference: simply reflect the vertical position of the point against the vertical axis, and read the output value that it corresponds to by using the vertical scale. The corresponding output value can thus be unequivocally specified.

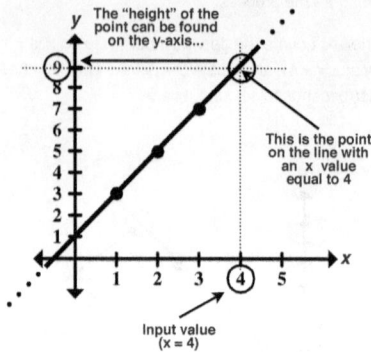

The "height" of the point can be found on the y-axis...

This is the point on the line with an x value equal to 4

Input value
(x = 4)

correctly read a graph (specify the input value and the output value of a given point that is part of the graph of the equation under consideration), you can unequivocally state that the coordinate of said point, of the form (input , output), represents a valid input/output pairing of the equation.

Let's revisit this same equation, and try to specify its output (the **y** value in this case) if the input is equal to **2.5** : using the two step process, we would first locate the input value of **2.5** , find the point that corresponds to that input value, and then determine its height (the **y** value).

By reading the graph correctly, we are able to determine the output value of this equation if the input value is **4** ... thus, the corresponding solution pair is: **(4 , 9)**. To confirm, let's replace replace **x** with **4** in the equation, and solve for the output variable, **y** :

$$y = 2(4) + 1$$

$$y = 8 + 1$$

$$y = 9$$

$$(4 , 9)$$

And the solution pair is confirmed!

What I am going to say once more is such an important principle of equations and their graphs that I want to ask you to read it carefully, over and over again, until it makes as much sense as going to sleep is when one is tired:

The graph of an equation is formed by all of the solution pairs of the equation. If you can

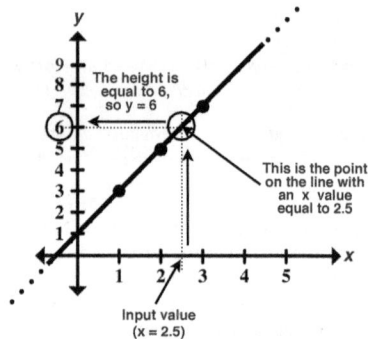

The height is equal to 6, so y = 6

This is the point on the line with an x value equal to 2.5

Input value
(x = 2.5)

As you can see from the graph above, the equation $y = 2x + 1$ has the solution pair **(2.5 , 6)**, which means that if you replace **x** with **2.5** , the equation's output will be **6**. Want to check it? Let's confirm our graph-reading abilities.

$$y = 2(2.5) + 1$$

$$y = 5 + 1$$

$$y = 6$$

$$(2.5 , 6)$$

And the solution pair is confirmed...

The principle we have been using above to find the output given the input value can also help us *find the input* given a specific *output* value. Can you see why? Can you deduce the two step process that would be involved?

To test the theory you may have come up with, try to specify the input value that produces an output value of **8** ; just remember, this number corresponds to the output, not the input value, which means it corresponds to the **y** variable in the case of this equation (and thus, the vertical axis). Another way to ask this same question is to simply present you with the incomplete coordinate (__ , **8**), and ask you find the missing value.

Did your method work out? Observe.

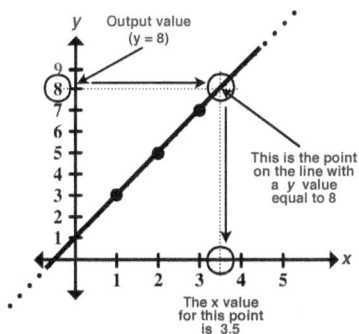

As you can see from the diagram, this time, because we are given the *output* value, **y = 8** , we find the point on the line that corresponds to that "height", using the vertical axis, and once the point is found, simply reflect the point against the horizontal axis to determine its horizontal displacement from the origin (lateral movement), which corresponds, of course, to the *input* value we were looking for. We can now revisit and complete the half-filled coordinate:

$$(\underline{} , \mathbf{8})$$

$$\downarrow$$

$$(\underline{\mathbf{3.5}} , \mathbf{8})$$

Care to prove it? Simple (note that all methods involve using the equation of the line, $y = 2x + 1$). We can either replace the **y** variable with **8** and solve for **x** (the input variable) to see if its value does, in fact, correspond to the value we read from the graph, or, we could "test" the coordinate to see if it is, in fact, a solution pair of the equation. The latter can be achieved in one of two ways: either by replacing the **x** variable with **3.5** and then solving for **y** , in which case we would see if this value corresponds to the **8** that was originally given for **y** , or, we could replace the **x** and the **y** variables with the values of the coordinate (**3.5** and **8** respectively), and check to see if the resulting equation is true. I will use all three methods so that you can observe first-hand how and why these options work.

Replacing y with 8 and solving for x ...

$$y = 2x + 1$$

replace **y** with **8** and solve for **x**...

$$8 = 2x + 1$$

$$8 - 1 = 2x + 1 - 1$$

$$7 = 2x$$

$$\frac{7}{2} = \frac{2x}{2}$$

$$3.5 = x$$

This confirms that we read the graph correctly.

Replacing the x variable with 3.5 and then solving for y ...

$$y = 2x + 1$$

replace x with **3.5** and solve for y...

$$y = 2(3.5) + 1$$

$$y = 7 + 1$$

$$y = 8$$

As you can see, this method can also be used to confirm that the value read from the graph is correct, since replacing the x with **3.5** produces an output of **8**, which corresponds to the output value provided in the prompt (the corresponding coordinate is (**3.5** , **8**), exactly what we set out to determine).

Replace the x and the y variables with the values of the coordinate (3.5 and 8 respectively), and check to see if the resulting equation is true...

$$y = 2x + 1$$

replace x with **3.5** and y with **8** ; notice the use of the question mark, since we are checking to see if the relationship is, in fact, equal...

$$8 \overset{?}{=} 2(3.5) + 1$$

$$8 \overset{?}{=} 7 + 1$$

$$8 \overset{\checkmark}{=} 8$$

This method also allows us to confirm our reading; since the equality remains unbroken (left side is equal to the right side), we can confirm that the coordinate is, in fact, a solution pair of the equation. Therefore, the (**3.5** , **8**) coordinate that we read form the graph is correct.

At this point you should be getting a sense of all the information that is contained in a graph of an equation. Because it is created by the solution pairs of the equation, it is a visual representation of the relationship that exists between the input and the output variables, and is thus able to provide the input and output value pairings.

Before moving on, I must clarify something I did earlier that if you were highly alert, should have caused alarm bells to ring inside your head. Not that I did anything wrong, of course, but simply that it must be validated.

Remember when I asked "what would happen if we joined the three points that we initially obtained form the equation $y = 2x + 1$ by using the input values of **1** , **2** , and **3** ? In other words, when we connected the coordinates (**1** , **3**), (**2** , **5**), and (**3** , **7**)? Did it occur to you to ask why this would be valid? Why not draw a curve instead? Or a series of points that are not immediately adjacent to each other? Or--err, you get the idea, right?

Well, the *only* reason why we can do this is that the equation is **LINEAR**, and that its domain is the entire set of **REAL** numbers (in other words, the equation can take on *any* REAL number input whatsoever: from $-\infty$ to ∞). This allows us to connect any two points that are part of the line using the shortest path possible within the plane, and since the plane is Euclidean (if you do not remember this from geometry, it simply means that the parallel postulate holds), we end up with a "***straight***" line, which, of course, can be extended indefinitely on either ends.

If it is unclear why, just think of it this way: a ***linear*** equation, as we said previously, maintains the exact same rate of change throughout all subsequent solution pairs: if you change, say, the input value **x** from the previous **x** value by a certain amount (keeping this amount constant), then the output will vary at the ***same rate*** with respect to the previous output. Just look at the three coordinates that we obtained from the equation we have been currently using: **(1 , 3)**, **(2 , 5)**, and **(3 , 7)**. If we start with the first coordinate (and use it as the basis of future changes), changing the **x** value by **1** unit will produce an output that will increase **2** units with respect to the previous output. Hence, an input of **2** gives us an output of **(3+2) = 5** , an input of **3** gives us an output of **(5+2) = 7** , an input of **4** gives us an output of **(7+2) = 9** , an input of **5** gives us an output of **(9+2) = 11**, etc... This rate of change can also be expressed as follows:

$$\text{Rate of change} = \frac{change\ in\ Output}{change\ in\ Input}$$

In the case of this equation, its rate of change is, using, say, coordinates **(1 , 3)** and **(2 , 5)**:

$$\text{Rate of change} = \frac{5-3}{2-1} = \frac{2}{1} = 2$$

Which makes sense: for every change in **x** that is **1** unit long, we end up with a change in the output value of **2** units, as we saw earlier. Look what happens if we use the coordinates **(2 , 5)** and **(3 , 7)**; do you think the rate will change?

$$\text{Rate of change} = \frac{7-5}{3-2} = \frac{2}{1} = 2$$

Of course it's the same rate of change! Since we increased the **x** value by **1** unit, we should find the same rate of change as before, and we do, in fact, find that same rate of change of **2** with respect to the ***previous*** output.

Watch what happens when we use two other coordinates, both **1** unit (**x**-wise) from each other: **(2.5 , 6)**, and **(3.5 , 8)**:

$$\text{Rate of change} = \frac{8-6}{3.5-2.5} = \frac{2}{1} = 2$$

The same rate of change! A rate of change of **2** for every **1** unit that is changed in the **x** value (***with respect to the previous x value***).

Now watch what happens when we use two coordinates whose **x** value differs by more than **1** unit (we are still using the same equation, hence, the same set of coordinates). Using a change in **x** of **2** units produces the following result:

$$\text{Rate of change} = \frac{7-3}{3-1} = \frac{4}{2} = 2$$

As if by some mysterious phenomena, we again obtain the same rate of change of **2** ! So what is happening here? Well, the reason we obtain the same rate is that the fraction $\frac{4}{2}$ simplifies to $\frac{2}{1}$ which is the **2** we have been finding all along. However, please note that if we were to, say, start at $(2,5)$ and we were told that the rate of change of the equation is 2, and then we were asked to figure out the output if the input value used was 4, we couldn't just add 2 units to the previous output (to 5 in this case) because that corresponds to a change in x of only 1 unit! Remember, a rate of change must specify the change of the output **based on** a specified change in input: and 2 is actually a $\frac{2}{1}$ which corresponds to a change of 2 in the output (numerator) for every change of 1 in the input (denominator). So in that case we would need to add 2 twice, because going from 2 to 4 implies a change in x of 2 units! Therefore, we would need to use the equivalent rate of change of $\frac{4}{2}$, which means that if $(2,5)$ is our starting point, and we want to know $(4,\_\_)$, we see that a change of 2 "x-units" implies changing the y value by adding 4 to the previous y value: since $5+4=9$, the coordinate is thus $(4,9)$.

Graphically, observe the rate of change in action going from the coordinate $(2,5)$ to $(3,7)$:

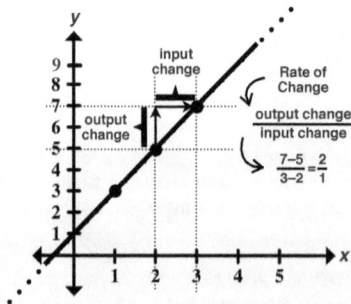

As you can see from the diagram, the rate of change quantifies the vertical change versus the horizontal change (output change, or change in y, divided by the input change, or change in x). In short, the rate of change is a **ratio**.

Thus, rate of change is also called **SLOPE** (the dictionary defines the word slope as "ground that slants upward or downward"... see the connection?). The big idea here is that the slope of the equation of a line is an indication of how steep its graph is on the **x-y plane**, or, in terms of its output, how large or small its output change is (between two different input values). Just remember that a rate of change must specify not only the change in y (numerator), but the change in x as well (a denominator). When the slope value is a whole number, this leads to confusion because **apparently**, only the change in y is specified, but this is **not** true. As we have reviewed numerous times throughout the book, a whole number can be written as a fraction simply by dividing the whole number by 1; thus, a slope of 2 is actually a slope of $\frac{2}{1}$. So why do we use 2 instead of $\frac{2}{1}$? Because it is generally easier to work with whole numbers, and to just remember (and know) that when considering slope, if the slope value is an **INTEGER**, we must be able to visualize the "dividing 1" which specifies the change in x, otherwise that piece of information would be missing. If the slope value is in fraction form, then there is no need to "visualize" any other dividing number: the change in x will be spelled out explicitly, as in $\frac{3}{5}$ or $\frac{2}{7}$, etc...

The example we have been working with happens to have an upward slope, which means that if we are using a "**left to right**" frame of reference (which we always use, unless otherwise stated), the line is said to be "rising", or "pointing upwards", or, mathematically speaking, has a **positive** slope. You see, the number **2** is positive (and so is $\frac{2}{1}$, the full-fledged slope value it represents); what this means is that if we change **x** (advance to the right **1** unit), we will change **y** by moving up **2** units on the **y-axis**.

⚓

As a rule, if the slope of a linear equation is positive, it means that its line graph must "rise" from left to right.

A nice way to remember this (and to imagine it as well) is to think of a person riding a bike on the line. If the slope is positive, the cyclist will have to exert effort to move from left to right, since it is going "up the hill"... So a positive slope is a "positive pedal". Observe this way of thinking about slope in the graph below.

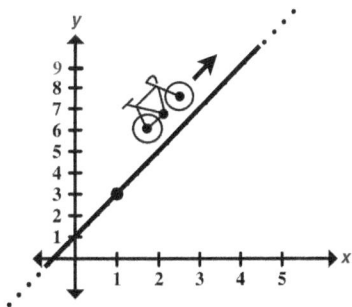

See the bicycle going "up the hill"? That is what happens whenever a line has a positive slope.

So what happens if the slope of a linear equation is negative? Any guesses? Well, in this case, continuing with the cyclist analogy, the cyclist will be coasting down a hill, no pedaling necessary! Let me show you an example of a linear equation with negative slope, and its corresponding graph.

$$y = -x + 4$$

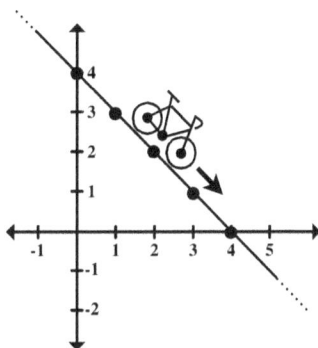

Can you see how the bicycle is coasting down the hill, absolutely no pedaling necessary? That's what happens whenever the slope of a line is negative.

The linear equation above has a slope of −1, as you can see from the coordinates that are clearly indicated on the graph: (**0** , **4**), (**1** , **3**), (**2** , **2**), (**3** , **1**), and (**4** , **0**); every input change (in other words, of **x**) of **1** unit (from left to right) implies a change in **y** of **−1** units with respect to the previous **y** value. If I write this slope fully and explicitly as $\frac{-1}{1}$, the output change (numerator, remember?) versus the input change (denominator) can be clearly seen.

And so we can write the rule that applies whenever the slope of a line is negative. It is as follows.

⚓

As a rule, if the slope of a linear equation is *negative*, it means that its line graph must "fall" from left to right.

That leaves us with two other possibilities. Can you see which ones they are? I have to warn you that one of the remaining two is not easy to envision, though the other one certainly is. Let's focus on the one that is much more obvious.

Let's see, so far we have talked about what happens if the slope is positive, and what happens if the slope is negative... Do you remember the **REAL** number system? How we defined positive numbers and negative numbers? If you do, you possibly (immediately?) thought of a third possibility: a slope that is equal to *zero*. Remember that positive and negative numbers are defined using **0** as the reference point: positives are to the right of **0** on the number line, while negatives are to the left of **0** on the number line, which leaves **0** in the middle, neither positive nor negative (since **0** isn't to the right or to the left of itself!). Thus, if we are saying that slopes can be positive and negative, then certainly we can talk about a slope that is equal to **0**. So what doe a slope equal to zero mean? What would you expect to see from the graph of a linear equation that has a slope of **0**? Well, you should expect to see a line that is flat, neither sloping upward nor sloping downward; in other words, parallel to the horizontal axis (the **x-axis**).

A cyclist on this line is therefore said to be riding on "flat terrain". Let me show you an example of a linear equation with slope of **0**, and its corresponding graph.

$$y = 0x + 2$$

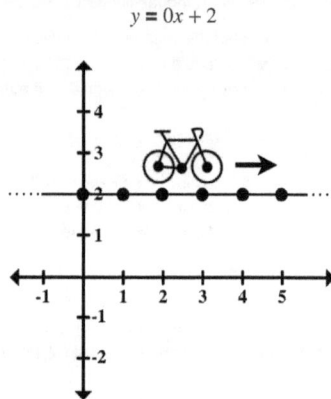

Can you see how the bicycle is traveling along "flat" terrain? That's what happens whenever the slope of a line is equal to zero.

The linear equation above has a slope of **0**, as you can see from the coordinates that are clearly indicated on the graph: **(0 , 2)**, **(1 , 2)**, **(2 , 2)**, **(3 , 2)**, **(4 , 2)**, and **(5 , 2)**; every input change (in other words, of **x**) of **1** unit (from left to right) implies a change in **y** of **0** units with respect to the previous **y** value. If I write this slope fully and explicitly as $\frac{0}{1}$, the output change (numerator, remember?) versus the input change (denominator) can be clearly seen. Looking back at the equation, note that the coefficient of the **x** variable is **0**.

And so we can write the rule that applies whenever the slope of a line is zero. It is as follows.

⚓

As a rule, if the slope of a linear equation is equal to zero, it means that its line graph is "flat", or parallel to the horizontal axis.

We are now left with one last possibility, the one that is the most elusive of the four. Instead of just blurting it out, I would rather you think about what it might be; I'll try to guide you towards the light (in other words, towards the answer!).

Let's revisit the meaning of slope and its fraction representation (remember that slope is equal to rate of change):

$$\text{slope} = \frac{change\ in\ Output}{change\ in\ Input} = \frac{rise}{run}$$

So far, we have explored **positive slopes** (which means that both the numerator and the denominator contain positive numbers), **negative slopes** (which means that either the numerator or the denominator contains a negative number), and a **slope equal to 0**, which means that the **numerator** is equal to **0** (as in $\frac{0}{1}$, or $\frac{0}{2}$, or $\frac{0}{50}$, or $\frac{0}{-4}$, or $\frac{0}{.746...}$, or $\frac{0}{-\pi}$, etc...). What other setup is possible here? If you still don't see it, think about where the **0** is located in the case of slopes that are equal to **0**; is it in the numerator or in the denominator of the fraction that represents the slope? The answer is that it is in the numerator, correct? So what would happen if

the **0** is not in the numerator but in the **denominator** instead? Do you remember what happens when a fraction has that particular setup, regardless of the numerator's value? As in:

$$\frac{change\ in\ Output}{change\ in\ Input} = \frac{anything}{0} = ???$$

I hope you remember that the answer is UNDEFINED. In other words, we may never divide any **REAL** number by **0** because the answer is not a **REAL** number, it is UNDEFINED. Thus, if we are exploring all possible scenarios regarding the value of slope, we must include the possibility that the change in input, or in **y**, is equal to **0**, in which case we end up with a slope that is UNDEFINED. Strange? Not really. Let's take this to the graph and see what it may mean...

Let's pick any location on the **x-y plane** and plot a point there. Let's see... **(3 , 1)** seems as good a spot as any. Alright, so suppose that you are standing on that point, and you encounter a rate of change value of $\frac{1}{0}$ (which we know is UNDEFINED, but never mind that for a minute; we are deciding to keep the fraction format so that we may access the change of output versus the change of input); so what is that slope value telling you to do? Well, if we extrapolate from what we have been doing with the rates of change so far, it would be directing us to move up **1** unit (the numerator's value), and move **0** units to the right; in other words, to stay put *laterally*. So, we do get to move vertically, but horizontally, our position remains the same.

Let's take this to the graph and see what happens.

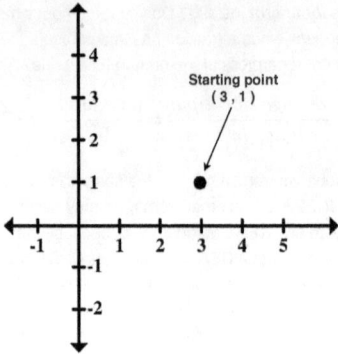

Starting point
(3 , 1)

Imagine standing on the point...

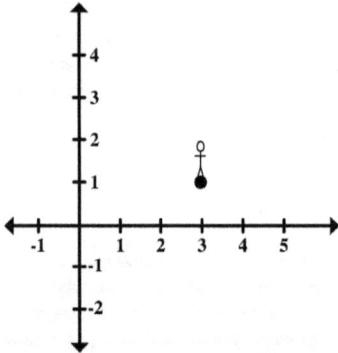

...then moving according to the "instructions" provided by the slope value of

$$\frac{change\ in\ Output}{change\ in\ Input} = \frac{1}{0} = \frac{rise}{run}$$

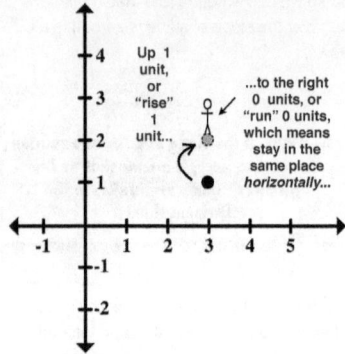

Up 1 unit, or "rise" 1 unit...

...to the right 0 units, or "run" 0 units, which means stay in the same place horizontally...

We can now draw the line of this unknown linear equation, since strictly speaking, remember that once you have the location of *two* points that belong to a *linear equation* (and its given line), you are able to draw its graph by simply connecting the points with a "straight" line, and extending it indefinitely at either ends:

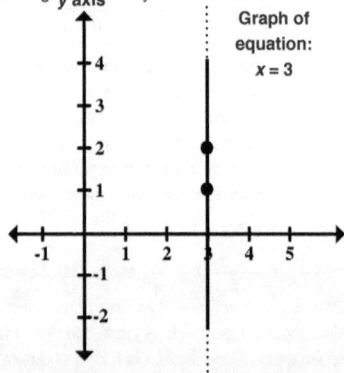

y axis

Graph of equation:

$x = 3$

So what kind of a slope is this? The apparently UNDEFINED slope?

486

Well, this slope is said to be infinite (or negative infinite) because the line is completely vertical, and so the rate of change can be expressed as "a change in the output that can be a value anywhere from $-\infty$ (negative infinity) to ∞ (positive infinity), with a change in the input of 0". Mathematically, the slope of this line can be expressed as a progressively larger (both positive-wise and negative-wise) numerator divided by 0 ; they are all equivalent (I am using **INTEGERS** to keep things simple, but any **REAL** number is valid):

$$\frac{-\infty}{0} = \frac{-999999}{0} = \frac{-1000}{0} = \frac{-20}{0} = \frac{-1}{0} \cdots$$

or, on the positive side of the number line, as

$$\cdots = \frac{1}{0} = \frac{20}{0} = \frac{1000}{0} = \frac{999999}{0} = \frac{\infty}{0}$$

I must clarify here that *negative infinity* and *positive infinity* is an idea, not a number, and therefore, the fraction "negative infinity divided by zero" is simply conceptual, and not a valid "number"; the same thing applies to the fraction "positive infinity divided by zero".

So how do we know that we are dealing with a linear equation that has an infinite slope? This one is extremely easy. All linear equations whose graphs are vertical lines (with infinite slopes) are of the following form:

$$x = \text{any } \textbf{REAL} \text{ number}$$

For example, the equation of the vertical line that we reviewed moments ago is:

$$x = 3$$

I know it seems a bit strange: the y variable is nowhere to be seen, and it looks more like the kind of statement that is used to state the value of a variable. However, *in the context of equations of lines*, it is, in fact, a linear equation, and it graphs to a vertical line. Do you remember what I said at the beginning of this chapter about how to identify linear equations? Well, context is everything. Unless you are told that $x = 3$ is a linear equation, who could blame you for thinking it is simply a statement saying "assign 3 to the variable x"? However, if you are told point-blank "graph the linear equation $x = 3$" then you would need to flip a mental switch and say "okay; this is an *equation of a line* and I must treat it as such...".

Now that we have seen all possible slope types, we can explore how to identify them simply by looking at the equation of a line; in fact, you may have already deduced how to do it by analyzing the four examples used in this section (if not, revisit the equations and their graphs and try to find a way to determine this).

The most convenient method to determine slope requires the linear equation to be of the form

$$y = m x + b$$

...in other words, solved for the variable y. This is called the "*slope-intercept*" form of a linear equation, for reasons that will shortly become very clear.

You may be wondering about the equation of the vertically-sloped line $x = 3$ that we worked with earlier, and I don't blame you. Because the y variable is not part of the equation, it cannot be solved for it (obviously!). Therefore, it is solved for the variable x instead. Remember that linear equations will **always** have either both variables as part of their equation or just one of them, but never a lack of both variables simultaneously.

So how can we know the type of slope that a linear equation possesses by looking at its slope-intercept form? Simple (just remember that the frame of reference is always "from left to right" unless otherwise stated). Here is the rule:

⚓

A linear equation of the form

$$y = mx + b$$

will have a slope-type that is determined by the value of the constant m, as follows:

(1) If m is positive: the slope "rises"

(2) If m is negative: the slope "falls"

(3) If m is 0: the slope is zero (horizontal line)

So what happened to the line that is completely vertical, whose slope is infinite (or negatively-infinite)? Well, it is a special case (and thus treated separately). As stated earlier, this happens whenever the equation of the line is of the form

(4) $x =$ any **REAL** number

Let me show you examples of each of the four cases listed above.

(1) Examples of a positive m value, with a slope that rises.

$$y = 2x + 1$$
$$y = 4x - 3$$
$$y = x + \pi$$
$$y = .5x - \sqrt{3}$$
$$y = x$$
$$y = \frac{2}{5}x - 100$$
$$y = \frac{x}{2}$$
$$y = \sqrt{5}x - 4$$
$$y = x - \frac{2}{5}$$

All of these equations produce a line with a rising slope (of course, the steepness will vary depending on the specific m value, as we will see later):

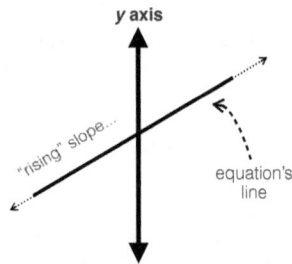

As you can see from the examples above, what's important is not the constant that is adding (or subtracting) the x term, but rather the **sign** of the x term's coefficient. The coefficients of the equations listed above, respectively, are: **2**, **4**, **1**, **.5**, **1**, $\frac{2}{5}$, $\frac{1}{2}$, $\sqrt{5}$, and **1**. And as you can appreciate, they are all **positive**, regardless of their number type (**INTEGER**, **RATIONAL** or **IRRATIONAL**).

I would like to stress that the coefficient of the x term in the third equation, $y = x + \pi$, for example, is **1** even though the number was not explicitly written (the x term appears by itself); as we have seen in earlier chapters of the book, it is implied (we can always think of any mathematical element as being multiplied by **1**, since multiplying by **1** does not change the element's value, properties, etc...). Thus, this equation could be written as $y = 1x + \pi$ and it would still represent the same linear relationship; however, you would never see it written like this since it would be simplified and the coefficient value of **1** would be removed, left to be implied by the person using the equation.

Another aspect worth reviewing is that of the coefficient of a term (the x term in this discussion) whenever a fraction value is involved, as in the sixth and seventh examples on the list above:

$$y = \frac{2}{5}x - 100$$
$$y = \frac{x}{2}$$

The coefficient of the x term on the first equation above is clearly $\frac{2}{5}$. However, it may be trickier to specify if the same equation had been written as follows:

$$y = \frac{2x}{5} - 100$$

The point here is to remember that the coefficient of a term is formed by any number multiplying **and** dividing it; in the form above, the **2** is clearly multiplying the x, and the **5** is clearly dividing it, so you must be able to "extract" the

coefficient of $\frac{2}{5}$. Saying that its coefficient is **2** is completely false, just as saying it is **5** or $\frac{1}{5}$.

The lesson is that sometimes you will see the coefficient completely **integrated** into the term as in the latter case ($y = \frac{2x}{5} - 100$), or separated from the term as in the former case ($y = \frac{2}{5}x - 100$); just keep in mind that **both** forms are **equivalent to each other**.

Moving on to the seventh example, $y = \frac{x}{2}$, I am certain that after reviewing in the detail the sixth example on the list you are able to clearly state why the x term's coefficient is $\frac{1}{2}$: although the x term is expressed as $\frac{x}{2}$, you should know that there is a **1** multiplying the x on the numerator, and therefore, when you extract the coefficient, you end up with $\frac{1}{2}$. Imagine if you did not see the **1** that is implied; would you then say that the coefficient was $\frac{0}{2}$, or $\frac{\square}{2}$, or **2**? Of course not, right? Those coefficients do not make any sense at all. The first choice, $\frac{0}{2}$, would imply **0**, and we can see clearly that the m value is not **0** in this case. The second choice, the number **2** dividing empty space is completely meaningless– mathematically speaking, at least–and therefore incorrect. And the third other option, **2**, does not apply, because if that was true then the equation would look like this: $y = 2x$ (clearly not the same as the equation $y = \frac{x}{2}$ which is what was given).

Let's continue with examples of the other slope-types.

(2) Examples of a negative m value, with a slope that "*falls*".

$$y = -2x + 1$$
$$y = -4x - 3$$
$$y = -x + \pi$$
$$y = -.5x - \sqrt{3}$$
$$y = -x$$
$$y = -\frac{2}{5}x - 100$$
$$y = \frac{-x}{2}$$
$$y = -\sqrt{5}x - 4$$
$$y = -x - \frac{2}{5}$$

All of these equations produce a line with a falling slope (of course, the steepness will vary depending on the specific m value, as we will see later):

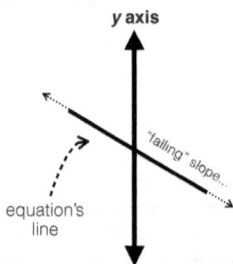

As you can see from the examples above, what's important is not the constant that is adding (or subtracting) the x term, but rather the **sign** of the x term's coefficient. The coefficients of the equations listed above, respectively, are: -2, -4, -1, $-.5$, -1, $-\frac{2}{5}$, $-\frac{1}{2}$ (or $\frac{-1}{2}$, or $\frac{1}{-2}$

which are equivalent, since the negative sign can be incorporated into the fraction's numerator or denominator as we have seen repeatedly throughout the book), $-\sqrt{5}$, and -1. And as you can appreciate, they are all **negative**, regardless of their number type (**INTEGER**, **RATIONAL** or **IRRATIONAL**).

I do not need to clarify anything regarding the fraction coefficients since this was already done on the positive slope list, and the same principles apply here.

I do, however, think it's appropriate to stress how to deal with the negative sign of a coefficient that is a fraction. Remember the general rule that states the following:

$$-\frac{a}{b} = \frac{-a}{b} = \frac{a}{-b}$$

where a and b represent any **REAL** number, except that $\mathbf{b} \neq \mathbf{0}$.

This rule means that when specifying the coefficient of the sixth equation on the list of negatively-sloped examples. we can either say that the slope is $-\frac{2}{5}$, or $\frac{-2}{5}$, or $\frac{2}{-5}$. It is very important to know that these are all equivalent to each other. As to the seventh equation on the list, its slope is thus $-\frac{1}{2}$, or $\frac{-1}{2}$, or $\frac{1}{-2}$.

Moving on to examples of equations of lines with a slope of 0 (and that produce a line that is completely horizontal, or "flat", with "zero slope"):

(3) Examples of an m value equal to 0, with a slope that is thus zero (or "flat").

$$y = 1$$
$$y = -3$$
$$y = \pi$$
$$y = \sqrt{3}$$
$$y = 0$$
$$y = -100$$
$$y = -\frac{2}{5}$$

All of these equations produce a line with a slope that is equal to **0** (which means that they all correspond to a line on the graph that is horizontal, or "flat"; of course, its specific location within the **x-y plane** will differ as we will explore shortly: they will be positioned at different "heights" using the **y-axis** as a reference):

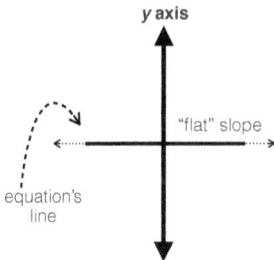

Observe how all of the examples are of the form "**y** is equal to a **REAL** number"; this effectively means that the **m** value is equal to **0**. Think about it for a minute: I already stated that linear equations of the form $y = mx + b$ that have a slope value of **0** (which means that the constant **m** is equal to **0**) correspond to a line on the graph with a "flat" slope (in other words, to a line that is horizontal). So, if you replace **m** with **0** you would end up with the following: $y = 0x + b$

in which case, the **x** term would completely disappear from the equation, since **0** times **anything** is equal to **0**. Thus, simplifying the right side of the equation when **m = 0** leaves us with the general equation $y = $ **b** :

Beginning with a linear equation that is in slope-intercept form...

$$y = mx + b$$

...and then considering all cases where **m = 0**:

$$y = 0x + b$$

which leads to:

$$y = 0 + b$$

or simply

$$y = b$$

What all of this essentially means is that *in the context of linear equations*, whenever you see the statement "$y = b$" you should know right away that you are dealing with a linear equation, that its slope value **m** is equal to **0** , and that if you were to graph the line that the equation specifies, you would end up with a completely horizontal line (a line with zero slope, or "flat"). The mystery of why these equations lack an **x** term should now be a thing of the past.

Finally, we can consider examples of linear equations that correspond to a line on the graph that has an infinite slope (or that is UNDEFINED).

(4) Examples of an **m** **value that is UNDEFINED, with a slope that is said to be infinite or completely vertical.**

$$x = 1$$
$$x = -3$$
$$x = \pi$$
$$x = \sqrt{3}$$
$$x = 0$$
$$x = -100$$
$$x = -\frac{2}{5}$$

All of these equations produce a line with a slope that is UNDEFINED or infinite (which means that they all correspond to a line on the graph that is vertical, or with an infinite rise; of course, its specific location within the **x-y plane** will differ as we will explore shortly: they will be positioned at different "lateral" or "horizontal" positions, using the **x-axis** as a reference):

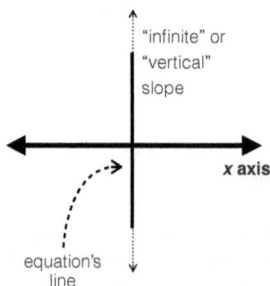

"infinite" or "vertical" slope

x axis

equation's line

Observe how all of the examples are of the form "**x** is equal to a **REAL** number"; this effectively means that the **m** value is UNDEFINED. So once again, *it is in the context of linear equations* that a statement of the form "**x** = a **REAL** number" corresponds to a linear equation that corresponds to a line on the graph

that is completely vertical, and therefore that has a slope that is UNDEFINED, or infinite.

We can now explore these linear equations in more detail.

Linear equations have slope, and as we have seen the slope may be either positive, negative, flat (or zero), or UNDEFINED (vertical or infinite). Is this the only aspect of a linear equation that is said to define it? No. There are three other characteristics that we must be able to state about any linear equation (all four slope-types considered earlier). These are:

a) The specific value of the slope (beyond weather it is positive, negative, zero, or UNDEFINED); it is a direct reflection of its rate of change.

b) The equation's **y-intercept**. This is the exact location (in coordinate form, typically) of where the line of the equation, when graphed, crosses the **y-axis**, if it does so at all.

c) The equation's **x-intercept**. This is the exact location (in coordinate form, typically) of where the line of the equation, when graphed, crosses the **x-axis**, if it does so at all.

As far as the first item on this list is concerned, we have already worked extensively with slope. Not only should you now know how to identify its sign, if it's zero, or if it is UNDEFINED, but you should also be able to specify it in terms of a rate of change, or "rise over run". Just remember that slope is equivalent to:

$$\textbf{slope} = \textit{m} = \frac{change\ in\ Output}{change\ in\ Input} = \frac{rise}{run}$$

We will use this often when graphing linear equations, since it offers a method of quickly finding additional points that belong to a given line (in other words, of solution pairs of its linear equation), with the only pre-requisite that point be known through which the line passes. More on this shortly.

On to the second aspect: the *y-intercept*. As the name suggests (and as I have already pointed out), the *y-intercept* is the point where a given line crosses the *y-axis*. In other words, it is the height of the point that is part of a linear equation's line that must be *exactly* on the *y-axis*. Observe these three lines and their corresponding (and differing) *y-intercept* points, as well as their corresponding equations. Try to determine how to find a line's *y-intercept* solely based on its *slope-intercept form* equation.

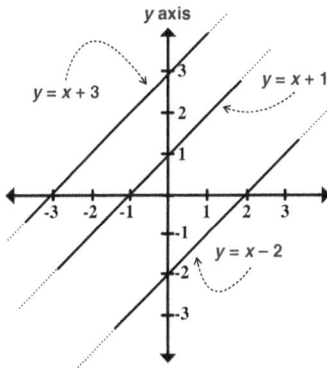

Can you identify what part of a linear equation (when written in slope-intercept form) specifies the *y-intercept*? If you can't see it, look at the specific location where each line crosses the *y-axis*:

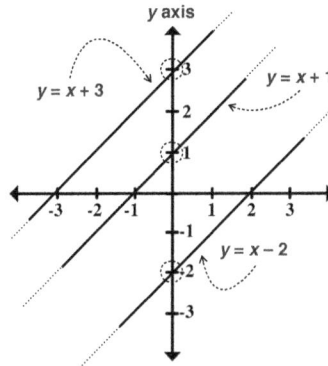

You should be able to specify the location pinpointed using the dotted circles in the above graph... They are:

$$\text{For } y = x + 3$$
$$(\, 0 \, , \, 3 \,)$$

$$\text{For } y = x + 1$$
$$(\, 0 \, , \, 1 \,)$$

$$\text{For } y = x - 2$$
$$(\, 0 \, , \, -2 \,)$$

Now can you see what part of the equation allows you to specify the *y-intercept* directly? Did you also notice that all *y-intercept* points **have an x value equal to 0**? This is a key characteristic of a *y-intercept* point, so obvious and yet at times a bit elusive (or easy to forget): since the *y-intercept* is, by definition, a point that must lie on the *y-axis*, it automatically follows that its *x* value will always be equal to **0**, because only then would a coordinate lie somewhere on the *y-axis*. Reread this part if you still can't make this connection.

The implications of the fact that a *y-intercept* point will *always* have an *x* value equal to **0** is that whenever you are trying to specify the *y-intercept* of a linear equation (and of its line), you automatically know, even before doing any work whatsoever, that you are looking for the coordinate with the format (**0** , **?**); it is thus the "question mark" that you are trying to specify, and nothing else, since by default the coordinate's *x* value *must be* equal to **0**, if it has one at all.

At this point I'm sure that if you have not yet been able to deduce how to find a given linear equation's *y-intercept* point solely form its slope-equation (if expressed in slope-intercept form), if you go back and revisit the previous two graphs again, keeping in mind that you are trying to specify (**0** , **?**)., you will be able to determine this for yourself.

Ready to see how it works? Observe.

⚓

A linear equation of the form

$$y = mx + b$$

has a *unique y*-intercept at:

$$(\, 0 \, , \, b \,)$$

For example:
y = 2x + 5 ---> (0 , 5)
y = 2x − 5 ---> (0 , − 5)

Notice that if the value of the constant **b** is being subtracted, then the *y* value of the *y-intercept's* coordinate will be negative. Furthermore, a linear equation *of this type* will have *only one y-intercept*.

As you can see from this rule, simply look at the independent term (called the constant *b* as per the slope-intercept general form of a linear equation, $y = mx + b$), extract it from the equation (as a positive if it is being added, but as a negative if it is being subtracted), and knowing that its *x* value pairing must be equal to **0**, you are ready to construct your *y-intercept* coordinate: it will *always* be (**0** , *b*).

Just check the three examples that were graphed earlier, look at where the lines cross the *y-axis*, and look at the corresponding coordinate; then, apply the rule above to each line's equation (which was given in slope-intercept form), and you will see that the rule does, in fact, work.

So why does the rule work? The answer is simple, and yet it requires a deep understanding of linear equations, their graphs, and what a coordinate that is part of the equation's line is to begin with. Remember that for a given linear equation, a coordinate that is part of its line is, by definition, a solution pair of said equation. This means that, for example, the equation $y = x + 3$, when graphed, produces a line that includes the point located at (**1** , **4**) because if you were to replace *x* with **1** and solve the equation, you would find that *y* is equal to **4**:

$$y = x + 3$$

replacing *x* with **1**

$$y = (1) + 3$$
$$y = 4$$

this may then be written in coordinate form as

$$(\, 1 \, , \, 4 \,)$$

So, if we are trying to determine the **y-intercept** of a linear equation, it means we are trying to determine $(0, ?)$, which means we are trying to determine for that specific equation the value of **y** when the value of **x** is equal to **0**. Since we are working with the so-called slope-intercept form, which is of the type $y = mx + b$, we may replace **x** with **0** in this general version of a linear equation, and see what it leads to:

$$y = mx + b$$

because we are interested in finding $(0, ?)$, replacing **x** with **0** leads to the following (in general, regardless of the specific equation we may actually be dealing with):

$$y = (0) + b$$
$$y = 0 + b$$
$$y = b$$

this may thus be written in coordinate form as

$$(0, b)$$

As you can see, whenever the variable **x** is to be replaced with **0** and we have the linear equation in slope-intercept form, we know that the **x-term** will disappear (since **0** times anything–whatever the **x-term's** coefficient may happen to be–is equal to **0**), leaving the variable **y** equal to the constant **b**. And so the mystery of why the rule works is solved.

So what do you do if the linear equation is not in slope-intercept form, but in general form, or any other form? Well, math, as you well know by now, offers many options. One of those is to solve the equation for the variable **y** so that you do have the slope-intercept form at your disposal. But another option, one which should be obvious to

you at this point, is to simply replace the variable **x** with **0** and solve for **y**, since you are interested in knowing the solution pair of the linear equation when **x = 0** (in other words, your aim is to find $(0, ?)$, remember?). Take, for example, the equation $-2x + 2y = 6$. To find the **y-intercept** we would have to apply one of the methods discussed above since the equation is not in slope-intercept form, and therefore, it is not possible to extract the **b** value directly (it is simply not present in any form other than the slope-intercept form). I will use both to show you how they work.

Option 1. Solve for **y**

$$-2x + 2y = 6$$

Adding $2x$ to both sides:

$$-2x + 2y + 2x = 2x + 6$$
$$2y = 2x + 6$$

then dividing both sides by **2** and simplifying:

$$\frac{2y}{2} = \frac{2x + 6}{2}$$
$$y = \frac{2x}{2} + \frac{6}{2}$$
$$y = x + 3$$

The equation is now in **slope-intercept** form (**y = mx + b**)

"**b**" value... therefore, the **y-intercept** is $(0, 3)$

Does the equation look familiar? It is the same equation as the first linear equation we graphed earlier. And so its **y-intercept** is $(0, 3)$ just as we specified earlier. Just remember to check whether the **b** value is being added or subtracted; if the former, it is positive, if the latter, it is negative (hence the dotted circle in the diagram above includes the addition symbol).

It is important to notice how we can't really extract the **3** from the equation's original form ($-2x + 2y = 6$), since the equation is *not* in the right format (the slope-intercept form).

Option 2. Replace **x** with **0** and solve for **y**

$$-2x + 2y = 6$$

Replacing **x** with **0** :

$$-2(0) + 2y = 6$$
$$0 + 2y = 6$$
$$2y = 6$$
$$\frac{2y}{2} = \frac{6}{2}$$
$$y = 3$$

Therefore, the solution pair is **(0 , 3)**, which corresponds to the equation's **y-intercept**.

As you can see from the method above, the same result is (of course) obtained. However, note that the variable **x** is not part of the final answer, as should be expected, since it was replaced by the number **0**. This option is a more explicit or direct way of obtaining the **y-intercept**, since you are essentially asking the equation to provide you with the value of **y** given an **x** value of **0**, as the **y-intercept** requires; however, the first option is perfectly valid as well. Which method you use will probably depend on whether you have the slope-intercept form or not. If you do, extracting the **b** value is the way to go; if not, I would recommend replacing **x** with **0**, since it will always eliminate the **x** term from the expression, leaving you with an easier simplification/solving process.

Now it's your turn. Try to specify the **y-intercept** of the following linear equations.

a) $-8x + 4y = 16$

b) $-5x + 5y = -5$

c) $-3x + y - 2 = 0$

d) $4x = 2y + 8$

e) $y = -x - 3$

f) $-y = 2x + 5$

g) $x = y$

h) $2 = -y$

Ready to check your work? Compare your answers:

a) $-8x + 4y = 16$

This equation is not in slope-intercept form. I will therefore use the second method described earlier: replace **x** with **0** and solve for **y** in order to find the **y-intercept**, **(0 , ?)**:

$$-8x + 4y = 16$$
$$-8(0) + 4y = 16$$
$$0 + 4y = 16$$
$$4y = 16$$
$$\frac{4y}{4} = \frac{16}{4}$$
$$y = 4$$

Therefore, the **y-intercept** is **(0 , 4)**

Problem solved. Let's move on to the next one.

b) $-5x + 5y = -5$

This equation is not in slope-intercept form. I will therefore use the second method described earlier: replace x with 0 and solve for y in order to find the **y-intercept**, **(0 , ?)**:

$$-5x + 5y = -5$$
$$-5(0) + 5y = -5$$
$$0 + 5y = -5$$
$$5y = -5$$
$$\frac{5y}{5} = \frac{-5}{5}$$
$$y = -1$$

Therefore, the **y-intercept** is **(0 , – 1)**

c) $-3x + y - 2 = 0$

This equation is not in slope-intercept form. I will therefore use the second method described earlier: replace x with 0 and solve for y in order to find the **y-intercept**, **(0 , ?)**:

$$-3x + y - 2 = 0$$
$$-3(0) + y - 2 = 0$$
$$0 + y - 2 = 0$$
$$y - 2 = 0$$
$$y - 2 + 2 = 0 + 2$$
$$y = 2$$

Therefore, the **y-intercept** is **(0 , 2)**

d) $4x = 2y + 8$

This equation is not in slope-intercept form. I will therefore use the second method described earlier: replace x with 0 and solve for y in order to find the **y-intercept**, **(0 , ?)**:

$$4x = 2y + 8$$
$$4(0) = 2y + 8$$
$$0 = 2y + 8$$
$$0 - 8 = 2y + 8 - 8$$
$$-8 = 2y$$
$$\frac{-8}{2} = \frac{2y}{2}$$
$$-4 = y$$

Therefore, the **y-intercept** is **(0 , – 4)**

e) $y = -x - 3$

The equation is in slope-intercept form. Therefore, all I need to do to find the **y-intercept** is "extract" the value of b from this particular equation. Because the b value is being subtracted (look at the **subtracting** 3), I must consider b to be a negative number; therefore, the **y-intercept** is **(0 , – 3)**:

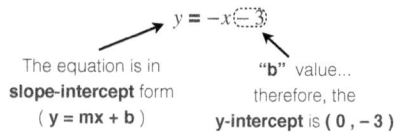

$$y = -x - 3$$

The equation is in **slope-intercept** form (**y = mx + b**)

"**b**" value... therefore, the **y-intercept** is **(0 , – 3)**

f) $-y = 2x + 5$

Be careful with this type of equation format. Although it appears as if the equation was, in fact, solved for y, it is not the case. Observe closely the left side of the equation, where the y variable is located: there is a negative sign attached to it that prevents us from saying that the equation is solved for y; it is, in fact, solved for $-y$ instead. Notice the difference? Therefore, we can either replace x with 0 as we have been doing, or, since the equation is *almost* solved for **y**, simply solve for **y** and extract the desired **b** value. I will do the latter, though, of course, the former leads to the same answer as well.

$$-y = 2x + 5$$

To get rid of the negative sign, I can multiply both sides of the equation
by **−1** and the simplify:

$$(-1)(-y) = (2x + 5)(-1)$$
$$y = (-1)(2x) + (-1)(5)$$
$$y = -2x + (-5)$$
$$y = -2x - 5$$

The equation is in **slope-intercept** form (**y = mx + b**)

"**b**" value... therefore, the **y-intercept** is (0 , −5)

Make sure you understand why we can't say that $-y = 2x + 5$ is solved for **y**; it is a very common misconception.

g) $x = y$

The equation is in slope-intercept form (don't be confused by the left-side/right-side placement of the variables; remember that $x = y$ is the same as $y = x$). Therefore, all I need to do to find the **y-intercept** is "extract" the value of **b** from this particular equation. Because the **b** value is "missing" (neither the left side nor the right side have an independent term adding or subtracting), I must consider **b** to be **0**, a perfectly valid value (it means that the line of this equation crosses the **y-axis** at the **x-y plane's** origin); therefore, the **y-intercept** is (0 , 0). The reason this works is because we could rewrite the equation as $x + 0 = y$ and we have not changed

it from the original. Therefore, **b = 0**. By the way, do not be thrown off by the left/right side switch.

$$x = y$$

is the same as...

$$x + 0 = y$$

"**b**" value... therefore, the **y-intercept** is (0 , 0)

The equation is in **slope-intercept** form (**y = mx + b**)

And so the major lesson here is that if a linear equation lacks an independent term (in other words, is defined in terms of an **x-term** and a **y-term** only), then the **b** value is **0**.

h) $2 = -y$

Once again, be very careful with this type of equation format. Although it appears as if the equation was, in fact, solved for **y**, it is not the case (just like problem **(f)**, remember?). Observe closely the right side of the equation, where the **y** variable is located: there is a negative sign attached to it that prevents us from saying that the equation is solved for **y**; it is, in fact, solved for **−y** instead. Notice the difference? Therefore, we can either replace **x** with **0** as we have been doing, or, since the equation is *almost* solved for **y**, simply solve for **y** and extract the desired **b** value. I will do the latter, though, of course, the former leads to the same answer as well. Observe the process.

➡

498

$$2 = -y$$

To get rid of the negative sign, I can multiply both sides of the equation by -1 and the simplify:

$$(-1)(2) = (-y)(-1)$$
$$-2 = y$$

"b" value... therefore, the **y-intercept** is (0 , – 2)

The equation is in **slope-intercept** form (**y = mx + b**)

In this example, the major lesson is that as far as the **y-intercept** is concerned, if the **x-term** is missing is irrelevant. The independent term is what defines the **y-intercept**.

Graphing linear equations

Before proceeding any further, it is a good idea to see the graphs of equations **(g)** $x = y$ (in other words, $y = x$) and **(h)** $2 = -y$ (in other words, $y = -2$) on the **x-y plane**. Observe.

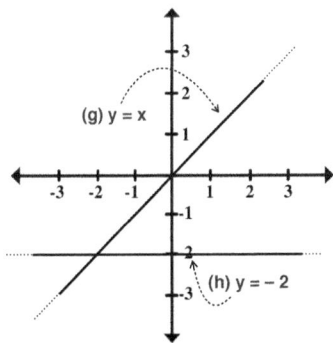

(g) y = x

(h) y = – 2

Note the slope or rate of change of equation **(g)** and of equation **(h)**; they are perfectly consistent with what we have learned so far about linear equations and the value of **m**. In the case of **(g)**, since the equation is $y = x$, you should be able to specify its **m** value: **1** (notice the **x-term's** coefficient?); then, since a rate of change requires a denominator, we write the number **1** in its equivalent fraction form as $\frac{1}{1}$. This now tells us that for this equation, a change of **1** unit in the output value corresponds to a change of **1** unit in the input value. We would next need to know a solution pair (in other words, a point that is part of this equation's line) to use as our starting point, from which this rate of change can be applied. Well, which point do you think is the easiest to determine? If you answered "The **y-intercept** point", then you are exactly right. Since this equation does not have an independent term, we know its **y-intercept** occurs at (**0 , 0**) because the constant **b** is equal to **0** . If you cannot see it, remember that the equation $y = x$ can be rewritten as $y = x + 0$ (in fact, we can rewrite it as $y = 1x + 0$ so that you can clearly see the **m** and **b** values explicitly).

Knowing that this equation has a solution pair that is (**0 , 0**) and that therefore, this coordinate is part of the equation's line allows us to plot a point at that precise location (in this case, the origin), and then use the rate of change that we determined above to locate a second point: simply "**rise**" **1** unit from the origin, and then "**run**" **1** unit from that temporary location. Finally, to draw the equation's line, since it is linear, we can simply connect the two points with a "**straight**" line, as we have done previously.

Observe this step by step process in the following diagrams:

Step 1. Plot the point **(0 , 0)** that corresponds to the equation's **y-intercept**.

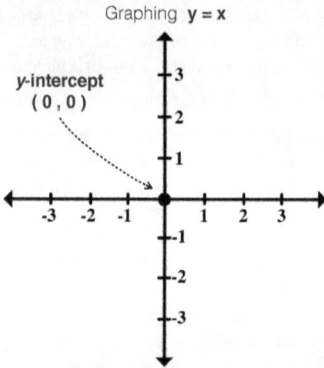

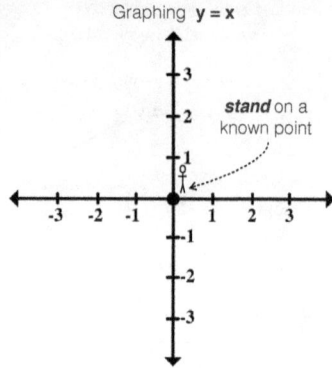

Graphing **y = x**

y-intercept (0 , 0)

and run values are positive, ***rise*** corresponds to moving up, while ***run*** to moving to the right.

Graphing **y = x**

stand on a known point

...and then ***rise*** **1** unit, and then ***run*** **1** unit...

Step 2. Use the rate of change, or slope, to find a second point that belongs to the line. In order to do this, as reviewed earlier in this chapter, imagine standing on the known point that is part of the line (in this case, the **(0 , 0)** that corresponds to the **y-intercept**), and then use the slope value of this equation: **1** or $\frac{1}{1}$. It is convenient to use the slope interpretation of "***rise over run***" since it quickly allows us to move from one known point of the equation to another point that can then be marked or plotted on the **x-y plane**. Thus:

$$\textbf{slope} = \frac{change\ in\ Output}{change\ in\ Input} = \frac{rise}{run} = \frac{1}{1}$$

The idea is to "rise" **1** unit and then "run" **1** unit ***from a known point*** of the equation (the **(0 , 0)** previously determined). Since the rise

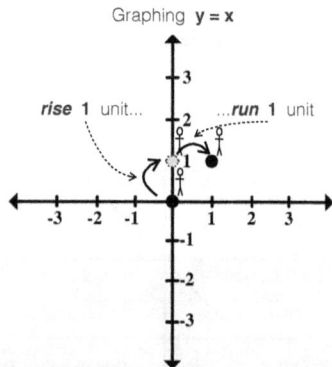

Graphing **y = x**

rise **1** unit...   ...***run*** **1** unit

The diagram illustrates the idea of imagining standing on the known point, then shifting up or down (as per the *rise* value specified by the slope: if *positive*, move up, if *negative*, move down), and then from that *temporary* resting place (a *stepping stone*, if you will), move laterally (as per the *run* value specified by the slope: if *positive*, shift to the right, if *negative*, shift to the left). The three stick figures correspond to:

- The initial location, or starting point, at (0 , 0).

- The temporary *stepping stone* after executing the *rise* portion of the slope, at (0 , 1).

- The final destination, or location of the point that is actually part of the line, after executing the *run* portion of the slope, at (1 , 1).

Once two points are plotted, as long as they are part of the line that is to be drawn, a "straight" line may be used to connect them, and they may be extended indefinitely at either ends (this only works for linear equations, however, so make sure to check for this). The linear equation has thus been graphed (into its corresponding *line*):

Graph of **y = x**

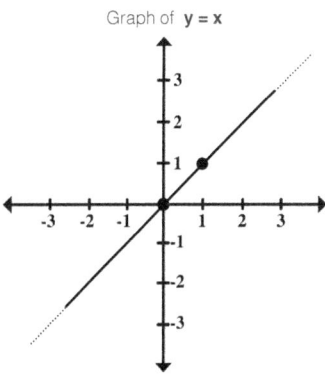

You should know that we could continue to find additional points (solution pairs) that are part of this equation's line, but because my only goal was to graph it, there really is no need. However, I do want to show you how easy it is to continue to find points that are part of the line, simply by using the slope (rise over run) information and starting out on a known point of a line. In fact, we can find points that are located both to the right of a known point (just as we found the (1 , 1) coordinate earlier), or to the left of a known point.

More points to the right of a known point...

As I did earlier, simply stand on (1 , 1), the last right-most known point that is part of the line, and use slope (in its rise over run interpretation) to find a third point, and a fourth point, etc..., *ad nauseam*, as illustrated below:

Finding more points of **y = x**

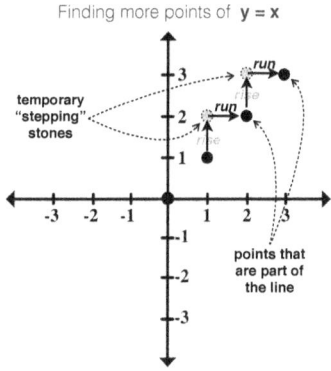

And so the equation $y = x$ contains, among its infinite solution set, the solution pairs (0 , 0), (1 , 1), (2 , 2), and (3 , 3).

But what about points to the left of a know point? How can you find them? If slope (and therefore, rise over run) is positive, how can you "**run**" to the left? The answer is actually simple: instead of considering the equation's slope to be $\frac{1}{1}$, you could think of it as being $\frac{-1}{-1}$. Is this a trick? I hope you know that it is not. Two negative **REAL** numbers dividing each other will always be equal to a positive **REAL** number, so we know that the slope $\frac{-1}{-1}$ is equivalent to the slope $\frac{1}{1}$, and they are both equivalent to the slope value of **1**. So, when applying this slope of $\frac{-1}{-1}$ we must read the **rise** value as **−1** which means we would have to move **down** 1 unit, and when reading the **run** value of **−1** we must move to the **left** 1 unit, as the diagram below illustrates:

Finding points of **y = x**

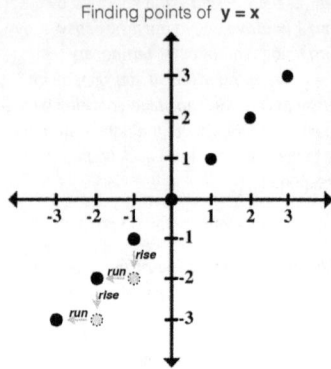

And we could, of course, keep using the slope value of $\frac{-1}{-1}$ in order to find more points to the left of a known point of the line, as the following diagram illustrates.

➡️

Finding points of **y = x**

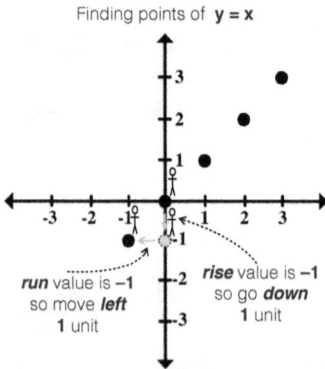

And so the equation $y = x$ contains, among its infinite solution set, the solution pairs $(-1,-1)$, $(-2,-2)$, $(-3,-3)$, as was determined by using the "tweaked" slope value of $\frac{-1}{-1}$.

Now, let's explore the graph of problem **(h)**, $2 = -y$ (or its equivalent version, $y = -2$). If we were asked to graph this equation, just as with the previous example, we could start by asking ourselves if we know a point that is part of this equation's line, because if we do, we could then apply this equation's slope **from that known point** to find a second point, and that would allow us to graph the equation correctly (linear equations can be graphed with a minimum of **two** known points; no more are necessary).

At first, the equation $y = -2$ seems odd: it does not have an **x**-term. But remember that this simply means that the slope is **0** (and therefore, that the **m** value is **0** in this specific equation). We could thus rewrite it as $y = 0x - 2$.

Regardless of this, it should be obvious by comparing it to the general form of a linear equation (in slope-intercept form)...

$$y = mx + b$$
$$y = 0x - 2$$

...that the **b** value is equal to -2 for this specific linear equation. Therefore, we can determine the **y-intercept** as long as we remember that for a linear equation in slope-intercept form, it is always given by (**0** , **b**).

$$y = 0x - 2$$
Since **b** $= -2$...
...the **y-intercept** is (**0** , -2)

This gives us the starting point that we need in order to use the equation's slope value (interpreted as **rise over run**) to find a second point. So what's the slope value? Once again, comparing the general form with the equation clearly gives us the answer:

$$y = mx + b$$
$$y = 0x - 2$$

So what do we do now that we know that **m** = **0** ? We follow the same process that we executed on the previous example: because we need a slope value in the form of a fraction (so that we can determine the **rise** and the **run** values), we simply divide the slope by **1** (as I have stated before, there is absolutely nothing wrong with this, since **0** divided by **1** is equal to **0**, so we have not arbitrarily changed the slope's value by doing this). Thus, we find that as far as this equation is concerned, the following holds:

$$\text{slope} = \frac{change\ in\ Output}{change\ in\ Input} = \frac{rise}{run} = \frac{0}{1}$$

Now, we are ready to fully graph the equation. Simply plot the **y-intercept** (**0** , -2) and from that location, apply the "**rise over run**" values of "**0** and **1**". Before you complain, a **rise** of **0** implies not moving up nor down (in other words, staying put vertically, from wherever you may be initially located). Observe.

Graphing **y** = -2

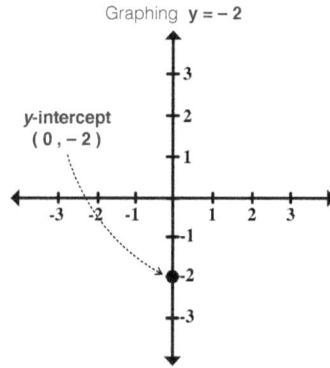

y-intercept
(**0** , -2)

Applying "rise over run" of **0** and **1** respectively...

Graphing **y** = -2

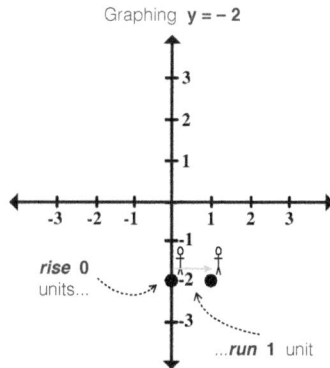

rise 0 units...

...run **1** unit

Now that we have two points that are part of this equation's line, we can connect the, with a "straight" line and extend them at either end indefinitely, completing the graphing process:

Graph of $y = -2$

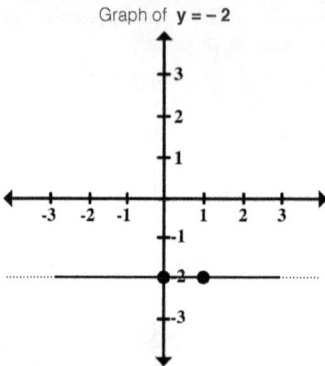

There is an important clarification that must be made at this point. Looking at the above graph, for example, you could get the wrong idea by the presence of the two points that were initially plotted (and used to draw the equation's graph, which corresponds to a line, since the equation is linear).Those two points are not different from the rest of the points that make up the line. What other points, you may ask? The infinite points that make up the line to begin with. As I stated earlier in the book, a line is formed by an infinite number of points, all of them "lined up" one after the other, without interruption or gaps. The two "thick points" that are drawn on the graph above are drawn that way in order to ease their identification. But do not attach any special significance to how they are shown on the graph. In fact, strictly speaking, the graph of the

equation should be drawn without those two points standing out the way they do, as follows:

Graph of $y = -2$

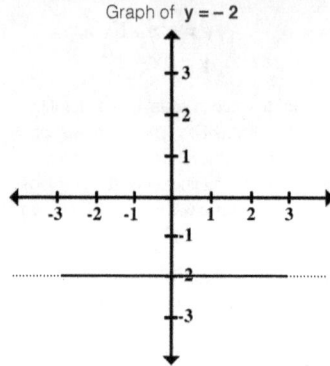

See the difference? In the graph above, no distinction is made between any of the points since that could be misleading.

Specifying solution pairs from a graph

So what exactly is the graph above communicating to us about the linear equation that produced it? Besides the obvious (that its slope is equal to **0** and that its **y-intercept** occurs at (**0** , **− 2**)), it allows us to determine as many solution pairs of the equation as we needed to or wanted to specify. Any coordinate of a point that is part of the line of the graph corresponds to a solution pair of the equation, such as those found on this list:

$$(-3 , -2), (-2 , -2), (-1 , -2), (0 , -2),$$
$$(1 , -2), (2 , -2), (3 , -2), (4 , -2),$$
$$(5 , -2), \text{etc...}$$

I am using **INTEGERS** for the *x* value, but other number types are part of the solution set as well:

$$(- 1.874... \,, - 2 \,), (- 1.25 \,, - 2 \,), (.5 \,, - 2 \,),$$
$$(\pi \,, - 2 \,), (\sqrt{17} \,, - 2 \,), (\frac{33}{2} \,, - 2 \,), \text{etc...}$$

What is particularly interesting about this specific linear equation is that it will always yield an output value of − **2** regardless of the input value used, precisely because its slope is **0**. Do you see the connection? Just remember that there are an *infinite* number of solution pairs to the equation, and therefore an infinite number of points that form its graph.

If at this point you are scratching your head, unsure how I was able to determine these coordinate values from the graph, just remember that a coordinate is of the form (**input** , **output**), which in this case corresponds to (*x* , *y*); therefore, looking at the graph, it is easy to see that for any input value used (*x* value), the output (or the value of *y*) is *always* equal to − **2** . For example, observe how finding the output when the input is equal to − **3** would work:

Graph of **y** = − **2**

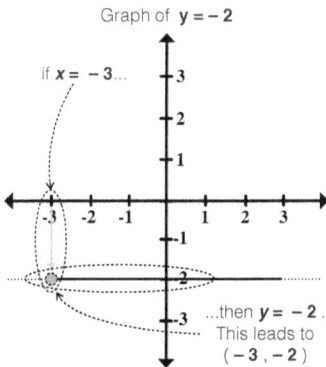

Or, observe how finding the output when the input is equal to − **2** would work:

Graph of **y** = − **2**

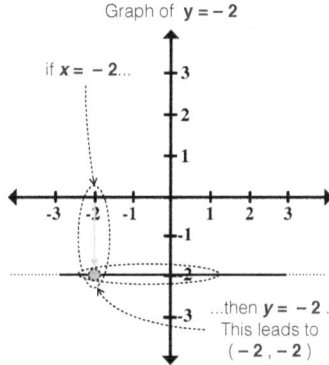

Or, observe how finding the output when the input is equal to − **1.5** would work:

Graph of **y** = − **2**

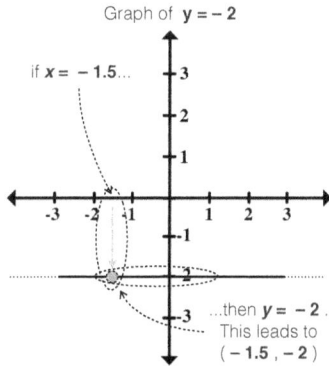

...and on and on and on. And so for any **REAL**-numbered input, the output will always be − **2** .

Using the same principle, try to specify at least five points that belong to the following equation by looking at its graph and extracting the coordinates that belong to its graphed line (I am saving you the trouble of having to graph it yourself, but you may wish to think about how you would do it if that had been the request).

Graph of $y = 3$

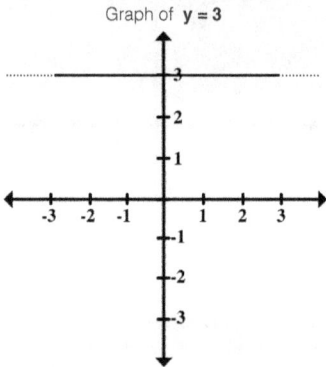

Ready to check your answer?

Well, this time, I can't possibly know which five coordinates you came up, so rather than list an infinite number of possibilities (which can't be done even if I tried), just check your coordinates to see if they all have the following format:

(*any REAL number* , 3)

Thus, points such as $(-\pi, 3)$, $(-1, 3)$, $(0, 3)$, $(1.5, 3)$, $(2.376..., 3)$, etc... , are all part of the line, and therefore, solution pairs of the equation. Easy, right?

And what about the following graph? This time, the slope is not equal to **0** , so you do have to be a bit careful reading the graph correctly. Furthermore, since the slope is not **0** , you can't really get away with trying to read the output for

an input such as $x = \pi$, because the graph's scale will not really allow you to express a reasonable estimate (in other words, to reasonably approximate the output). However, in this case you can easily determine the output of **INTEGER** inputs, or inputs that are halfway between two **INTEGERS** (such as -1.5 or $.5$, etc...). Try to state six coordinates that are part of the following equation's line:

Graph of $y = -x - 1$

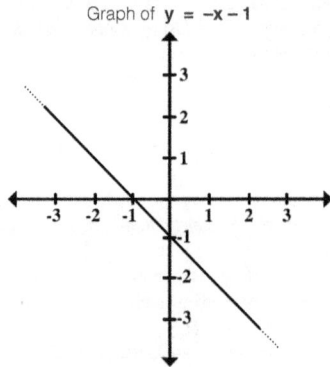

Ready? Did you specify all six? Check to see if they are among the following list of coordinates that belong to points that are part of the equation's line:

$(-3, 2)$, $(-2.5, 1.5)$, $(-2, 1)$,
$(-1.5, .5)$, $(-1, 0)$, $(-.5, -.5)$, $(0, -1)$,
$(.5, -1.5)$, $(1, -2)$, $(1.5, -2.5)$,
$(2, -3)$, etc...

These would be the most reasonable coordinates to specify given the graph above (considering the scale). If you wanted to know any other coordinates using input values such as **.3** or $-\pi$, etc.., you would have to use the equation itself (replace **x** with the desired input, and then solve to find its corresponding output value, the *y*). This would then allow you to specify that coordinate, using the template **(*x* , *y*)**.

Linear equations of the form *x* = any REAL number

Remember the graph of *x* = 3 that we worked on earlier in the chapter? Allow me to revisit its graph:

Graph of equation:
x = 3

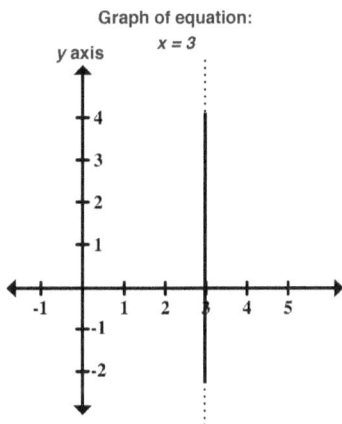

Notice the missing **_thick points_** that were originally used to draw it? Remember: they can be misleading so I am not using them except when plotting points or in the process of drawing a graph, but not as part of the final equation's graph.

The reason I am revisiting this equation and its graph is the following: what would you say if I asked you to specify this equation's (and therefore, its line's) **_y_-intercept**? Can you provide a coordinate for this **_y_-axis** crossing point? Don't

jump to conclusions... think about what the **_y_-intercept** is, observe the graph of *x* = 3 and then try to come up with a mathematically valid answer to the question.

Ready for the semi-obvious? The correct answer is that this particular linear equation **_does not_** have a **_y_-intercept**. Just check the graph if you don't believe me. Do you see a location where this equation's line would pass through the **_y_-axis**? Impossible, right? After all, the line is perfectly vertical, or perpendicular to the **_x_-axis**, or parallel to the **_y_-axis**; therefore, it will never cross the **_y_-axis** at all (this is based on the parallel postulate that applies to Euclidian geometry; remember that the cartesian plane, or **_x_-_y_ plane**, is constructed within this type of geometry. If you are wondering how this postulate could possibly fail to hold, think of a curved surface; for example, that of our planet: two parallel lines will eventually intersect–where? At the poles, of course, if they are initially draw parallel to each other at the equator...).

So we find that not all linear equations have a **_y_-intercept**. But let me ask you this: do all equations of the form *x* = any REAL number lack a **_y_-intercept**? Once again, do not jump to conclusions. Think about all of the possible vertical lines that may be drawn on the **_x_-_y_ plane**, and try to correctly answer the question.

Ready for the (again) semi-obvious? At first glance, the answer might be that if the equation is of the form *x* = **any REAL number** it will never have a **_y_-intercept**, except, perhaps, for one specific linear equation of this type (or form). Can you guess which might be? If you answered "the equation *x* = 0" then you are absolutely on the right path. You see, this particular equation may be said to have an infinite number of **y-axis** crossing points, or **y-intercepts**, simply because it will lie exactly where the **y-axis** is located to begin with, and so you would be justified in assuming that all of its solution pairs correspond to a **y-intercept**. Except, of course, if you want to interpret the word "***intercept***" as "a point where a "***crossing***" occurs", in which case, we can't really say that it applies to any solution pair of $x = 0$ because there is never a "***crossing***" to begin with: all points lie *along* the **y-axis**, never to its right or left, so a "***crossing***", strictly-speaking, never occurs! Observe the graph of $x = 0$.

Graph of $x = 0$

Equation's line...

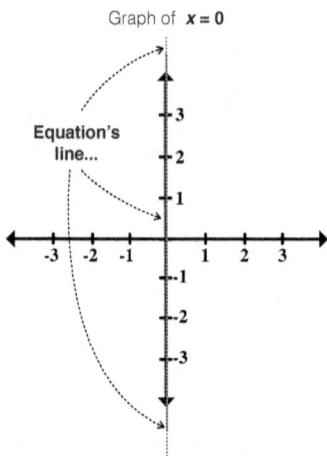

Do you see how the line is located exactly where the y-axis is located? All of its solution pairs will have the form **(0 , any REAL number)**, such as:

$$(0 , -2), (0 , -.333...), (0 , 0),$$
$$(0 , \sqrt{3}), (0 , 5), \text{ etc...}$$

A "***y-axis crossing***" thus never takes place, so strictly speaking (formally, that is), this equation does not have *a* (note the singular version of this article, as in *one*) **y-intercept**.

Everything we've seen so far regarding linear equations and the **y-intercept** is included in the following diagram.

⚓

A linear equation of the form

$$y = mx + b$$

has a *unique y-*intercept at:

(0 , b)

For example:
$$y = 2x + 5 \dashrightarrow (0 , 5)$$
$$y = 2x - 5 \dashrightarrow (0 , -5)$$

Notice that if the value of the constant **b** is being subtracted, then the **y** value of the **y-intercept's** coordinate will be negative. Furthermore, a linear equation *of this type* will have *only one y-*intercept.

A linear equation of the form

x = any REAL number

does not have a **y-intercept**.

We can now move on to review the concept of a linear equation's **x-intercept**.

The x-intercept of a linear equation

After reviewing the concept of a linear equation's **y-intercept**, I am sure you can figure out for yourself the meaning of a linear equation's **x-intercept** . It is, of course, the precise location where its graphed line crosses the **x-axis**, and it will always have the form:

(any REAL number , 0)

Please note that this time, it is the **y** value that is equal to **0** , as is to be expected, since any point that is located somewhere along the **x-axis** will have a "*height*" of **0** , as the following diagram indicates.

y axis

Any point located along the
x-axis has a "**height**" equal to **0**

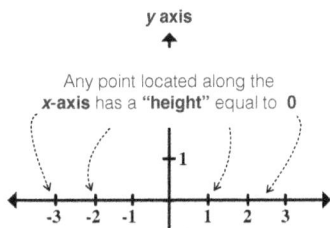

For example:
(− 3 , 0) , (− 2 , 0) , (1 , 0) , etc...

↓

It is relevant to point out that the idea of thinking about the **y** value of a coordinate as its "*height*" is appealing because most of us can relate to it by thinking of, for example, a building. Each floor is typically numbered (above ground level) as **1** , **2** , **3** , **4** , etc..., and if we were to go down below ground level (towards the "basement"), we could label those floors as − **1** , − **2** , − **3** , etc... I have personally been in elevators that have such a numbering system, so I can attest to the fact that they do exist. In this scenario, the

ground level would be considered floor "**0**" . Let me illustrate this with the following diagram:

Graph of **x = 0**

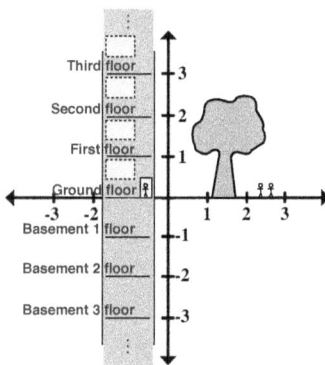

Keeping this in mind, you can now probably see what I mean when I say that a point that lies on the **y-axis** has a "*height*" of "**0**" .

So back to the **x-intercept**. If the **x-intercept** occurs when **y** is equal to **0** , then all we need to do to find the **y-intercept** of a linear equation is replace the variable **y** with **0** (the dependent variable, that is, otherwise known as the output), and then solve for the other variable, **x** (the independent, or input variable). Of course, if you happen to have the graph laying around, then all you would have to do is extract the coordinate from the graph by locating the precise point where the equation's line crosses the **x-axis**. This of course, assuming the scale is appropriate and that the point where the line is crossing the **x-axis** can be determined either precisely or with a margin of error that is acceptable.

This can be summed up as follows:

⚓

A linear equation of the form

$$y = mx + b$$

has a *unique* x-intercept at:

$$(\frac{-b}{m}, 0)$$

This is derived from the process of solving $y = mx + b$ for x, after assigning 0 to the y variable.

For example:

$$y = 2x + 6 \dashrightarrow (\frac{-6}{2}, 0) \dashrightarrow (-3, 0)$$

$$y = 4x - 8 \dashrightarrow (\frac{8}{4}, 0) \dashrightarrow (2, 0)$$

A linear equation *of this type* will have *only up to one* x-intercept.

Note that if $m = 0$, then the equation's line will not have an x-intercept, since the equation's line is in this case parallel to the x-axis, and therefore, it will either never cross the x-axis at all, or, in the case of $y = 0$, it will lie exactly on the x-axis and so a "crossing" never occurs...

I do not recommend trying to memorize the "formula" version of the x-intercept. Instead, just know that if you want to know the precise location where a linear equation's line crosses the x-axis, simply replace the variable y with 0 and solve for x. Those two values will then become the values of the x-intercept's coordinate.

For example, say you wish to know the x-intercept of the following lines:

a) $y = 5x + 20$

b) $y = 3x - 18$

c) $y = 4x$

d) $y = -x$

e) $y = 5$

f) $x = 4$

Before reading on, I strongly suggest you try to work it out on your own. Be careful with those special cases where the line is parallel to the x-axis.

Ready to check your answers? Compare your work.

a) $y = 5x + 20$

Since we want to know the x-intercept, in other words, (? , 0), simply replace y with 0 and solve for x...

$$y = 5x + 20$$

First, replace y with 0 ...

$$0 = 5x + 20$$

then solve for x ...

$$0 - 20 = 5x + 20 - 20$$

$$-20 = 5x$$

$$\frac{-20}{5} = \frac{5x}{5}$$

$$-4 = x$$

We now know that if we assign a value of **−4** to **x**, the output will be **0**, and hence, **y** will equal **0**. Let me prove it to you by replacing **x** with **−4** this time, and finding the value of **y**:

$$y = 5x + 20$$

Replacing **x** with **−4** ...

$$y = 5(-4) + 20$$

then solving for **y** ...

$$y = -20 + 20$$

$$y = 0$$

Do you see how it does work out?

In conclusion, this equation's **x-intercept** is located at:

$$(-4, 0)$$

b) $y = 3x - 18$

$$y = 3x - 18$$

First, replace **y** with **0** ...

$$0 = 3x - 18$$

then solve for **x** ...

$$0 + 18 = 3x - 18 + 18$$

$$18 = 3x$$

$$\frac{18}{3} = \frac{3x}{3}$$

$$6 = x$$

This equation's **x-intercept** is therefore located at:

$$(6, 0)$$

c) $y = 4x$

$$y = 4x$$

First, replace **y** with **0** ...

$$0 = 4x$$

then solve for **x** ...

$$\frac{0}{4} = \frac{4x}{4}$$

$$0 = x$$

This equation's **x-intercept** is therefore located at:

$$(0, 0)$$

In other words, this line passes through the origin, and will therefore have a **y-intercept** in that location as well.

d) $y = -x$

$$y = -x$$

First, replace **y** with **0** ...

$$0 = -x$$

then solve for **x** ...

$$\frac{0}{-1} = \frac{-x}{-1}$$

$$0 = x$$

Just like the previous exercise, this equation's **x-intercept** is also located at:

$$(\,0\,,\,0\,)$$

In other words, this line passes through the origin, and will therefore have a **y-intercept** in that location as well.

e) $y = 10$

Be careful with this equation: if you try to replace **y** with **0** you end up with a meaningless (or false) statement:

$$y = 5$$

First, replace **y** with **0** ...

$$?\;0 \overset{?}{=} 5\;?$$

This is obviously false: zero is *not* equal to ten.

What you need to remember is that this is the equation of a horizontal line, since its slope value is **0** (we could rewrite the equation as $y = 0x + 10$ so that you can clearly see its **m** value, remember?). Therefore, it is parallel to the **x-axis** and it will never cross it.

Conclusion: this equation *does not* have an **x-intercept**.

f) $x = 4$

This time, you may be wondering how to find this equation's **x-intercept** if it doesn't have a **y** variable to begin with... Well, remember that this

is the equation of a vertical line, and therefore, all of its solution pairs are of the form:

(4 , any REAL number)

Since we are interested in the solution pair that has a **y** value of **0**, this knowledge allows us to conclude that the line will therefore cross the **x-axis** at:

$$(\,4\,,\,0\,)$$

Observe all graphs of the equations from this exercise above, and how the answers we found are consistent with the point where the lines cross the **x-axis** (when there is such a solution, because as we already stated, problem **(e)** does not have an **x-intercept**).

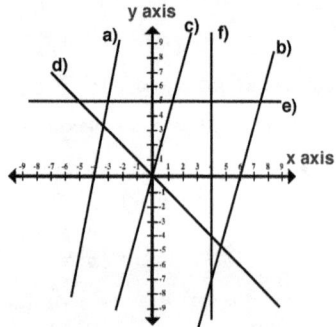

a) $y = 5x + 20$; **x-intercept** @ $(-4\,,\,0)$

b) $y = 3x - 18$; **x-intercept** @ $(\,6\,,\,0\,)$

c) $y = 4x$; **x-intercept** @ $(\,0\,,\,0\,)$

d) $y = -x$; **x-intercept** @ $(\,0\,,\,0\,)$

e) $y = 5$; does not have an **x-intercept**

f) $x = 4$; **x-intercept** @ $(\,4\,,\,0\,)$

Before we end this part of the review, I should clarify that as encountered previously, not all linear equations are presented to you in slope-intercept form. If that is the case, remember that you have options:

1) Solve the linear equation for **y** –*if possible* (remember that a linear equation may be of the form **x = any REAL number** , in which case, solving for **y** is meaningless). This essentially switches the linear equation *to* slope-intercept from; having done this, you may extract the slope value, **m** , the **y-intercept**, graph the equation's line, and figure out the **x-intercept** by following the steps we reviewed earlier in this chapter.

> *** If the equation does not have the variable **y** (or its dependent equivalent, or output, because remember that other letters may be used in place of the **x** and **y**), then you should identify it as the equation of a vertical line. Simply solve for **x** if not already in that state, and extract the information using the principles reviewed earlier in this chapter.

2) If you do not wish to solve the equation for **y** (assuming said variable is present in the equation), you could find the **y-intercept** by replacing **x** with **0** and solving for **y** ; then, simply use **x = 0** and the **y** value found to specify the point where the line crosses the **y-axis**. Similarly, for the **x-intercept**, replace **y** with **0** and solve for **x** . You can then use these two points on the graph to draw the linear equation's line. If you used this method, and you need to determine the equation's slope, you would then have to use a slope formula that will be reviewed shortly.

There are, of course, other methods (for example, using the slope value, if known, and a known point of the equation's line, and finding more points that belong to the line; from this graph version, specify the **y-intercept**, **x-intercept**, or specific solution pairs that may be needed, etc...). However, these two methods are the most practical and therefore the most commonly used.

The only special case worth mentioning is this: if you have the linear equation's graph, you can quickly determine the equation's slope, **y-intercept**, and **x-intercept**, so long as its scale is appropriate and if the graphs contains tick-marks on both axes, allowing you to precisely read the desired coordinates.

To check your mastery of the concepts reviewed so far, try to solve the following prompt by yourself. You can check the examples that we worked on earlier, but avoid peeking at the answer before you try to do all you can to solve it on your own.

Graph the following linear equations, and specify, whenever possible, their slope, **y-intercept**, and **x-intercept** (the latter two in coordinate form). Hint: because I am asking you to specify slope, it is ideal to express the linear equations in slope-intercept form.

a) $y = 2x - 4$

b) $y = -x + 3$

c) $-2x + 2y - 2 = 0$

d) $4 - y = 0$

e) $-2x - 6 = 0$

f) $-x + y - 3 = -3$

Ready to check your answers? Observe.

a) $y = 2x - 4$

This equation is already in slope-intercept form. Therefore, we know the following:

$$\text{Slope} = m = 2 = \frac{2}{1}$$

y-intercept @ (0 , – 4)

x-intercept @ (? , 0) ; to find it, simply replace *y* with **0** and solve for *x* :

$$y = 2x - 4$$
$$0 = 2x - 4$$
$$0 + 4 = 2x - 4 + 4$$
$$4 = 2x$$
$$\frac{4}{2} = \frac{2x}{2}$$
$$2 = x$$

Therefore: *x*-intercept @ (2 , 0)

Finally, to graph this linear equation, you can either use the **y-intercept** on the **x-y plane**, plot a point there knowing that this equation's line must pass through this point, and then use the slope value (as rise over run) from this point to find a second point, and then draw a "straight" line passing through these two points. Or, you could simply use the **y-intercept** and **x-intercept** and draw a "straight" line passing through them. Since I have already found these two "special" points that are part of the linear equation, I will use this method. Just remember to extend the line at either ends, and to avoid drawing these two points thicker than the rest, especially if you think doing so could mislead you regarding the linear equation's true graph form.

➡

Graph of y = 2x – 4

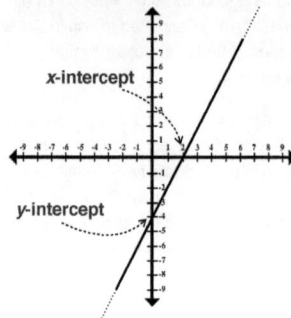

Observe how I point at the intercepts instead of using "thick dots" in order to keep drilling the fact that all points that belong to a given linear equation's line are equally relevant, and that they all form the line to begin with.

b) $y = -x + 3$

This equation is already in slope-intercept form. Therefore, we know the following:

$$\text{Slope} = m = -1 = \frac{-1}{1}$$

y-intercept @ (0 , 3)

x-intercept @ (? , 0) ; to find it, simply replace *y* with **0** and solve for *x* :

$$y = -x + 3$$
$$0 = -x + 3$$
$$0 - 3 = -x + 3 - 3$$
$$-3 = -x$$
$$\frac{-3}{-1} = \frac{-x}{-1}$$
$$3 = x$$

Therefore: **x-intercept** @ **(3 , 0)**

I will use the same method that I used on the previous problem to graph this linear equation:

Graph of $y = -x + 3$

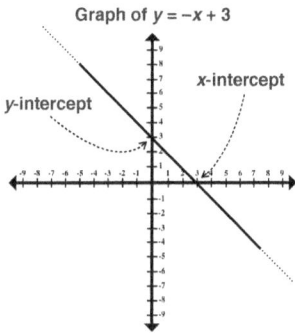

y-intercept

x-intercept

At this point, you should be able to reaffirm what we discussed earlier in the chapter regarding the value of **m** and the equation's slope: if **m** is positive, the equation's line will rise from left to right; if it is negative, it will fall from left to right. And as you can see, the two equations graphed so far allow you to verify that the rule, of course, is valid for all linear equations.

c) $-2x + 2y - 2 = 0$

This equation is **not** in slope-intercept form. Therefore, I am going to solve it for **y** so that it is in slope-intercept form, and then proceed from there:

$$-2x + 2y - 2 = 0$$
$$-2x + 2y - 2 + 2x + 2 = 0 + 2x + 2$$
$$2y = 2x + 2$$

$$\frac{2y}{2} = \frac{2x + 2}{2}$$
$$y = x + 1$$

$$\textbf{Slope} = \textbf{\textit{m}} = 1 = \frac{1}{1}$$

y-intercept @ **(0 , 1)**
x-intercept @ **(? , 0)** ; to find it, simply replace **y** with **0** and solve for **x** :

$$y = x + 1$$
$$0 = x + 1$$
$$0 - 1 = x + 1 - 1$$
$$-1 = x$$

Therefore: **x-intercept** @ **(− 1 , 0)**

I will use the same method that I used on the previous problem to graph this linear equation:

Graph of $y = x + 1$

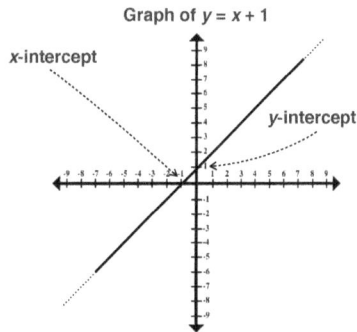

x-intercept

y-intercept

Please note that the original equation, $-2x + 2y - 2 = 0$, and the equation that was found during the process of solving for **y** , $y = x + 1$, are exactly the same. I used the latter version in the graph's heading because it is in slope-intercept form and can thus be used to verify the slope and the **y-intercept** values. Just remember that I could have used the other version as well, since they are interchangeable.

d) $4 - y = 0$

This equation is **not** in slope-intercept form. Therefore, I am going to solve it for **y** so that it is in slope-intercept form, and then proceed from there:

$$4 - y = 0$$
$$4 - y - 4 = 0 - 4$$
$$0 - y = -4$$
$$-y = -4$$
$$\frac{-y}{-1} = \frac{-4}{-1}$$
$$y = 4$$

Slope $= m = 0 = \dfrac{0}{1}$

y-intercept @ (0 , 4)

x-intercept @ (? , 0) ; since this linear equation's graph is a horizontal line, it **does not** have an **x-intercept**.

To graph this linear equation, simply plot the **y-intercept** and draw a **horizontal** line passing through that point; no additional points required:

Graph of y = 4

y-intercept

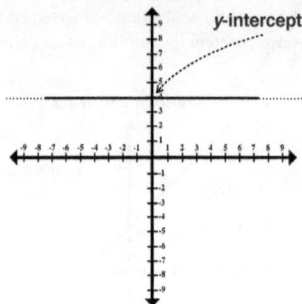

As expected, the graph of this linear equation **never** crosses the **x-axis**, precisely because its slope value is equal to **0** .

One last comment regarding this equation: you should remember that another way of expressing this equation is to write it as follows:

$$y = 0x + 4$$

This clearly allows you to identify it as the equation of a line that is horizontal, since the value of **m** is equal to **0** .

e) $-2x - 6 = 0$

This equation is **not** in slope-intercept form; furthermore, the equation lacks the variable **y**. Therefore, I am going to solve it for **x** so that I can extract its **x** value from the **x-intercept** and graph it. Note that even before we do this, we know that this linear equation corresponds to a vertical line with an infinite slope (UNDEFINED).

$$-2x - 6 = 0$$
$$-2x - 6 + 6 = 0 + 6$$

$$-2x = 6$$
$$\frac{-2x}{-2} = \frac{6}{-2}$$
$$x = -3$$

Slope $= m =$ **UNDEFINED**

y-intercept @ *does not have one*

x-intercept @ (− 3 , 0)

Please note that we did not have to do any additional work to figure out the **x-intercept**; whenever the linear equation is of the form **x = any REAL number** or " **x = d** ", we automatically expect the **x-intercept** to be located at (**0 , d**) . To graph this linear equation, simply plot the **x-intercept** and draw a **vertical** line passing through that point:

Graph of x = − 3

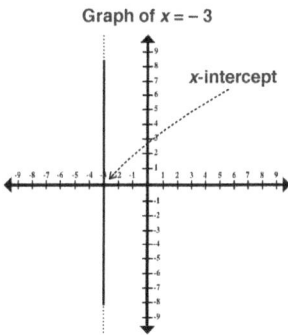

x-intercept

f) $-x + y - 3 = -3$

This equation is **not** in slope-intercept form. Therefore, I am going to solve it for **y** so that it is in slope-intercept form, and then proceed from there:

$$-x + y - 3 = -3$$
$$-x + y - 3 + x + 3 = -3 + x + 3$$
$$y = x$$

Slope $= m = 1 = \dfrac{1}{1}$

y-intercept @ (0 , 0)

x-intercept @ (? , 0) ; to find it, simply replace **y** with **0** and solve for **x** :

$$y = x$$
$$0 = x$$

Therefore: **x-intercept** @ (0 , 0)

You should note that since the **y-intercept** is (**0 , 0**), we know that this equation's line passes through the origin. And if it passes through the origin, we know that its **x-intercept** will also be (**0 , 0**). Therefore, we could have saved ourselves the trouble of replacing **y** with **0**, since once the **y-intercept** was known, the **x-intercept** could have been thus determined.

Graphing this equation may be a bit trickier because the **y-intercept** and the **x-intercept** occurs at the exact same location: at the origin. Therefore, I will use the "**plot a point, then use the slope to find a second point**" method to graph this linear equation.

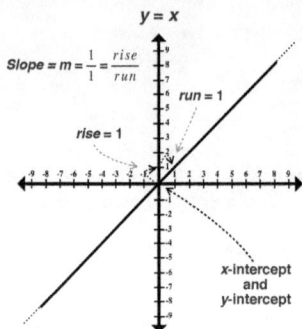

$y = x$

Slope $= m = \dfrac{1}{1} = \dfrac{rise}{run}$

run $= 1$

rise $= 1$

x-intercept
and
y-intercept

As you can see from the graph above, I first located the *y*-intercept (which also corresponds to the *x*-intercept, since they are the same for this specific linear equation), **(0 , 0)**, and then using the equation's slope, and starting at said *y*-intercept, moved **1** unit up and then **1** unit to the right to find a second point that is part of the equation's line: in other words, the coordinate **(1 , 1)**. Having located those two points, the equation's line can now be graphed by drawing a "straight" line that passes through them.

So far, we have worked with linear equations. In other words, we have been given the equation to work with, whether in slope-intercept form or not, and with the equation at our disposal, specified its slope, *y*-intercept, *x*-intercept, as many solution pairs as may be needed, and used this information to graph it.

The logical question to ask ourselves is this: what happens if all we have is the graph of a line? Can we determine its equation (in other words, *reverse-engineer it*)? Or, if we know for a fact that a line passes through two specific points

(and we know the coordinates in question), how can we determine the equation of that specific, unique line? Finally, what if all we know about a linear equation is that its line passes through a given point, and one of the following?

1) The linear equation's slope

2) That the equation' line whose equation we are trying to determine is parallel or perpendicular to another line, call it **Line B**, and we have the equation of **Line B** available (or we are given enough information to figure this **Line B's** equation).

These processes are important because many real world applications involve linear relationships. When modeling a real world phenomenon, it is very common to obtain data from the field (by performing measurements), plot the points obtained (in the form **(input , output)** or **(independent var , dependent var)**, where var stands for variable), and then to try to find a linear equation that best fits said measurements so that predictions can be made and causal relationships established.

Well, to deal with situations such as these, we can proceed as follows.

Determining the slope of a linear equation when two points through which it passes are known

Let's explore the following scenarios.

Scenario A Suppose we know that a line passes through the point **(1 , 3)**. We know absolutely nothing else. Can we specify a *unique* linear equation with this information?

Think about it and do not jump to conclusions. A graph representing this scenario may help you answer this question correctly. Observe.

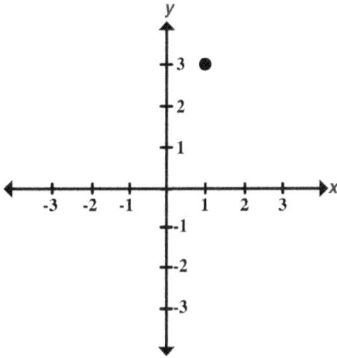

The graph above contains the plot for the only point specified: **(1 , 3)**. Once again, is there a *unique* line that passes through it? Or can many different lines (belonging to as many different linear equations) pass through that point? Lines that will have different slopes, different **y-intercepts**, and **x-intercepts**, and different sets of solutions pairs?

The answer should be obvious. Let me show you a few lines that pass through the given point and that belong to different linear equations:

a) The line of the equation $y = 3$

b) The line of the equation $x = 1$

c) The line of the equation $y = 2x + 1$

d) The line of the equation $y = -x + 4$

e) The line of the equation $y = 3x$

etc...

As you can see, there will be many different lines that can be drawn on the graph that pass through the point **(1 , 3)** and that belong to just as many different linear equations. There are, in fact, an infinite number of them, believe it or not. Observe the graphs of the equations that I used to illustrate this:

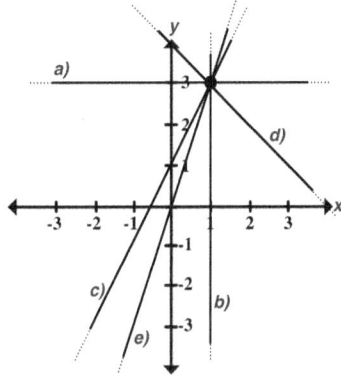

By now, I am sure you can see why we would never finish drawing different lines belonging to different linear equations that pass through the coordinate **(1 , 3)** ; simply by changing the slope value slightly, we can change the equation but still ensure that it passes through the given point, and this simply because there are an infinite number of slope (or **m**) values available (because the **REAL** number system contains an infinite number of elements).

So *Scenario A* is a dead-end street. We cannot specify a *unique* equation for the line that passes through only one point. We can, however, specify infinitely many. More on how to do this later.

Scenario B Suppose we know that a line passes through the point **(1 , 3)** and through the point $(0, 1)$. We know absolutely nothing else. Can we specify a ***unique*** linear equation with this information? This time, as we have mentioned earlier, we can. Why? Because in ***euclidean geometry*** (which is the geometry we are working with when graphing linear equations on the ***x-y plane*** or Cartesian plane), there can only be one unique ("straight") line that passes through any given two points.

Therefore, we should be able to specify this line's equation. But how?

First, let's plot the two given points:

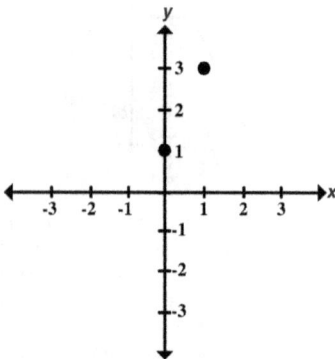

In this specific example, we are given the ***y-intercept*** point $(0, 1)$; did you notice it? Being able to identify the points as either the ***x-intercept*** or the ***y-intercept*** can be helpful, though this will not always be the case (this is one of those cases where ***it will*** help us).

Okay... so our goal is to define the equation of the line that passes through the two given points. Which line? Well, let's draw it:

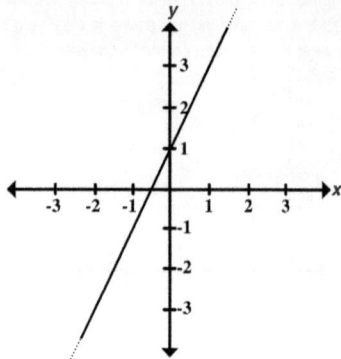

So how can we figure out its equation? Or rather, what linear equation, when graphed, produces that exact line, that passes through the two given points–$(1, 3)$ and $(0, 1)$–, and therefore, has the slope that can be visually seen on the graph above? We can already determine just by looking at the line above that its slope must be positive because it rises from left to right, and we can also see that it is considerably steep.

Well, it might occur to you that we could find the slope value by reverse-engineering the process with which we graph an equation when we have a point through which it passes through and use the known slope to find a second point... You see, in this case, we already have two known points, so maybe if we start from the left-most point, and try to determine the "rise" and "run" values that would lead to the second known point, we could specify the lines's slope this way. Let's give this a try.

Bringing back the graph that contains the two known points only, we can imagine standing on

(0 , 1) and then trying to figure out the rise and run values that would land you on (1 , 3) :

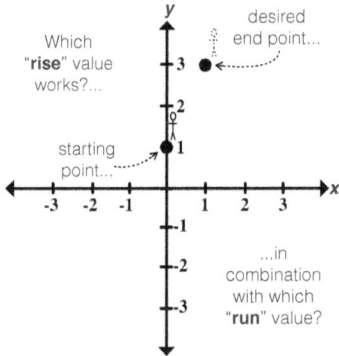

Well, if the stick figure is currently located at (0 , 1) , and we want it to end up at (1 , 3) , it should be fairly obvious to you that to achieve this, it needs to "*rise*" **2** units (using the building analogy, it is currently on the first floor; to get to the third floor, it needs to go up **2** floors; hence **2** units). But that only resolves the vertical movement; what about the lateral or horizontal movement? Moving up **2** units is not enough because it would leave our stick figure stranded at (0 , 3) , and we need it to make it all the way to (1 , 3) . Well, this is where the "*run*" value comes in: simply advance it **1** unit to the right, and our goal is achieved.

So, to recap, starting out at (0 , 1) and applying a slope value of "*rise*" **2** units and "*run*" **1** unit, we are able to obtain the graph of the exact same line for which we are trying to define its equation.

Well, if we know the slope, and we also know the **y-intercept**, we should certainly be able to specify the corresponding linear equation, specifically in the form...

$$y = mx + b$$

...the so-called slope-intercept form. Why? Because we now know the values of **m** and **b**, all that is needed to "create" a unique linear equation. The slope value **m** is:

$$\textbf{slope} \ = \ m \ = \frac{change \ in \ Output}{change \ in \ Input} = \frac{rise}{run} = \frac{2}{1}$$

In other words, the equation must have an **m** value equal to **2** .

Because the **y-intercept** is located at (0 , 1) , we also know the value of **b** (we could have specified this at the beginning of this exercise, but without **m**, it was pointless to do so). A quick comparison between the general version of the **y-intercept** when the linear equation is in slope-intercept form and the coordinate we have for it in this case is the source of the value we need:

(0 , **b**)

(0 , **1**)

Therefore: **b** = 1

Replace the values of **m** and **b** that we were able to determine, and we will have found the linear equation of the line that passes through (0 , 1) and (1 , 3) :

$$y = mx + b$$

$$y = 2x + 1$$

And we have found our answer!

But what if neither one of the two points that are known (or given) correspond to the **y-intercept**? What would we do then?

To understand this potential problem, let's consider the following scenario.

Scenario C   Suppose we know that a line passes through the point **(1 , 2)** and through the point **(2 , 3)**. We know absolutely nothing else. Can we specify a *unique* linear equation with this information? As we have mentioned earlier, and as we determined on the previous scenario, we can.

First, let's plot the two points:

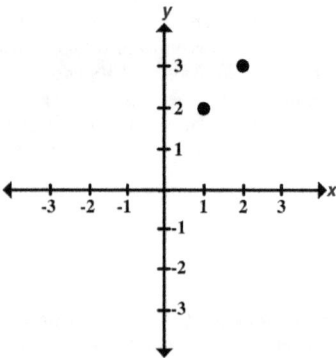

In this specific example, we are **not** given the **y-intercept** point. This will be a problem later on, after we have found the slope. But first things first...

Okay; so our goal is to define the equation of the line that passes through the two given points. Let's draw the line:

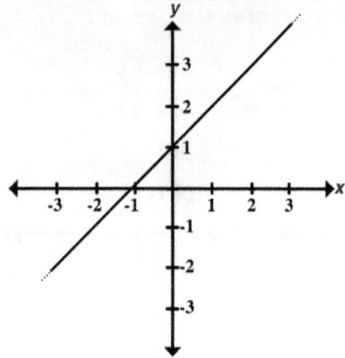

So how can we figure out its equation? Or rather, what linear equation, when graphed, produces that exact line, that passes through the two given points—**(1 , 2)** and **(2 , 3)**—, and therefore, has the slope that can be visually seen on the graph above? We can already determine just by looking at the line on the graph that its slope must be positive because it rises from left to right, and we can also see that it is less steep that the previous line (therefore, we should expect a slope value, **m** that should be less than **2** , the previous scenario's line slope value).

Remember how we obtained the previous slope value? Well, we do the exact same thing,

Bringing back the graph that contains the two known points only, we can imagine standing on **(1 , 2)** and then trying to figure out the rise and run values that would land you on **(2 , 3)** :

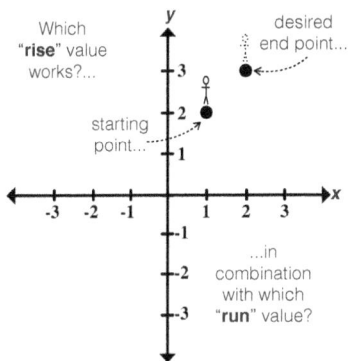

Which "**rise**" value works?...

desired end point...

starting point...

...in combination with which "**run**" value?

Well, if the stick figure is currently located at (1 , 2), and we want it to end up at (2 , 3), it should be fairly obvious to you that to achieve this, it needs to "*rise*" 1 unit (using the building analogy, it is currently on the second floor; to get to the third floor, it needs to go up 1 floor; hence 1 unit). But that only resolves the vertical movement; what about the lateral or horizontal movement? Moving up 1 unit is not enough because it would leave our stick figure stranded at (1 , 3), and we need it to make it all the way to (2 , 3). Well, this is where the "*run*" value comes in: simply advance it 1 unit to the right, and our goal is achieved. Thus, $m = 1$.

So, to recap, starting out at (1 , 2) and applying a slope value of "*rise*" 1 unit and "*run*" 1 unit, we are able to obtain the graph of the exact same line for which we are trying to define its equation.

Well, if we know the slope, and–wait... this is where we encounter a dead-end. What do we do next? After all, unlike the previous problem, we were not given the **y-intercept**, so do we give

up? Or do we try to find the **y-intercept**, perhaps by using the slope value we found, and going back (from right to left, just as we did in an earlier exercise)? The problem with this method is that although it will work **sometimes**, there are many times where we will not be able to pinpoint the **y-intercept** precisely by using the graph. Why is this? Let me illustrate the answer with the following scenario: consider an equation that has a **y-intercept** of (0 , .333...) ; this is a perfectly valid coordinate for the **y-axis** crossing, and yet how could we be expected to read the decimal number .333... , which contains an infinite number of decimal digits, from a graph, considering its scale? How do we know it isn't .3332 ? Or .33335 instead? Or any other possibility? There must be a precise way to find the equation of the line if we know the slope and if we know a point through which it passes. How?

Well, let's first explore the formula that allows you to find the slope of a linear equation when you know at least two points through which its line passes, as in the problem we are currently working on. Revisiting the method we used to find the slope of the line earlier will allow us to figure out the general slope formula ourselves, which can then be used whenever we need to find the slope of a line as long as we have two points through which it passes.

If you recall, we knew that our starting point was (1 , 2) and that we wanted to end up at (2 , 3). Next, we can think of the "*rise*" and "*run*" as a "*change in vertical position*" and a "*change in horizontal position*" respectively. This means that if we think of any two coordinates in general, and use (x_1 , y_1) and (x_2 , y_2) to represent them (in this case, x_1 would correspond to the first coordinate's *x* value: **1** , and y_1 would correspond to the first coordinate's *y* value: **2**, while x_2 would correspond to the second coordinate's *x* value: **2** , and y_2 would correspond to the second coordinate's *y* value: **3**), we can use a mathematical statement to express the process we used to pinpoint the rise and run values earlier. Just keep in mind the following definition of slope:

$$\textbf{slope} = m = \frac{rise}{run} = \frac{change\ in\ vertical}{change\ in\ horizontal}$$

Just keep in mind the following relationship that we established above:

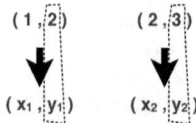

The rise value

To find the rise value, we thought of a stick figure standing on the second floor (or in this case, the y_1 value); then, we asked ourselves the following question: "How many floors must the stick figure go up (positive rise value) or go down (negative rise value) in order to reach the desired floor, in this case, the third floor of the "building"?" In other words, how can we get the stick figure to the desired level specified by y_2? It should be obvious to you that what we computed was essentially this: **3 – 2** , which led to the answer that we were looking for: "the *rise* value is equal to **1** unit". Well, for any two coordinates in general, we should be able to establish the following:

$$\textit{rise} = y_2 - y_1$$

Do you agree? Do you truly understand why it works out? This should not feel like an imposition on my part, nor like magic, pulling formulas out of thin air... All we did here was think of any two coordinates by using variables instead of actual values (using subscripts–the numbers that appear next to the *x* and *y* variables–which allow us to distinguish them) and then come up with the mathematical statement that matches the process we used to intuitively find the rise value.

If you are still in doubt as to whether this formula works or not, let's test it.

To find the rise value of the equation of the line that passes through the points:

(1 , 2)   (2 , 3)

(x_1 , y_1)   (x_2 , y_2)

...we can use the rise formula we found

$$\textit{rise} = y_2 - y_1$$

...by replacing y_2 with **3** and y_1 with **2** :

$$\textit{rise} = 3 - 2 = 1$$

Which is exactly the value we found earlier. If you are wondering what happens if we were to switch the coordinates, like this:

(2 , 3)   (1 , 2)

(x_1 , y_1)   (x_2 , y_2)

...the answer is that our rise value will be different. Observe:

$$\textit{rise} = 2 - 3 = -1$$

524

So what does this mean? Is the answer wrong? No. The answer is perfectly valid. You just need to truly understand the setup: if you call the "first" coordinate (2 , 3), it means you are positioning the stick figure on that point, and that your goal is to move it to the "second" coordinate, (1 , 2) :

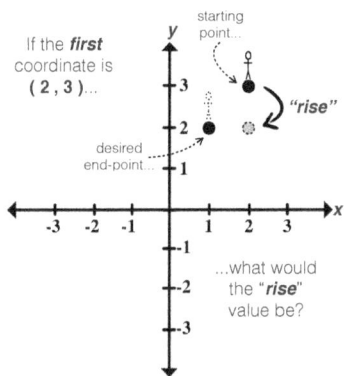

If the **first** coordinate is (2 , 3)...

starting point...

desired end-point...

"rise"

...what would the "rise" value be?

As you can probably appreciate from the graph above, the "**rise**" value would have to correspond to going **down** 1 floor, since in this set up (determined by the fact that the "first" coordinate was deemed to be (2 , 3)) that's what it would take to eventually land on the "second" point, the coordinate (1 , 2). So, if an **upwards vertical movement** is said to be a positive number, it stands to reason that a **downwards vertical movement** must correspond to a negative number. And now you see why the rise value that the formula gave us is **negative**: it is **automatically** considering the starting position and providing the correct "rise" value that would take us to the second coordinate's height, or **y** value. If this is a bit confusing, it might be because of the word "**rise**": our brains automatically assume it means upwards, which

matches the meaning of the word in everyday use, but in the context of slope, it actually means "vertical displacement"; thus, it may be positive (if the vertical displacement is upwards, or negative if downwards.

Now, on to the "**run**" value, or the horizontal displacement; similarly to all that we established for the "**rise**" value will apply to the "**run**" value as well. Just keep in mind the definition of slope that we have been using:

$$\text{slope} = m = \frac{rise}{run} = \frac{change\ in\ vertical}{change\ in\ horizontal}$$

And keep in mind the following relationship that we established previously:

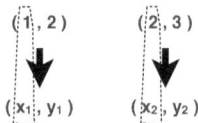

$$(1 , 2) \qquad (2 , 3)$$

$$(x_1 , y_1) \qquad (x_2 , y_2)$$

The run value

To find the run value, earlier we thought of a stick figure standing on the **x** position of **1** (or in this case, the x_1 value); then, we asked ourselves the following question: "How many steps must the stick figure move to the right **laterally** (positive run value) or left **laterally** (negative run value) in order to reach the desired ending **x** value, in this case, **2** ?" In other words, how can we get the stick figure to the desired **x** position specified by x_2? It should be obvious to you that what we computed was essentially this: **2 − 1** , which led to the answer that we were looking for: "the **run** value is equal to **1** unit". Well, for any two coordinates in general, we should be able to establish the following:

$$run = x_2 - x_1$$

Once again, all we did here was think of any two coordinates by using variables instead of actual values (using subscripts–the numbers that appear next to the x and y variables–which allow us to distinguish them) and then come up with the mathematical statement that matches the process we used to intuitively find the run value.

Let's test it.

To find the run value of the equation of the line that passes through the points:

$$(1,2) \qquad (2,3)$$

$$(x_1, y_1) \qquad (x_2, y_2)$$

...we can use the run formula we found

$$run = x_2 - x_1$$

...by replacing x_2 with 2 and x_1 with 1 :

$$run = 2 - 1 = 1$$

Which is exactly the value we found earlier. If you are wondering what happens if we were to switch the coordinates, like this:

$$(2,3) \qquad (1,2)$$

$$(x_1, y_1) \qquad (x_2, y_2)$$

...the answer is that our run value will be different. Observe:

$$run = 1 - 2 = -1$$

So what does this mean? Is the answer wrong? No. Once again, just like what we found when we did the same thing to the *rise* calculation, the answer is perfectly valid. If you call the "first" coordinate $(2,3)$, it means you are positioning

the stick figure on that point, and that your goal is to move it to the "second" coordinate, $(1,2)$:

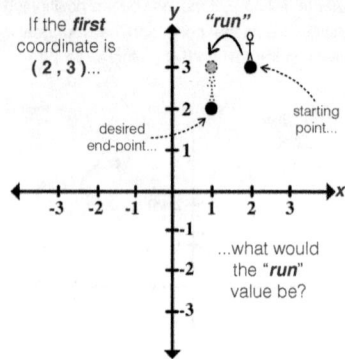

As you can probably appreciate from the graph above, the "*run*" value would have to correspond to moving *to the left* 1 unit, since in this set up (determined by the fact that the "first" coordinate was deemed to be $(2,3)$) that's what it would take to eventually land on the "second" point, the coordinate $(1,2)$. So, if a *rightwards horizontal movement* is said to be a positive number, it stands to reason that a *leftwards horizontal movement* must correspond to a negative number. And now you see why the run value that the formula gave us is *negative*: it is automatically considering the starting position and providing the correct "*run*" value that would take us to the second coordinate's horizontal position, or x value.

Slope formula

Finally, we are in a position to come up with the slope formula ourselves! We have already established that

$$\text{slope} = m = \frac{rise}{run} = \frac{change\ in\ vertical}{change\ in\ horizontal}$$

...and the following as well: for any line that passes through the two points

$$(x_1 , y_1) \qquad (x_2 , y_2)$$

...the following holds:

$$rise = y_2 - y_1$$

$$run = x_2 - x_1$$

So, bringing everything together, we can express slope as the following:

$$\text{slope} = m = \frac{y_2 - y_1}{x_2 - x_1}$$

Even though it took us a while to get here, if you followed the process closely, you should now know why this formula works, where it comes from, and how to derive it all by yourself, if necessary.

Let's finish what we started, and verify that the slope formula will in fact give us the correct slope value for the equation that passes through the two points $(1 , 2)$ and $(2 , 3)$, which we used to determine the slope formula to begin with.

We assign the following correspondence:

$$(1 , 2) \qquad (2 , 3)$$
$$(x_1 , y_1) \qquad (x_2 , y_2)$$

Then, applying the slope formula, we find that:

$$m = \frac{y_2 - y_1}{x_2 - x_1} = \frac{3 - (2)}{2 - (1)} = \frac{1}{1} = 1$$

And we find that the slope formula, of course, works perfectly! Furthermore, you may remember that we explored what would happen to the rise and run values if we assigned the (x_1 , y_1) and the (x_2 , y_2) correspondence with the given points the other way around. We found that both the rise and the run values would end up being different. But doesn't this contradict what I have stated regarding the slope of a line that passes through a pair of given points, mainly that the slope is and should be unique, always the same for a line passing through those exact same given points? So what's going on here? Observe:

If we assign the correspondence as follows:

$$(2 , 3) \qquad (1 , 2)$$
$$(x_1 , y_1) \qquad (x_2 , y_2)$$

Then, applying the slope formula, we find that:

$$m = \frac{y_2 - y_1}{x_2 - x_1} = \frac{2 - (3)}{1 - (2)} = \frac{-1}{-1} = 1$$

As if by some mysterious force, we do obtain the same *slope value* in the end! Well, it's not really a mystery; you see, switching the correspondence creates a different perspective, as we saw during the rise and run calculations. But since slope is the *ratio* of the rise and the run values, the sign difference in the individual rise and run values takes care of the switch since *both* values are changed when the correspondence is switched.

So there you have it. The famous slope formula of a linear equation that passes through two given points, completely (I hope) demystified.

To sum up:

⚓

The slope of a linear equation that passes through the two points

$$(x_1 , y_1) \text{ and } (x_2 , y_2)$$

can be determined using the slope formula

$$m = \frac{y_2 - y_1}{x_2 - x_1}$$

The slope formula will even tell you if the equation of the line that passes through the given two points corresponds to a linear equation that graphs a vertical or a horizontal line–although to be fair, this should be obvious to you simply by analyzing the coordinates of the two given points, since a *horizontal line* will always have points with equal *y* values (*equal heights*) while *vertical lines* will always have equal *x* values (*equal horizontal positions*).

Let's explore this using the following three examples.

Find the slope of the linear equations whose graphs (their lines, that is) pass through:

a) $(-2 , -3)$ and $(4 , -6)$
b) $(-6 , -4)$ and $(-2 , -4)$
c) $(-3 , -2)$ and $(-3 , 5)$

Before looking at the answers, give these a try by yourself. Just be careful with the negative numbers. Remember what we reviewed earlier in the book regarding replacing variables with a specific number: always use a set of parentheses and use the sign table where necessary.

Where you able to determine the slopes of these three equations? Compare your answers to the correct responses below.

a) $(4 , -6)$ and $(-2 , -3)$

A quick analysis of these two coordinates tells us that the equation *must* have a negative slope since the left-most coordinate that's given (that's $(-2 , -3)$) is "*higher*" than the right-most coordinate given (that's $(4 , -6)$). Observe:

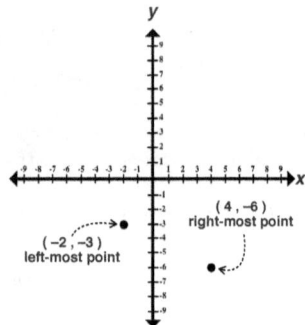

Do you see how the point on the left is higher than the point on the right? That's what I mean by the analogy "the left point is "higher" than the right point"; this implies that "*rock bottom*" (against which we may determine which point is higher than another, in this frame of reference) occurs at "negative infinity", or that the highest point is found at "positive infinity"... If this seems a bit confusing, think about it; read it a couple of times and I'm sure you'll get it. By the way, remember that both negative infinity and positive infinity are concepts, not actual locations on the graph. So when I say that "**rock bottom**" occurs at "**negative infinity**" I mean "the idea of **rock bottom**, since we can never actually reach **rock bottom**". Think about it...

So, back to this problem. Let's use the slope formula to see if we do, in fact, end up with a negative-valued slope. If we don't we would be able to detect that an error was made during the use of the formula, because the slope value **must** be negative. The quick slope-sign analysis we performed at the beginning of the solution process thus serves as a partial error-detecting device. Of course, we could still make a computational mistake and end up with a negative slope but with the incorrect negative number, so it isn't a foolproof device.

So who do we call (x_1, y_1) and who do we call (x_2, y_2)? Remember, this is literally irrelevant as far as the slope value is concerned. So I will just assign the following correspondence, matching the order in which the coordinates were initially given to us:

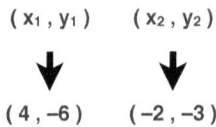

$$(x_1, y_1) \qquad (x_2, y_2)$$
$$\downarrow \qquad\qquad \downarrow$$
$$(4, -6) \qquad (-2, -3)$$

Replacing the variables with the corresponding values extracted from the coordinates above yields the following:

$$m = \frac{y_2 - y_1}{x_2 - x_1} = \frac{-3 - (-6)}{-2 - (4)}$$

$$= \frac{-3 + 6}{-2 - 4} = \frac{3}{-6} = \frac{1}{-2} = \frac{-1}{2} = -\frac{1}{2}$$

The answer we obtained matches our expectation: the slope should be negative. I have listed explicitly all three forms of a negative fraction so that you remember that they are all equivalent to each other.

A slope value m of $\frac{1}{-2}$ would imply, from a given starting point, a rise value of "**up 1** unit" and a run value of "to the **left 2** units". A slope value m of $\frac{-1}{2}$ would imply, from a given starting point, a rise value of "**down 1** unit" and a run value of "to the **right 2** units".

But please note that a slope value of $-\frac{1}{2}$ cannot actually be used as we used the slope values above, because the negative sign is to the side of the fraction; to fix it, either rewrite the fraction as any one of the two other forms, the preferred choice, or, in what turns out to be a much more complicated possibility, consider the entire slope value of $-\frac{1}{2}$ as the **rise** portion of the slope, and dividing this by **1**, use said dividing **1** as the "**run**" value:

$$m = -\frac{1}{2} = \frac{-\frac{1}{2}\ \cdots\text{ rise}}{1\ \cdots\text{ run}}$$

...which would then imply a **rise** value of "**down** one thirds of a unit, or **−.5** " and a **run** value of "**to the right** one unit". See why this is much more complicated?

I therefore strongly recommend rewriting a slope that is expressed in terms of a negative fraction so that the negative sign is incorporated into either the numerator or the denominator of the fraction, as I illustrated in this example.

Before moving on to the next problem, try to find on the **x-y plane** additional points using all three versions of the slope value that we found for this exercise, using as the starting point any of the two coordinates that were given as part of the original prompt, and check to see if you do, in fact, "land" on the line as you move either to the left or to the right of your chosen point (this will depend on the version of the slope that you use: if you are using $\frac{1}{-2}$ then you will find points to the left of your starting point; if you use $\frac{-1}{2}$ then you will find points to the right of your starting point. The version of the slope written as $-\frac{1}{2}$ would correspond to the second type, since you would have to divide it by **1** to have a usable slope format (rise over run values explicitly written), which then instructs you to move down and to the right of the starting point.

Ready? For example, if we use the point **(4 , – 6)** , we would locate the following points:

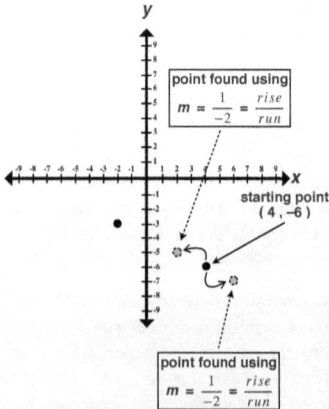

It is crucial that you understand that even though using the two different versions of slope ($\frac{1}{-2}$ and $\frac{-1}{2}$) will lead to the location of different points (the former to the left of the starting point, the latter to the right of it), they are all, of course, part of the linear equation's graph; in other words, the points have coordinates that are solution pairs of the linear equation. Observe:

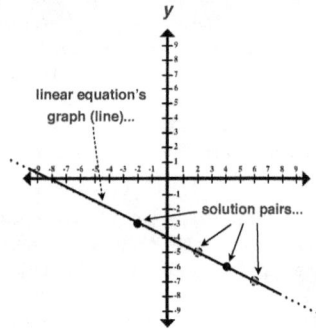

Can you see how all points are part of the linear equation's line, since they are all part of its graph?

I would like to point out that we have not yet reviewed how to figure out the linear equation that corresponds to the graphed line above (in other words, the linear equation whose graph passes through the two given initial coordinates: **(4 , –6)** and **(–2 , –3)** ; we will see this shortly...

Let's move on to the next problem of the set we are currently working with.

b) $(-6, -4)$ and $(-2, -4)$

A quick analysis of these two coordinates tells us that the equation *must* have a slope equal to **0** since the left-most coordinate that's given (that's $(-6, -4)$) is o at the same "height" as the right-most coordinate given (that's $(-2, -4)$). In other words, their y values are the same. Observe:

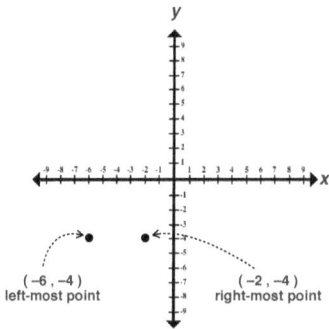

$(-6, -4)$
left-most point

$(-2, -4)$
right-most point

Do you see how the point on the left is at the same height as the point on the right?

Let's use the slope formula to see if we do, in fact, end up with a slope that is equal to **0**. If we don't we would be able to detect that an error was made during the use of the formula, because the slope value *must* be **0**, regardless of which coordinate we call the (x_1, y_1) and which one the (x_2, y_2).

So who do we call (x_1, y_1) and who do we call (x_2, y_2)? Remember, this is literally irrelevant as far as the slope value is concerned. So I will just assign the following correspondence, matching

the order in which the coordinates were initially given to us:

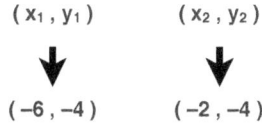

$$(x_1, y_1) \qquad (x_2, y_2)$$

$$\downarrow \qquad\qquad \downarrow$$

$$(-6, -4) \qquad (-2, -4)$$

Replacing the variables with the corresponding values extracted from the coordinates above yields the following:

$$m = \frac{y_2 - y_1}{x_2 - x_1} = \frac{-4 - (-4)}{-2 - (-6)}$$

$$= \frac{-4 + 4}{-2 + 6} = \frac{0}{4} = 0 = \frac{0}{1} = \frac{0}{-1}$$

The answer we obtained matches our expectation: the slope is equal to **0**. I have written the slope of **0** in fraction form as well, both using a positive denominator and a negative denominator, so that we may use them to find additional points on the line, both to the right and to the left of a starting point respectively. But remember: **0** is equal to $\frac{0}{1}$ and is also equal to $\frac{0}{-1}$. A slope value *m* of $\frac{0}{1}$ would imply, from a given starting point, a rise value of "*up (or down)* **0** units" (in other words, don't move vertically), and a run value of "to the **right** 1 unit".

A slope value *m* of $\frac{0}{-1}$ would imply, from a given starting point, a rise value of "*up (or down)* **0** units" (in other words, don't move vertically), and a run value of "to the **left** 1 unit".

At this point, try to find on the *x-y* plane additional points using both versions of the slope value that we found for this exercise, using as the starting point any of the two coordinates that were given as part of the original prompt, and check to see if you do, in fact, "land" on the line as you move either to the left or to the right of your chosen point.

Ready? For example, if we use the point (–6 , –4) , we would locate the following points:

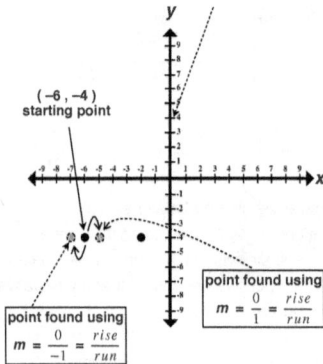

It is crucial that you understand that even though using the two different versions of slope ($\frac{0}{1}$ and $\frac{0}{-1}$) will lead to the location of different points (the former to the right of the starting point, the latter to the left of it), they are all, of course, part of the linear equation's graph; in other words, the points have coordinates that are solution pairs of the linear equation.

To prove this, observe the graph of this linear equation (the line that passes through the points given by the original prompt):

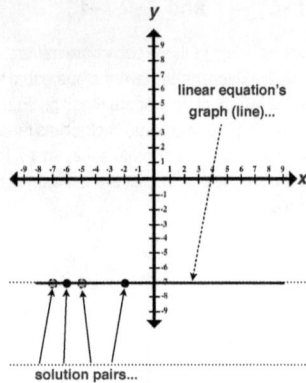

solution pairs...

Once again, can you see how all points are part of the linear equation's line, since they are all part of its graph?

As opposed to the previous example (we did not establish the linear equation itself), this linear equation is very easy to define. Since its graph corresponds to a completely horizontal line, we know it is of the following type:

$$y = b$$

Where **b** is the **y** value of the **y-intercept**, the latter defined as (0 , **b**) , remember? Looking at the coordinates we see that **y** always equals –4 in this specific line; therefore, the **y-intercept** must occur at (0 , – 4) . Therefore, this linear equation is:

$$y = - 4$$

The lesson here is that it is very easy to determine the linear equation of a horizontal line: simply look at any of the coordinate's **y** value!

c) $(-3, -2)$ and $(-3, 5)$

A quick analysis of these two coordinates tells us that the equation *must* have a slope that is UNDEFINED or *infinite* since the given coordinates have the same x value; in other words, both points are located on the exact same lateral position, one "higher up" than the other one. Observe:

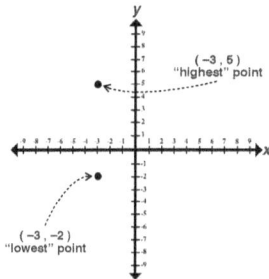

Do you see how we can't state that one of the points is to the left of the other one? That's because both coordinates have an x value that is the same: -3. Therefore, we know right away that we are dealing with a linear equation that when graphed will yield a vertical line. Furthermore, this tells us that we are dealing with an equation of the form

$$x = \text{any } \textbf{REAL} \text{ number}$$

At this point, you should know that it would be pointless to use the slope formula since we expect an UNDEFINED value (in this case, a fraction with a denominator equal to 0). But let's assume that we miss this and use the slope formula anyway. What would happen? Let's check it out.

So who do we call (x_1, y_1) and who do we call (x_2, y_2)? Remember, this is literally irrelevant as far as the slope value is concerned. So I will just assign the following correspondence, matching the order in which the coordinates were initially given to us:

$$(x_1, y_1) \qquad (x_2, y_2)$$
$$\downarrow \qquad\qquad \downarrow$$
$$(-3, -2) \qquad (-3, 5)$$

Replacing the variables with the corresponding values extracted from the coordinates above yields the following:

$$m = \frac{y_2 - y_1}{x_2 - x_1} = \frac{5 - (-2)}{-3 - (-3)}$$

$$= \frac{5 + 2}{-3 + 3} = \frac{7}{0} = UNDEFINED$$

The answer we obtained matches our expectation: the slope is UNDEFINED.

At this point, there is very little left to do. Since we know the type of linear equation that we are dealing with:

$$x = \text{any } \textbf{REAL} \text{ number}$$

...and we know that the x value is always equal to -3 in this specific case, we are able to quickly come up with the linear equation that passes through the two given points:

$$x = -3$$

To graph this line, simply connect the two points with a vertical line. If we needed more coordinates (or solution pairs), simply create them using the template

$$(-3, y)$$

...and assign any value to y that you wish. For example:

$(-3, -8)$, or $(-3, \pi)$, or $(-3, 9.85)$, etc...

Observe its graph fully drawn:

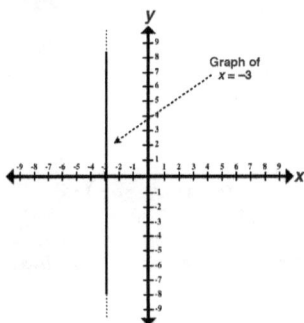

We can now go back to the problem we were working on before exploring this idea of the slope formula; the prompt was:

"**<u>Scenario C</u>** Suppose we know that a line passes through the point $(1, 2)$ and through the point $(2, 3)$. We know absolutely nothing else. Can we specify a **unique** linear equation with this information?"

If you go back to this problem, you will see that we plotted the two points, graphed the line, and then derived the slope formula ourselves, which allows us to determine the slope of any linear equation as long as we know two points through

which it passes. Let me bring back the important elements that we worked on...

The plot of the two points:

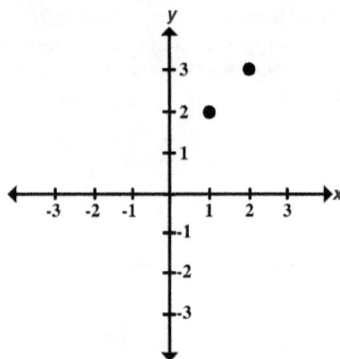

The line itself:

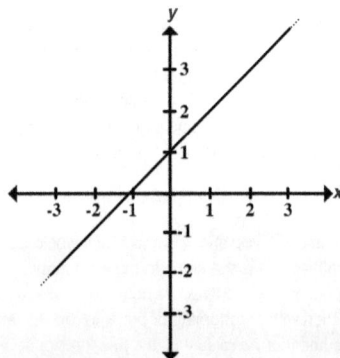

Using the idea that slope is equal to "rise over run", we found this line's slope to be equal to **1** :

534

We also corroborated that the slope formula works:

Assigning the following correspondence...

$$(1, 2) \qquad (2, 3)$$

$$(x_1, y_1) \qquad (x_2, y_2)$$

...and applying the slope formula, we found that:

$$m = \frac{y_2 - y_1}{x_2 - x_1} = \frac{3 - (2)}{2 - (1)} = \frac{1}{1} = 1$$

You may be wondering what was the point of finding the slope formula to begin with considering the fact that we were able to come up with the slope of this line by using the "rise over run" method. Well, if you recall, after finding the slope we got stuck. Remember that our goal is to find the equation of the line, and since we do not have the y-intercept we can't specify b which means we cannot specify the equation in the form $y = mx + b$. Well, this is where the slope formula will really pay off (besides allowing us to figure out the slope of any line, as long as we are given two coordinates through which it passes, regardless of the x and y values of the coordinates that are provided).

What I am going to do next may seem a bit confusing, but if you read it carefully, and if you try to follow along by referring to the graph that serves to illustrate the process, you should be able to make the conceptual connections necessary for a deep understanding of linear equations, the formulas typically associated with them, and their graphs.

First, let's agree that the slope formula allows us to find the slope of a line that passes through any two of the infinite number of points that it is

formed by (or that it "passes through"). The key part of this is the "any" word: for a given line, I can use the slope formula to figure out the line's slope, regardless of which two points I choose to use. The only requirement is that the points **must be part** of the line (in other words, the two coordinates used **must be solution pairs** of the equation).

Based on the above, we can imagine standing on any point that is part of the line, let's call that point, in general, (x_1, y_1), just as we did before, but this time, as for the second point, let's call it (x, y); we will then use the slope formula to find the line's slope. Please note that the only thing that I am changing at this point from what we did earlier when we found the slope formula is opting to **name** one of the points as (x, y) instead of what we originally used: (x_2, y_2). Why the switch? You'll see it in a minute.

So, let's use these two points on the slope formula, and pretend we wish to know the line's slope. Remember, I said I am standing on (x_1, y_1) and I wish to land on (x, y). Observe:

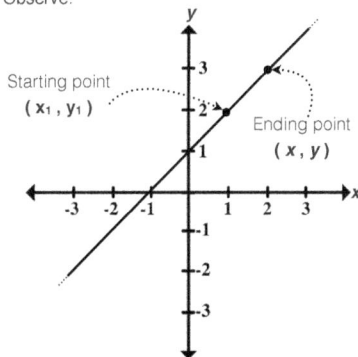

I am not going to consider the actual values of the coordinates given, because we simply want to treat this as the general case.

Alright, so based on the coordinates specified, the slope of the line would have to be:

$$m = \frac{y - y_1}{x - x_1}$$

Do you agree? Based on the chosen starting point, (x_1, y_1), and the ending point, (x, y), the equation above follows.

So what now? Well, we know that the equation of a line, when expressed using the variables x and y as its input/output variables, must be expressed, in general, in terms of said x and y (think of the slope-intercept formula: $y = mx + b$, see how it is written or expressed, in general, in terms of the variables x and y?); therefore, I am going to simplify the above formula, and to try to rearrange its sides so that I end up with an equation that contains the variable y on the left side, and the variable x on the right side. Just keep in mind that we are trying to consider any possible line, which is why I am not replacing the variables with any specific values.

Simplifying the expression and moving the variables around leads to the following:

$$m = \frac{y - (y2)}{x - (x2)}$$

Simplify...

$$m = \frac{y - y2}{x - x2}$$

Multiply both sides by the right side's denominator...

$$(x - x1)(m) = \frac{y - y1}{x - x1} \ (x - x1)$$

Cancel the right side's denominator with the multiplying parentheses introduced in the previous step:

$$(x - x1)(m) = \ y - y1$$

Exchange sides to keep the y variable on the left side of the equation...

$$y - y1 \ = \ (x - x1)(m)$$

Place the (m) to the left of the multiplying parenthesis set:

$$y - y1 \ = \ (m)(x - x1)$$

Remove unnecessary sets of parentheses (surrounding m)...

$$y - y1 \ = \ m(x - x1)$$

What did we come up with here? It's called the *point-slope* formula of a linear equation. Its name clearly tells us that this equation contains the slope value m explicitly indicated, and unlike the slope-intercept formula, a specific point that is part of the equation's line (in other words, a solution pair): the (x_1, y_1) coordinate we used to derive the formula to begin with. If we keep the variables x and y intact, without replacing them for any actual values (points of the line or solution pairs of the equation), we now have a way of determining the equation of the line if we know the equation's slope and a solution set of the equation (which is to say, the slope of the equation's line and a single point through which it passes through).

Let's try it on the problem we have been working on. We were told that the line passes through the coordinates $(1,2)$ and $(2,3)$; using these two points, we were able to determine the line's slope (which is to say, the equation's slope): $m = 1$. Now, if I we are stating that a linear equation can be specified using the point-slope formula...

$$y - y1 = m(x - x1)$$

...it means that if we replace the m above with a value of 1, which is the value of this specific line's slope, and replace (x_1, y_1) with any one of the two coordinates that were originally given, $(1,2)$ or $(2,3)$, we should be able to specify the equation precisely. Let's try it.

$$m = 1$$

$$(x_1, y_1) \text{ ---> } (1,2)$$

which means that

$$x_1 = 1$$

$$y_1 = 2$$

We can replace these values into the slope formula, as follows:

$$y - y1 = m(x - x1)$$

$$y - (2) = (1)(x - (1))$$

Then, simplifying, we obtain:

$$y - 2 = (x - 1)$$

And we have found the equation of the line that passes through the two given points. If we wanted to, we could express it in slope-intercept form. Can you guess how? If you answered "...by solving for y", then you are absolutely correct. Observe:

$$y - 2 = (x - 1)$$

First, remove the unnecessary parentheses set that is on the right side...

$$y - 2 = x - 1$$

...then, add 2 to both sides of the equation:

$$y - 2 + 2 = x - 1 + 2$$

$$y = x + 1$$

And there it is! The equation of the line in the familiar-looking slope-intercept format.

To prove to you that this is, in fact, the equation of the line that passes through the points $(1,2)$ and $(2,3)$, let's replace the input values of these two points, and check to see if the equation we obtained gives us the correct output (or y) values.

Linear equation to check:

$$y = x + 1$$

Let's try the first coordinate: $(1,2)$. Replacing x with 1 and verifying that the equation yields the correct output (y value) of 2:

$$y = x + 1$$

$$y = (1) + 1$$

$$y = 2$$

Correct! But what about the second point? Will it also work? Remember, if it does, we know for a fact that the equation is the **one and only** equation of the line that passes through the given points, since a **unique** line passes through any two specific points.

Linear equation:

$$y = x + 1$$

Checking the second coordinate: **(2 , 3)**. Replacing **x** with **2** and verifying that the equation yields the correct output (**y** value) of **3**:

$$y = x + 1$$

$$y = (2) + 1$$

$$y = 3$$

It's official! The equation that we found is, in fact, the linear equation of the line that passes through the two given points **(1 , 2)** and **(2 , 3)**.

So what does this mean? Well, it means that all of the following equations are equivalent:

$y - 2 = (x - 1)$ --> **point-slope form**

$y = x + 1$ --> **slope-intercept form**

$-x + y - 1 = 0$ --> **general form**

The general form doesn't really provide any "explicit" information about the equation or the line that it graphs to, because it isn't solved for the variable **y** nor for the variable **x**; it is therefore the least useful format of all. However, many problems will provide you with a linear equation in general form and ask you to graph it, or to find the **y-intercept**, or the **x-intercept**, etc... To solve such prompts correctly, you will have to either take the general form and solve it

for the variable **y**, or you will have to apply the other strategies that we reviewed in this chapter, such as replacing **x** with **0** and solving for **y** to find the **y-intercept**, or setting **y** to **0** and solving for **x** to find the **x-intercept**, etc...

Let's test this point-slope formula that we were able to derive ourselves once more by applying it to the problem we solved before this one. It read:

"**Scenario B** Suppose we know that a line passes through the point **(1 , 3)** and through the point **(0 , 1)**. We know absolutely nothing else. Can we specify a **unique** linear equation with this information?"

To solve it, we found the slope using the slope formula (**m = 2**), and then used the **y-intercept** point given (that's the coordinate **(0 , 1)**) to extract the **b** value (which led to **b = 1**); finally, replacing these **m** and **b** values into the slope-intercept form, we were able to determine this line's equation:

$$y = mx + b$$

Since **m = 2** and **b = 1**

$$y = 2x + 1$$

But what if we used the point-slope formula instead? The slope value is going to be the same, of course, since we would use the same slope formula in both cases. So **m = 2** . But now, let's use the second point given, **(1 , 3)**, and check to see if the point-slope formula that we derived would give us the correct equation of the line as well. Please know that we could use the first point given, **(0 , 1)**, but I rather use the other point because it makes it that more obvious that we are using different formulas with different values, and yet we should still end up with the exact same linear equation.

$$m = 2$$

$$(\,x_1\,,\,y_1\,)\ ---\!\!>\ (\,1\,,\,3\,)$$

which means that

$$x_1\ =\ 1$$

$$y_1\ =\ 3$$

We can replace these values into the slope formula, as follows:

$$y - y1\ =\ m(x - x1)$$

$$y - (3)\ =\ (2)(x - (1))$$

Then, simplifying, we obtain:

$$y - 3\ =\ 2(x - 1)$$

So, this is supposed to be the exact same equation as the version "$y = 2x + 1$" except that it is in point-slope form. Well, to check it, let's solve it for the variable y.

$$y - 3\ =\ 2(x - 1)$$

First, distribute the number 2 that is multiplying the parentheses on the right side of the equation...

$$y - 3\ =\ 2x - 2$$

...then, add 3 to both sides of the equation and simplify...

$$y - 3 + 3\ =\ 2x - 2 + 3$$

$$y\ =\ 2x + 1$$

And there it is! The exact same equation. If you think that choosing the point $(\,1\,,\,3\,)$ was perhaps the trick that led to a matching equation, observe what would have happened if we had chosen the first point given instead: $(\,0\,,\,1\,)$.

$$m = 2$$

$$(\,x_1\,,\,y_1\,)\ ---\!\!>\ (\,0\,,\,1\,)$$

which means that

$$x_1\ =\ 0$$

$$y_1\ =\ 1$$

We can replace these values into the slope formula, as follows:

$$y - y1\ =\ m(x - x1)$$

$$y - (1)\ =\ (2)(x - (0))$$

Then, simplifying, we obtain:

$$y - 1\ =\ 2(x - 0)$$

So, this is supposed to be the exact same equation as the version "$y = 2x + 1$" except that it is in point-slope form. Well, to check it, let's solve it for the variable y.

$$y - 1\ =\ 2(x - 0)$$

First, distribute the number 2 that is multiplying the parentheses on the right side of the equation...

$$y - 1\ =\ 2x - 0$$

...then, add 1 to both sides of the equation and simplify...

$$y - 1 + 1 = 2x - 0 + 1$$

$$y = 2x + 1$$

And there it is! As expected, we obtained the same equation of the line that passes through the given points.

If you are wondering how this is possible (that using two different points will lead to the same linear equation), think of it this way: when we used the coordinate $(1, 3)$, the x value was 1 and the y value was 3. Using these values to replace their respective variables in the point-slope formula led to the linear equation $y = 2x + 1$. Then, when we used a different point (but that is also part of the line), the coordinate $(0, 1)$, the x value was 0 and the y value was 1. **Both numbers changed**... and that's why using that other coordinate ultimately led to the same linear equation: yes, the x value may have changed from a 1 to a 0, but the y value also changed: from a 3 to a 1. What is happening here is that the pair of values, *working together*, lead to the same linear equation, precisely because the equation is linear and because the points being used are part of the same line (in other words, both are solution pairs of the linear equation).

So there you have it. The point-slope formula of a linear equation:

$$y - y1 = m(x - x1)$$

Let's try a few more problems that can be easily solved using the point-slope formula before moving on.

Try to determine the equation of the line that passes through the following points:

a) $(3, 5)$ **and** $(6, 17)$
b) $(-4, 1)$ **and** $(-6, 7)$
c) $(-5, -8)$ **and** $(-4, -18)$

Before trying to solve them by yourself, just remember to use sets of parentheses when replacing the variables in the slope formula and in the point-slope formula, otherwise, your answers will be wrong.

Ready to check your answers? Let's solve them using the same process that we used earlier.

a) $(3, 5)$ **and** $(6, 17)$

First, we need to determine the slope of the linear equation. I will establish the following correspondence:

$$(3, 5) \qquad (6, 17)$$

$$(x_1, y_1) \qquad (x_2, y_2)$$

...and applying the slope formula, we find that

$$m = \frac{y_2 - y_1}{x_2 - x_1} = \frac{17 - (5)}{6 - (3)} = \frac{12}{3} = 4$$

Next, we choose one of the points given, trying to use whichever coordinate contains the easiest pair of values to work with (in this case, the first point would be the logical choice), and replace the variables m, x_1, and y_1 in the point-slope formula with their respective values:

$$m = 4 \; ; \; x_1 = 3 \; ; \; y_1 = 5$$

$$y - y1 = m(x - x1)$$

$$y - (5) = (4)(x - (3))$$

Then, simplifying, we obtain:

$$y - 5 \;=\; 4(x - 3)$$

The equation of the line, as requested.

But should we leave it in the point-slope form, or should we solve for y and express it in slope-intercept form? Well, it all depends on the prompt; if the problem explicitly requests the slope-intercept form, you will have to solve for y; or, if you are taking a multiple-choice test, and you notice that all of the choices given are in slope-intercept form, then you know you have to solve for y; otherwise, you can leave the equation in point-slope form, since it is ultimately a perfectly valid equation of the line that passes through the given points. On these three problems, I will rewrite the equation in slope-intercept form, just to help you practice this skill. But do note that you do not have to do this every-time.

Let's solve the equation in point-slope form for the variable **y** ...

$$y - 5 \;=\; 4(x - 3)$$

First, distribute the number **4** that is multiplying the parentheses on the right side of the equation...

$$y - 5 \;=\; 4x - 12$$

...then, add **5** to both sides of the equation and simplify...

$$y - 5 + 5 \;=\; 4x - 12 + 5$$

$$y \;=\; 4x - 7$$

And there we have it. The line that passes through the two points given has the following equation:

In **point-slope form**:

$$y - 5 \;=\; 4(x - 3)$$

In **slope-intercept form**:

$$y \;=\; 4x - 7$$

Just remember, these two linear equations are exactly equivalent, interchangeable, and therefore, when graphed, correspond to the same line! They are expressed using a different format, but they represent the same relationship.

I will go above and beyond and I will write the equation in general form as well. Starting with the slope-intercept form, we can send both terms that are on the right side of the equation to the left side, as follows:

$$y \;=\; 4x - 7$$

$$y - 4x + 7 \;=\; 4x - 7 - 4x + 7$$

Then, rearranging the terms on the left side...

$$-4x + y + 7 \;=\; 0$$

And there you have it. The same linear equation, but now in general form! At the risk of being too repetitive, let me present to you, one last time, this linear equation, in the three forms that are typically used when working with linear equations. Remember, these are all equal to each other!

$$y - 5 \;=\; 4(x - 3) \quad \text{-->} \quad \text{point-slope form}$$
$$y \;=\; 4x - 7 \quad \text{-->} \quad \text{slope-intercept form}$$
$$-4x + y + 7 \;=\; 0 \quad \text{-->} \quad \text{general form}$$

It is important to be able to "read" these forms correctly, in order to be able to extract the appropriate information from each of them. The general form, however, does not provide any information regarding the equation's slope, *x-intercept*, or *y-intercept*, and is thus the least useful of the three forms.

b) (−4 , 1) and (−6 , 7)

First, we need to determine the slope of the linear equation. I will establish the following correspondence:

$$(-4 , 1) \qquad (-6 , 7)$$

$$(x_1 , y_1) \qquad (x_2 , y_2)$$

...and applying the slope formula, we find that

$$m = \frac{y_2 - y_1}{x_2 - x_1} = \frac{7 - (1)}{-6 - (-4)} = \frac{6}{-2} = -3$$

Using one of the points given and the slope value, we can now specify the linear equation...

$$m = -3 \; ; \; x_1 = -4 \; ; \; y_1 = 1$$

$$y - y1 = m(x - x1)$$

$$y - (1) = (-3)(x - (-4))$$

Then, simplifying, we obtain:

$$y - 1 = -3(x + 4)$$

Please note that the x_1 value appears as an *adding* term on the point-slope form (inside the parentheses set). Why? Because the *x* coordinate was a negative number, which led to a set up where we were "subtracting a negative number", which as we have seen throughout the book, can be simplified as "an addition of the positive version of the number", as the diagram illustrates:

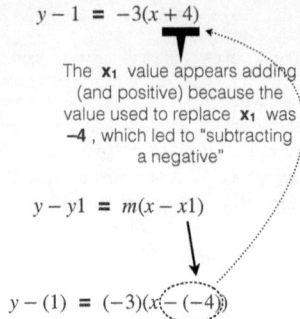

$$y - 1 = -3(x + 4)$$

The x_1 value appears adding (and positive) because the value used to replace x_1 was **−4** , which led to "subtracting a negative"

$$y - y1 = m(x - x1)$$

$$y - (1) = (-3)(x - (-4))$$

This aspect of how the x_1 and the y_1 appears on the point-slope form is crucial when reverse-engineering the process (in other words, when you are provided with the linear equation in point-slope form, and you are asked to extract the coordinate that it contains, as well as its slope value, as we will explore in a moment).

Now, to find the other forms, simply solve for **y** for the slope-intercept form, and send all terms to the left side (set to zero) for the general form:

$$y - 1 = -3(x + 4)$$
$$y - 1 = -3x + (-3)(4)$$
$$y - 1 = -3x + (-12)$$
$$y - 1 = -3x - 12$$
$$y - 1 + 1 = -3x - 12 + 1$$
$$y = -3x - 11$$

And we now have the slope-intercept form. As to the general form:

$$y = -3x - 11$$
$$y + 3x + 11 = -3x - 11 + 3x + 11$$
$$3x + y + 11 = 0$$

And so we find that the three forms of this linear equation are:

$$y - 1 = -3(x + 4) \rightarrow \text{point-slope form}$$
$$y = -3x - 11 \rightarrow \text{slope-intercept form}$$
$$3x + y + 11 = 0 \rightarrow \text{general form}$$

c) (–5 , –8) and (–4 , –18)

First, we need to determine the slope of the linear equation. I will establish the following correspondence:

$$(-5 , -8) \qquad (-4 , -18)$$
$$(x_1 , y_1) \qquad (x_2 , y_2)$$

...and applying the slope formula, we find that

$$m = \frac{y_2 - y_1}{x_2 - x_1} = \frac{-18 - (-8)}{-4 - (-5)} = \frac{-10}{1} = -10$$

Using one of the points given and slope value, we can now specify the linear equation...

$$m = -10 \ ; \ x_1 = -5 \ ; \ y_1 = -8$$

$$y - y1 = m(x - x1)$$

$$y - (-8) = (-10)(x - (-5))$$

Then, simplifying, we obtain:

$$y + 8 = -10(x + 5)$$

Please note that this time both the x_1 and the y_1 values appear as *adding* terms on the point-slope form, for the same reason that we explored on the previous problem.

Now, to find the other forms, simply solve for **y** for the slope-intercept form, and send all terms to the left side (set to zero) for the general form:

$$y + 8 = -10(x + 5)$$
$$y + 8 = -10x + (-10)(5)$$
$$y + 8 = -10x + (-50)$$
$$y + 8 = -10x - 50$$
$$y + 8 - 8 = -10x - 50 - 8$$
$$y = -10x - 58$$

And we now have the slope-intercept form. As to the general form:

$$y = -10x - 58$$
$$y + 10x + 58 = -10x - 58 + 10x + 58$$
$$10x + y + 58 = 0$$

And so we find that the three forms of this linear equation are:

$$y + 8 = -10(x + 5) \rightarrow \text{point-slope form}$$
$$y = -10x - 58 \rightarrow \text{slope-intercept form}$$
$$10x + y + 58 = 0 \rightarrow \text{general form}$$

I want to do one last example before moving on. Let's work on it together. Let's call it exercise **(d)**.

d) What is the equation of the line that passes through the following two points?

(–5 , –2) and (–3 , –7)

First, we need to determine the slope of the linear equation. I will establish the following correspondence:

$$(-5 , -2) \qquad (-3 , -7)$$
$$(x_1 , y_1) \qquad (x_2 , y_2)$$

...and applying the slope formula, we find that

$$m = \frac{y_2 - y_1}{x_2 - x_1} = \frac{-7 - (-2)}{-3 - (-5)} = \frac{-5}{2} = -\frac{5}{2}$$

Using one of the points given and slope value, we can now specify the linear equation...

$$m = \frac{-5}{2} \; ; \; x_1 = -5 \; ; \; y_1 = -2$$

$$y - y1 = m(x - x1)$$

$$y - (-2) = (-\frac{5}{2})(x - (-5))$$

Then, simplifying, we obtain:

$$y + 2 = -\frac{5}{2}(x + 5)$$

Now, to find the slope-intercept and the general forms of the equation, I would proceed as follows...

$$y + 2 = -\frac{5}{2}(x + 5)$$

First, rewrite the negative fraction by including the negative sign in the numerator. It will make the process of simplifying a lot easier and reduce the likelihood of making a mistake...

$$y + 2 = \frac{-5}{2}(x + 5)$$

Next, distribute (multiply the fraction by the terms contained in the set of parentheses)...

$$y + 2 = \frac{-5}{2}x + \frac{-5}{2}(5)$$

$$y + 2 = \frac{-5x}{2} + \frac{-25}{2}$$

Now, subtract **2** to both sides of the equation...

$$y + 2 - 2 = \frac{-5x}{2} + \frac{-25}{2} - 2$$

$$y = \frac{-5x}{2} + \frac{-25}{2} - \frac{2}{1}$$

$$y = \frac{-5x}{2} + \frac{(-25)(1) - (2)(2)}{(2)(1)}$$

$$y = \frac{-5x}{2} + \frac{-25 - 4}{2}$$

$$y = \frac{-5x}{2} + \frac{-29}{2}$$

$$y = -\frac{5x}{2} - \frac{29}{2}$$

And we now have the equation in slope-intercept form. As to the general form...

$$y = -\frac{5x}{2} - \frac{29}{2}$$

$$y + \frac{5x}{2} + \frac{29}{2} = -\frac{5x}{2} - \frac{29}{2} + \frac{5x}{2} + \frac{29}{2}$$

$$\frac{5x}{2} + y + \frac{29}{2} = 0$$

But is this truly in general form? The answer is no. Let me remind you how the general form (in general) is expressed:

$$\mathbf{Ax + By + C = 0}$$

(Keep in mind that the **B** in this general from has nothing to do with the **b** from the slope-intercept form, which corresponds to the **y-intercept** value; believe me, if I had my say in

544

this, I would change all math text books so that other letters were used to avoid any confusion).

So what is yet to be done? Well, the expectation is that the general form should be *fraction-less*. To achieve it, simply multiply both sides of the equation by the denominators that may be present; in this case, both denominators are the same (equal to **2**), so multiplying both sides of the equation times **2** will take get rid of both denominators simultaneously. But, if they were different, then you would want to multiply by the smallest number evenly divisible by both denominators.

$$\frac{5x}{2} + y + \frac{29}{2} = 0$$

$$(2)(\frac{5x}{2} + y + \frac{29}{2}) = (0)(2)$$

$$5x + 2y + 29 = 0$$

It is at this stage of the process that we can claim victory: we have the general form of the linear equation as would be expected: without any fraction coefficients.

And so we find that the three forms of this linear equation are:

$$y + 2 = -\frac{5}{2}(x + 5) \quad \text{--> point-slope form}$$

$$y = -\frac{5x}{2} - \frac{29}{2} \quad \text{--> slope-intercept form}$$

$$5x + 2y + 29 = 0 \quad \text{--> general form}$$

As of this moment, if you were able to solve the examples we worked on by yourself, you are perfectly capable of finding the equation of a line that passes through two given points, and to express the linear equation in any one of the three form that are defined, as the following box specifies.

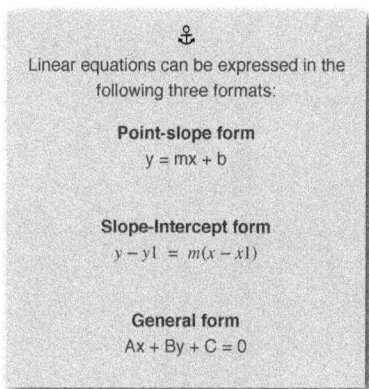

⚓

Linear equations can be expressed in the following three formats:

Point-slope form

$y = mx + b$

Slope-Intercept form

$y - y1 = m(x - x1)$

General form

$Ax + By + C = 0$

This, in combination with the slope formula, allows you to solve most problems that deal with linear equations and their graphs.

We must now explore the following two basic ideas regarding linear equations and their graphs.

Parallel lines

Parallel lines, as we stated earlier, in Euclidean Geometry, are lines that will never cross each other since they are said to have **equal slopes**. Train tracks, for example, are supposed to be perfectly parallel; otherwise, a train riding on them would eventually derail. The big idea, though, is what I wrote in bold: parallel lines *must have* equal slopes! Why? Because it is the only way to guarantee that two lines will never cross each other: they are rising or falling at the exact same rate, unit by unit, and thus can be called parallel to each other. Observe the following graph for an illustration of this important principle.

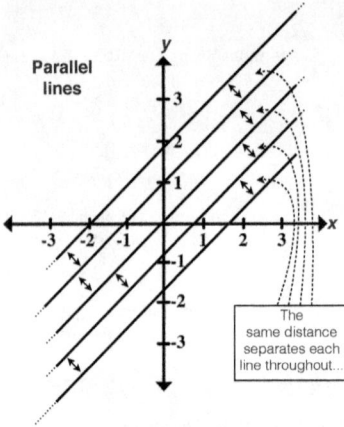

Parallel lines

The same distance separates each line throughout...

Observe how all of these lines are parallel to each other: they maintain the same distance between each other as you move from left to right...

Well, since parallel lines have equal slopes, we have a way of easily dealing with scenarios such as the following.

Scenario E   What is the equation of a line that passes through (**4 , 3**) and that is *parallel* to the line **4x + 2y − 8 = 0** ? I would imagine that at this stage of the chapter, based on what I said above, you can solve this problem on your own. Try it...

Ready to check your answer?

First, let's try to understand what we know about the the linear equation we are trying to specify. We are told that the point (**4 , 3**) is part of the equation's line (when graphed); in other words, that it is a solution pair of the equation. That by itself, as we saw earlier in **Scenario A**, is not enough to specify a unique line. The problem is that they do not specify a second point with which we can figure out the slope, and they do not give us directly the slope of this line either. So what can we do? Looking at the following equations of lines, we would seem to be stuck, because we only have partial information at best:

Point-slope form
$$y = mx + b$$

Slope-Intercept form
$$y - y1 = m(x - x1)$$

General form
$$Ax + By + C = 0$$

The known point could be used on the point-slope formula, but we would not be able to figure out the equation because we would be missing the slope (**m**) value. The slope-intercept is a brick wall, because we do not know the **y-intercept** (and therefore we can't know **b**). And as to the general form, that is useless in this case because we do not know the **A** , **B** , and **C** values.

However, remember that they told us that the linear equation we are looking for, the one that has the coordinate (**4 , 3**) as one of its solution pairs, corresponds to a line that is *parallel* to 4x + 2y − 8 = 0 . So, if we could find the slope of this second line, we could simply use that value, together with the given coordinate, and find the equation we are looking for, using the point-slope form.

So how can we find the slope of this other equation, the one that is expressed in general form?

$$4x + 2y - 8 = 0$$

Simple! Solve for **y** and it will be in slope-intercept form; then, extract the coefficient of the **x** term, and we will have found the slope of both lines!

Observe.

Step 1) Express the general form linear equation in slope-intercept form by solving for **y** in order to extract the slope value...

$$4x + 2y - 8 = 0$$
$$4x + 2y - 8 - 4x + 8 = 0 - 4x + 8$$
$$2y = -4x + 8$$
$$\frac{2y}{2} = \frac{-4x + 8}{2}$$
$$y = -2x + 4$$

And there we have it! The other line in slope-intercept form. Now we can specify both lines' slopes, since they must be equal to each other if they are to be parallel lines...

$$m = -2$$

Step 2) Using the slope value found, together with the coordinate given, **(4 , 3)**, replace the corresponding variables on the point-slope formula...

$$m = -2 \; ; \; x_1 = 4 \; ; \; y_1 = 3$$
$$y - y1 = m(x - x1)$$
$$y - (3) = -2(x - (4))$$

Then, simplifying, we obtain:

$$y - 3 = -2x - (-2)(4)$$
$$y - 3 = -2x - (-8)$$
$$y - 3 = -2x + 8$$
$$y - 3 + 3 = -2x + 8 + 3$$
$$y = -2x + 11$$

And there it is! The equation of the line that passes through **(4 , 3)** and that is parallel to **4x + 2y − 8 = 0** . I am going to graph both lines on the same **x-y plane** so that you can visually verify it.

Equation given: $y = -2x + 4$

Equation found: $y = -2x + 11$

Please note that both equations have the same slope: this automatically means that they must be parallel. Also, you can see how they have different **y-intercept** points: the given equation crosses the **y** axis at **(0 , 4)** while the equation that we found at **(0 , 11)** ...

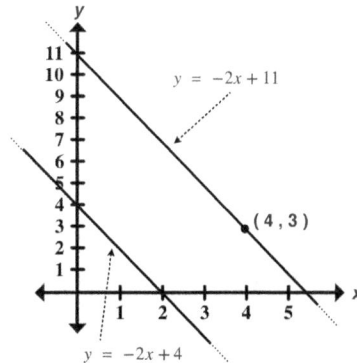

As you can see from the graph, the equation that we found does, in fact, contain the point $(4, 3)$ as a solution pair, and has the same slope (and thus is parallel to) the given equation.

At this stage of the chapter, you should be able to find the equation of a line that passes through a given point, and that is parallel to a second line.

But what if the lines are perpendicular instead? Take a look at the following scenario as an example of this type of problem.

Scenario F   What is the equation of a line that passes through $(3, 2)$ and that is *perpendicular* to the line $3x + y - 2 = 0$?

A graphical representation of this prompt will help you fully understand the question (and the set up it is describing).

We are looking for the equation of a line that has a solution pair of $(3, 2)$, which means that its graphed line should pass through the point given by that solution pair, $(3, 2)$ ...

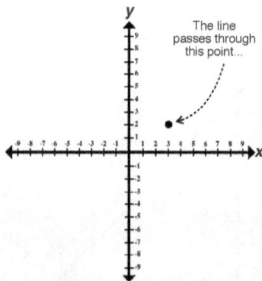

The line passes through this point...

However, as we saw earlier in this chapter, there are an infinite number of lines that pass through that point. This is where the other bit of information that is given becomes relevant: the

line must be perpendicular to the line specified by the equation $3x + y - 2 = 0$. So the obvious follow-up question is this: what does it mean for a line to be perpendicular to another line? Well, in Euclidean Geometry, on a "flat" surface (which is what we use when we graph linear equations on the *x-y plane* or the *cartesian plane*), it means that the lines will form a ninety degree angle between them (for example, the *x-axis* and the *y-axis* are perpendicular to each other). Observe examples of pairs of lines that are perpendicular to each other:

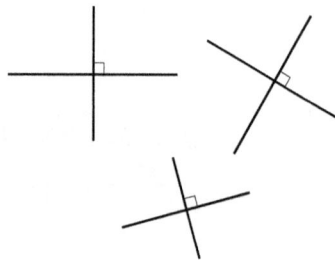

The "square element" that is placed between the lines is used to explicitly indicate that the lines are perpendicular to each other. They are **not** to be considered part of the lines. These same lines, perpendicular to each other, without that extra element, look as follows:

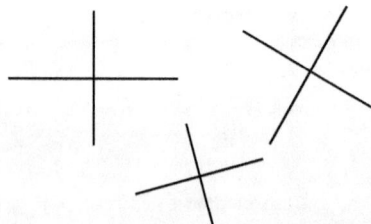

So how do we deal with this problem? Can we know the slope of the equation we are looking for, knowing that it is perpendicular to the given equation? The answer, of course, is yes.

In order for two lines to be perpendicular to each other, they *must have* "opposite reciprocal" slopes. Sounds complicated, but it's actually very simple. Let me break it down:

opposite...

Stands for "opposite-signed", as in: if one slope is positive, the other *must* be negative.

...*reciprocal*

Means that the slope values, in fraction form, must have their numerators/denominators switched.

Just remember that it is a *two part* process. Let me show you examples of slope values that are opposite reciprocals of each other:

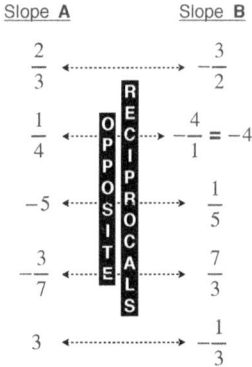

Slope A	Slope B
$\frac{2}{3}$	$-\frac{3}{2}$
$\frac{1}{4}$	$-\frac{4}{1} = -4$
-5	$\frac{1}{5}$
$-\frac{3}{7}$	$\frac{7}{3}$
3	$-\frac{1}{3}$

OPPOSITE RECIPROCALS

What this diagram illustrates is this: if you look at the first slope value on the top left (under the column labeled "Slope A"), which is $\frac{2}{3}$, and you want to determine its "opposite reciprocal", you would first need to change its sign (since $\frac{2}{3}$ is positive, its **opposite reciprocal** *must be* negative), and then you would need to "*flip*" the fraction so that the numerator become the denominator, and the denominator becomes the numerator; you have thus found the **opposite reciprocal** of $\frac{2}{3}$, which is $-\frac{3}{2}$. Observe:

To find the **opposite reciprocal** of

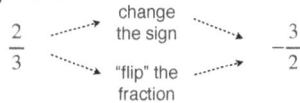

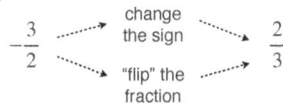

$\frac{2}{3}$ ⟶ change the sign ⟶ ⟶ $-\frac{3}{2}$
⟶ "flip" the fraction ⟶

Easy, right? Observe the other opposite reciprocal pairs and how they adhere to this rule illustrated above.

It is important to note that if the opposite reciprocal of $\frac{2}{3}$ is $-\frac{3}{2}$ then the reverse is also true: the opposite reciprocal of $-\frac{3}{2}$ must be equal to $\frac{2}{3}$. See the logic? Observe:

To find the **opposite reciprocal** of

$-\frac{3}{2}$ ⟶ change the sign ⟶ ⟶ $\frac{2}{3}$
⟶ "flip" the fraction ⟶

See how it works out both ways? It must, because if a first number is said to have a second number as its opposite reciprocal, then it follows that the second's number opposite reciprocal is the first number as well. It's a two way street!

I would like to explain further one of the examples shown on the diagram from the previous page: the opposite reciprocal of $\dfrac{1}{4}$. We expect the opposite reciprocal to be a negative number, right? That much is clear. But watch what happens when we "flip" the fraction to obtain the "value" part of the opposite reciprocal:

$$\frac{1}{4} \quad \text{<--->} \quad -\frac{4}{1} = -4$$

See how the opposite reciprocal ended up being an **INTEGER**? The reason is simple: when we "flip" the fraction one-fourth, we obtain the fraction "four divided by one"; and this, of course, simplifies to **4** . Taking into account the sign change, we thus find that the opposite reciprocal of $\dfrac{1}{4}$ is equal to -4 . This "**INTEGER conversion**" will happen whenever one of the numbers in an opposite-reciprocal pair has a numerator that is equal to **1** .

Another interesting case is the opposite reciprocal of -5 . The "sign-part" of the analysis is easy enough: this number is negative, so its opposite reciprocal must be positive. However, when it comes time to "flip" the fraction, you may be left standing there scratching your head wondering "what fraction? Negative five is an **INTEGER**!..."; well, although I sympathize with your confusion, it actually has an easy "fix". The number -5 can be written as the fraction $\dfrac{-5}{1}$,

and now you know how I was able to "flip" the **INTEGER**. Observe.

$$-5 = \frac{-5}{1} \quad \text{<--->} \quad +\frac{1}{5} = \frac{1}{5}$$

The "positive sign" is not necessary at all, which is why the final answer is simply written as $\dfrac{1}{5}$.

There are two other "special" cases that we should explore in detail.

What is the opposite reciprocal of **1** ? Sound innocent enough, but many make mistakes with this prompt. Try it on your own first.

Ready? Observe.

$$1 = \frac{1}{1} \quad \text{<--->} \quad -\frac{1}{1} = -1$$

Think I cheated? No. I did flip the fraction. Observe:

$$\frac{1}{1} \quad \text{<--->} \quad -\frac{1}{1}$$

What is happening here is easy to explain: flipping the "value" portion $\dfrac{1}{1}$ yields the exact same "value" portion, $\dfrac{1}{1}$, which is why the opposite reciprocal of **1** is equal to -1. It is only the sign that changes in this very specific case.

The last set-up that we need to discuss in detail is this: what is the opposite reciprocal of **0** ? Observe:

$$0 = \frac{0}{1} \quad \text{<--->} \quad -\frac{1}{0} = \textbf{UNDEFINED}$$

Did you predict that "flipping" $\frac{0}{1}$ would lead to an UNDEFINED statement? So what does this mean? In the context of linear equations, if I know that a line has a slope of **0** (in other words, that it is horizontal, such as $y = 4$), does it mean that it does not have a perpendicular line to it? Observe:

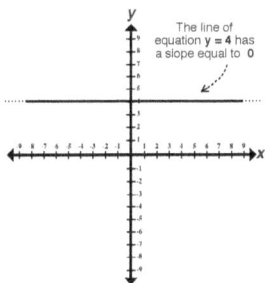

Clearly, there are an infinite number of lines that are perpendicular to this one. Observe:

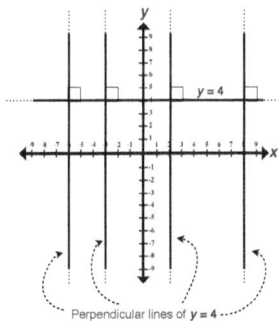

So what do all these perpendicular lines of **y = 4** have in common? Can you see it? That their slope is precisely UNDEFINED, exactly what we found to be the opposite reciprocal of **0** ! It

makes perfect sense from the point of view of linear equations and their graphs: if we are interested in finding a perpendicular line of a horizontal line (a line with a slope of **0**), then we expect the perpendicular line to be a vertical line, which we know has a slope that is UNDEFINED, or infinite.

So, the opposite reciprocal of **0** is UNDEFINED, but in the context of linear equations and their slopes, we now know that if a line has a slope that is equal to **0** we can state that a perpendicular line to it must have an infinite or UNDEFINED slope, and vice-versa.

Let's sum up everything we learned about parallel and perpendicular lines.

⚓

****

Parallel lines have equal slopes

$$m_1 \ = \ m_2$$

****

Perpendicular lines have *opposite-reciprocal* slopes

$$\text{if } m_1 \ = \ \frac{a}{b} \ \longleftrightarrow \ m_2 \ = \ -\frac{b}{a}$$

Two special cases to consider

Case 1)
$$\text{If } m_1 \ = \ 1 \ \longleftrightarrow \ m_2 \ = \ -1$$

Case 2)
$$\text{If } m_1 \ = \ 0 \ \longleftrightarrow \ m_2 \ = \ \text{UNDEFINED}$$

Just remember that if the slope of a given equation is not in fraction format–in other words, if it is an **INTEGER** or an **IRRATIONAL** such as $\sqrt{3}$ or π–, you can express it as a fraction (so that you know what values to "*flip*" when executing step two of finding the opposite reciprocal) simply by dividing it by one. Observe, for example, the following opposite reciprocals, and how dividing the original number by 1 when said number is not in fraction format allows you to easily "flip" the fraction (and don't forget to change the sign as well):

$$2 = \frac{2}{1} \quad \text{<--->} \quad -\frac{1}{2}$$

$$-7 = \frac{-7}{1} \quad \text{<--->} \quad +\frac{1}{7} = \frac{1}{7}$$

$$\sqrt{3} = \frac{\sqrt{3}}{1} \quad \text{<--->} \quad -\frac{1}{\sqrt{3}}$$

So, are you ready to tackle some problems on your own involving the concept of parallel and perpendicular lines? Try to solve them before looking at my answers, although you may certainly look at the previous examples that we worked on earlier and use them as a guide. Try to graph each problem as well: graph both lines and the point that is given. Visually check your answer (check that the lines that you graph are, in fact, parallel or perpendicular to each other, as each problem indicates).

a) What is the equation of a line that passes through $(1, 4)$ and that is *parallel* to the line $9x + 3y - 9 = 0$?

b) What is the equation of a line that passes through $(-2, 3)$ and that is *perpendicular* to the line $-x + y + 2 = 0$?

c) What is the equation of a line that passes through $(0, 5)$ and that is *parallel* to the line $-2x - y - 3 = 0$?

d) What is the equation of a line that passes through $(-5, 3)$ and that is *perpendicular* to the line $y = 2x - 4$?

e) What is the equation of a line that passes through $(-2, 0)$ and that is *parallel* to the line $x = 5$?

f) What is the equation of a line that passes through $(3, 2)$ and that is *perpendicular* to the line $y = -7$?

g) What is the equation of a line that passes through $(-4, 3)$ and that is *perpendicular* to the line $y = \frac{2}{3}x + 1$?

Ready for the answers? Problems **(a)** , **(b)**, and **(d)** were very similar to those that we worked on during the review; however, for problems **(c)**, **(e)** and **(f)** you had to carefully analyze the set up and use your common sense to come up with the answer, as you will see in a minute. Problem **(g)** involved the use of fractions, but other than that, the process was similar to those that we worked on during the review.

Time to check your work...

a) What is the equation of a line that passes through $(1, 4)$ and that is *parallel* to the line $9x + 3y - 9 = 0$?

First, I would solve the given equation for y because the equation of the line we are looking for is parallel to that given equation's line (which means that they will share the same slope). Know the given equation's line slope, know the slope of the equation we are trying to determine.

Step 1) Express the given linear equation in slope-intercept form by solving for y in order to extract its slope value...

$$9x + 3y - 9 = 0$$
$$9x + 3y - 9 - 9x + 9 = 0 - 9x + 9$$
$$3y = -9x + 9$$
$$\frac{3y}{3} = \frac{-9x + 9}{3}$$
$$y = -3x + 3$$

And there we have it! The other line in slope-intercept form. Now we can specify both lines' slopes, since they must be equal to each other if they are to be parallel lines: $m = -3$.

Visually, this is what we have so far:

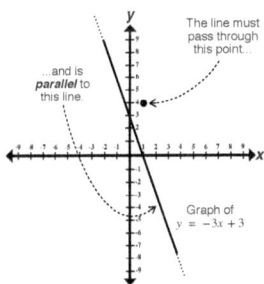

The line must pass through this point...

...and is **parallel** to this line.

Graph of $y = -3x + 3$

As you can see from the graph, we are trying to determine the equation of the line that passes through the given point, and that is **parallel** to the line of the given equation. We now know that the slope of the equation we are after must be equal to -3.

The next step involves using the slope value we found together with the point that was specified in the prompt: we know that the equation's line that we are trying to define passes through that point, and that it must have a slope value of $m = -3$. We can use the point-slope formula for this step.

Step 2) Using the slope value found, together with the coordinate given, (1 , 4), replace the corresponding variables on the point-slope formula.

$$m = -3 \; ; \; x_1 = 1 \; ; \; y_1 = 4$$

$$y - y1 = m(x - x1)$$

$$y - (4) = -3(x - (1))$$

Then, simplifying, we obtain:

$$y - 4 = -3(x) - (-3)(1)$$

$$y - 4 = -3x - (-3)$$

$$y - 4 = -3x + 3$$

$$y - 4 + 4 = -3x + 3 + 4$$

$$y = -3x + 7$$

And there it is! The equation of the line that passes through (1 , 4) and that is parallel to **9x + 3y − 9 = 0** . Although we could have left the equation in point-slope form, as $y - (4) = -3(x - (1))$, I solved it for y to rewrite it in slope-intercept form so that you can practice that skill. Just remember, they are equivalent equations:

slope-intercept: $y - 4 = -3(x - (1))$

is the same as

point-slope: $y = -3x + 7$

Now, if we graph the equation that we found above, we should end up with a line that passes through the given point and that is also parallel to the line of the given equation (the latter is self-evident: the equation we found has a slope equal to **-3**, the same slope as the given equation; thus, they are parallel).

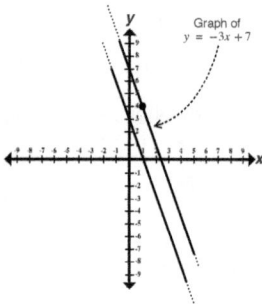

Graph of $y = -3x + 7$

Problem solved!

b) What is the equation of a line that passes through **(-2, 3)** and that is **perpendicular** to the line **-x + y + 2 = 0** ?

The solution process will be similar to the previous prompt, except that this time, once we have the slope of the given equation, before we use it as the slope of the equation we are trying to define, we will have to find its opposite reciprocal, since the lines are perpendicular to each other. Observe.

Step 1) Express the given linear equation in slope-intercept form by solving for **y** in order to extract its slope value...

$$-x + y + 2 = 0$$
$$-x + y + 2 + x - 2 = 0 + x - 2$$
$$y = x - 2$$

And there we have it! The other line in slope-intercept form. The slope of this line is: **m = 1**. Therefore, the slope of the equation we wish to define must be the opposite reciprocal of this value (the circled-value below):

$$1 = \frac{1}{1} \qquad \text{Opposite} \atop \xleftarrow{\text{-------->}} \atop \text{Reciprocals} \qquad -\frac{1}{1} = (-1)$$

Visually, this is what we have so far:

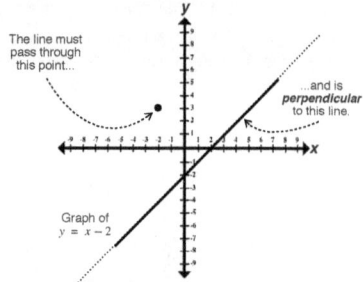

The line must pass through this point...

...and is **perpendicular** to this line.

Graph of $y = x - 2$

As you can see from the graph, we are trying to determine the equation of the line that passes through the given point, and that is **perpendicular** to the line of the given equation. We now know that the slope of the equation we are after must be equal to **-1** (the opposite reciprocal of the given equation's slope).

The next step involves using the slope value we defined for the equation we are looking for, together with the point that was specified in the prompt: we know that the equation's line that we are trying to define passes through that point, and that it must have a slope value of $m = -1$. We can use the point-slope formula for this step.

Step 2) Using the slope value found, together with the coordinate given, $(-2, 3)$, replace the corresponding variables on the point-slope formula.

$$m = -1 \ ; \ x_1 = -2 \ ; \ y_1 = 3$$

$$y - y1 = m(x - x1)$$

$$y - (3) = -1(x - (-2))$$

which is written as...

$$y - 3 = -(x + 2)$$

To express it in slope-intercept form, we can solve it for y as follows:

$$y - 3 = -(x) + (-(2))$$

$$y - 3 = -x - 2$$

$$y - 3 + 3 = -x - 2 + 3$$

$$y = -x + 1$$

And there it is! The equation of the line that passes through $(-2, 3)$ and that is perpendicular to $-x + y + 2 = 0$.

Now, if we graph the equation that we found above, we should end up with a line that passes through the given point and that is also perpendicular to the line of the given equation (the latter is self-evident: the equation we found has a slope equal to -1, the opposite reciprocal

of the slope of the given equation; thus, they are perpendicular to each other).

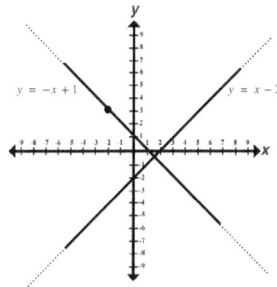

Problem solved! You can see how the line of the equation that we defined is, in fat, **perpendicular** to the line of the given equation, and that it passes through the point that was specified.

c) What is the equation of a line that passes through $(0, 5)$ and that is **parallel** to the line $-2x - y - 3 = 0$?

The solution process will be similar to the first problem of this series. However, we will not need to use the point-slope formula in order to define the equation we are looking for (can't see why? I'll explain shortly).

Step 1) Express the linear equation given in slope-intercept form by solving for y in order to extract its slope value...

$$-2x - y - 3 = 0$$

$$-2x - y - 3 + 2x + 3 = 0 + 2x + 3$$

$$-y = 2x + 3$$

To get rid of the negative sign "**attached**" to **y** simply multiply both sides by **−1** ...

$$(-1)(-y) = (2x + 3)(-1)$$
$$y = -2x - 3$$

And there we have it! The other line in slope-intercept form. Now we can specify both lines' slopes, since they must be equal to each other if they are to be parallel lines: $m = -2$.

Visually, this is what we have so far:

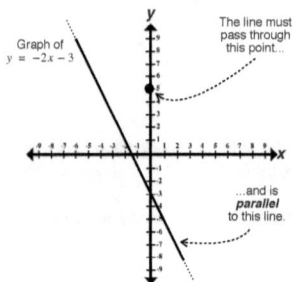

Graph of $y = -2x - 3$

The line must pass through this point...

...and is **parallel** to this line.

As you can see from the graph, we are trying to determine the equation of the line that passes through the given point, and that is **parallel** to the line of the given equation. We now know that the slope of the equation we are after must be equal to **−2**.

So what's next? We know the equation we are after must have a slope of **−2** and that it has a solution pair of **(0 , 5)** ; how can we define its equation?

If you think we need to use the point-slope formula again, you are correct, but you are embarking on a path that takes longer to

traverse. If you analyze this prompt carefully, you should be able to see that the given point for the equation's line that we are trying to define is a "special" point: remember what it is?

(0 , 5) is of the form **(0 , b)**

What about now? If you said "It's the **intercept point**..." then you are absolutely correct. And since it is the intercept point, we know that the equation's **b** value (when using the slope-intercept form) must be **5** . This means that we do not have to use the point-slope formula at all. Simply take the general slope-intercept form of a linear equation, and since we know that $m = -2$ and that $b = 5$ we can specify the linear equation, as requested, as follows:

$$y = mx + b$$

$$y = -2x + 5$$

And we are done! Of course, if the prompt had specifically requested the point-slope form (or if you were solving a multiple choice test and all of the choices were in point-slope form), then you would have to proceed as follows:

$$m = -2 \; ; \; x_1 = 0 \; ; \; y_1 = 5$$

$$y - y1 = m(x - x1)$$

$$y - (5) = -2(x - (0))$$

which may be written as...

$$y - 5 = -2(x - 0)$$

You can solve the point-slope form version for the variable **y** and you will get the slope-form we found earlier. I suggest you try it.

The lesson here is that it is always a good idea to stay alert when working on problems such as these. In this occasion, we were able to quickly come up with the linear equation by identifying the given point as the intercept point. Finally, let me to show you the graph of both equations:

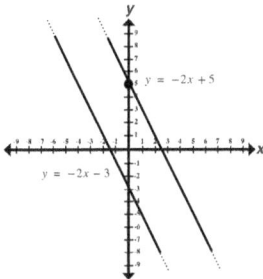

d) What is the equation of a line that passes through $(-5, 3)$ and that is **perpendicular** to the line $y = 2x - 4$?

For this problem, the given equation is in slope-intercept form: this means that we can extract the slope value directly, without having to do any additional work (unlike the previous three prompts where we had to solve the given equation for **y** in order to find the slope value). Thus, the slope of the given equation is $m = 2$. Therefore, the slope of the equation we wish to define must be the opposite reciprocal of this value (the circled-value below):

$$2 = \frac{2}{1} \quad \longleftrightarrow \quad \boxed{-\frac{1}{2}}$$

Visually, this is what we have so far:

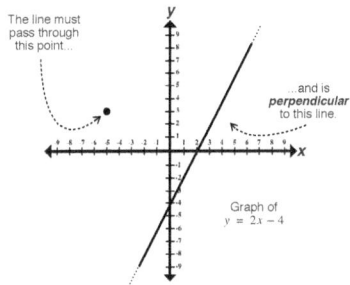

As you can see from the graph, we are trying to determine the equation of the line that passes through the given point, and that is **perpendicular** to the line of the given equation. We now know that the slope of the equation we are after must be equal to $-\frac{1}{2}$ (the opposite reciprocal of the given equation's slope).

The next step involves using the slope value we defined for the equation we are looking for, together with the point that was specified in the prompt. We can use the point-slope formula for this step.

Step 2) Using the slope value found, together with the coordinate given, $(-5, 3)$, replace the corresponding variables on the point-slope formula.

$$m = -\frac{1}{2} \; ; \; x_1 = -5 \; ; \; y_1 = 3$$

$$y - y1 = m(x - x1)$$

$$y - (3) = -\frac{1}{2}(x - (-5))$$

which may be written as...

$$y - 3 = -\frac{1}{2}(x + 5)$$

To express it in slope-intercept form, we can solve it for **y**. However, the first thing I will do to solve for **y** is change the form in which the fraction is written. Remember that I recommend incorporating a negative sign that may be part of a fraction into the numerator as opposed to leaving it "on the side":

$$y - 3 = \frac{-1}{2}(x + 5)$$

Now, I can continue as usual...

$$y - 3 = \frac{-1}{2}(x + 5)$$

$$y - 3 = (\frac{-1}{2})(x) + (\frac{-1}{2})(5)$$

$$y - 3 = (\frac{-1}{2})(\frac{x}{1}) + (\frac{-1}{2})(\frac{5}{1})$$

$$y - 3 = \frac{-x}{2} + (\frac{-5}{2})$$

$$y - 3 + 3 = \frac{-x}{2} + \frac{-5}{2} + 3$$

$$y = \frac{-x}{2} + \frac{-5}{2} + \frac{3}{1}$$

$$y = \frac{-x}{2} + \frac{(-5)(1) + (3)(2)}{(2)(1)}$$

$$y = \frac{-x}{2} + \frac{-5 + 6}{2}$$

$$y = \frac{-x}{2} + \frac{1}{2}$$

And there it is! The equation of the line that passes through **(-5 , 3)** and that is perpendicular to $y = 2x - 4$.

Now, if we graph the equation that we found above, we should end up with a line that passes through the given point and that is also perpendicular to the line of the given equation (the latter is self-evident: the equation we found has a slope equal to $-\frac{1}{2}$, the opposite reciprocal of the slope of the given equation; thus, they are perpendicular to each other).

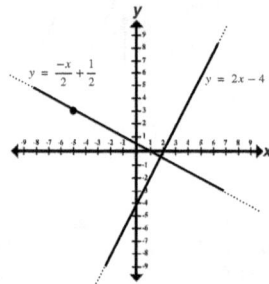

Problem solved! You can see how the line of the equation that we defined is, in fat, **perpendicular** to the line of the given equation, and that it passes through the point that was specified.

e) What is the equation of a line that passes through **(-2 , 0)** and that is **parallel** to the line $x = 5$?

This problem is actually very easy to solve. Let's start with a visual representation of this prompt.

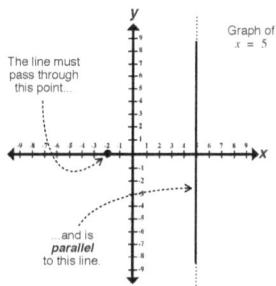

The line must pass through this point...

...and is *parallel* to this line.

Graph of $x = 5$

So, we are trying to determine the equation of a line that is vertical (we know this because the line must be parallel to a vertical line, the line given by $x = 5$, which means that they have the same–undefined or infinite–slope). You should remember that all vertical lines have the form

$$x = d$$
where d = any REAL number

...and that all of the solution points will be of the form

(d , any REAL number)

In this case, since the equation we are looking for must contain the solution pair $(-2, 0)$, it means that $d = -2$, which means that the equation must be

$$x = -2$$

This equation, when graphed, yields a vertical line that passes through $(-2, 0)$; all of its solution pairs (and therefore, all of the points through which the line passes) must have an

x-coordinate equal to -2, such as $(-2, -1)$ $(-2, 0)$, $(-2, 1)$, $(-2, 2)$, etc...

We were thus able to solve this prompt without doing much work. We simply had to realize (recognize) that the given equation was a vertical line and then apply the knowledge which we have reviewed regarding vertical lines.

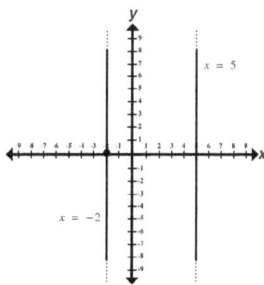

$x = -2$ $x = 5$

f) What is the equation of a line that passes through $(3, 2)$ and that is **perpendicular** to the line $y = -7$?

This prompt is very similar to the previous prompt; if you were unable to solve it the first time you attempted this series of problems on your own, try it again now.

Ready?

Let's create a visual representation of the prompt. Observe in the following graph the horizontal line defined by the given equation, and the location of the point that the line of the equation we are trying to define must pass through.

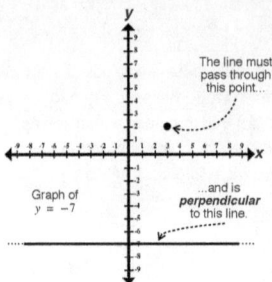

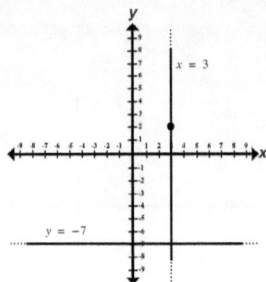

This time, the line of the given equation is a horizontal line, and if we are attempting to define the equation of a line that is perpendicular to it, then that equation must correspond to a vertical line. And if it's a vertical line, it must be of the form

$$x = d$$

where d = any REAL number

...and all of its solution points should thus be of the form

(d , any REAL number)

In this case, since we need the line to pass through (3 , 2) then it means that $d = 3$. Therefore, the equation we are looking for is:

$$x = 3$$

Simple, right? Most problems involving horizontal and/or vertical lines and lines that are parallel or perpendicular to them are relatively easy to solve as long as you remember the key characteristics of these two types of lines.

Observe the final graph, with the line of the equation that we defined.

g) What is the equation of a line that passes through (−4 , 3) and that is **perpendicular** to the line

$$y = \frac{2}{3}x + 1 ?$$

Step 1) As before, the first step is to define the slope of the equation we are trying to come up with, let's call it **line A**. Since this equation's line must be perpendicular to the line given by the equation $y = \frac{2}{3}x + 1$, let's call it **line B**, all we need to do is extract the slope from **line B**, and since perpendicular lines have opposite-reciprocal slopes, find the opposite-reciprocal of **line B's** slope and assign it as **line A's** slope.

Because the given equation is in slope-intercept form, we can easily specify line B's slope: $\frac{2}{3}$.

Next, we find the opposite reciprocal of this value, and assign the result to line B's slope.

$$\frac{2}{3} \quad \longleftrightarrow \quad -\frac{3}{2}$$

Thus, we know that the slope of the equation we are trying to define must be $-\frac{3}{2}$.

Next, since **line A** must pass through $(-4, 3)$, we can use the point-slope formula to define its equation now that we know its m value.

Step 2) Using the slope value found, together with the coordinate given, $(-4, 3)$, replace the corresponding variables on the point-slope formula.

$$m = -\frac{3}{2} \; ; \; x_1 = -4 \; ; \; y_1 = 3$$

$$y - y1 = m(x - x1)$$

$$y - (3) = -\frac{3}{2}(x - (-4))$$

Which may be written as...

$$y - 3 = -\frac{3}{2}(x + 4)$$

Remember that this is a perfectly valid representation (or form) of the linear equation we were asked to solve: it is in point-slope form. I am also finding its slope-intercept form to help you see how its done and to practice the skill. To do this, we solve it for y ...

$$y - 3 = -\frac{3}{2}(x + 4)$$

$$y - 3 = \frac{-3}{2}(x + 4)$$

$$y - 3 = (\frac{-3}{2})(x) + (\frac{-3}{2})(4)$$

$$y - 3 = \frac{-3x}{2} + (\frac{-12}{2})$$

$$y - 3 = \frac{-3x}{2} - 6$$

$$y - 3 + 3 = \frac{-3x}{2} - 6 + 3$$

$$y = \frac{-3x}{2} - 3$$

And there it is! The equation of the line that passes through $(-4, 3)$ and that is perpendicular to $y = \frac{2}{3}x + 1$:

In slope-intercept: $\quad y - 3 = -\frac{3}{2}(x + 4)$

In point-slope: $\quad y = \frac{-3x}{2} - 3$

Observe the graph of both equations, and how the line of the equation that we specified does pass through the point $(-4, 3)$.

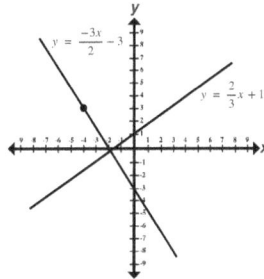

Before ending the chapter, I would like to review an important skill involving the point-slope form of a linear equation.

Extracting information from a linear equation's point-slope form.

If you know the equation of a line and it is in point-slope form, you should be able to extract its slope and the "point" that it specifies; this information can then be used to graph the line, for example, or to do anything else that involves the use of the equation's slope or the point that it specifies.

Although we have used the slope of a line and a known point through which it passes in order to define the equation of the line in point-slope form, the task of extracting the values is typically more confusing. Let's see why.

Example 1

Let's work with a line that has a slope value of **−4** and that passes through the point **(−3 , 2)**. We could easily specify its equation, using, of course, the point-slope form:

$$m = -4 \; ; \; x_1 = -3 \; ; \; y_1 = 2$$

$$y - y1 \; = \; m(x - x1)$$

$$y - (2) \; = \; -4(x - (-3))$$

Which may be written (simplified) as...

$$y - 2 \; = \; -4(x + 3)$$

Okay... so we now know the equation of the specified line. Now what? Well, let's suppose that you challenge a friend to "extract" from the equation that you just found its line's slope value and the point it specifies. In other words, you write the point-slope form equation on a piece of paper, and ask him to tell you its slope and the point it specifies:

$$y - 2 \; = \; -4(x + 3)$$

$$m = ? \; ; \; (x , y) = ?$$

Let's also assume that your friend happens to enjoy challenges such as this one, and doesn't laugh maniacally, walking away, crumpling the piece of paper and throwing it in the nearest wastebasket, wondering why you two are friends to begin with.

In order to do this challenge correctly, you could proceed as follows:

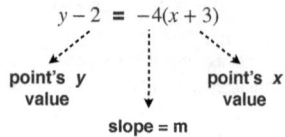

$$y - 2 \; = \; -4(x + 3)$$

point's *y*   point's *x*
value   value

slope = m

It is easy enough to see that the slope value for this equation is **−4** , since its position within the point-slope form is clear-cut, and even if the value is negative, this can easily be determined.

So, assuming your friend knows this particular math concept, would write the following:

$$y - 2 \; = \; -4(x + 3)$$

$$m = -4 \; ; \; (x , y) = ?$$

Now, your friend would need to determine the x and the y values of the point's coordinate, as specified by the equation. And this is where things get interesting.

Unless your friend has mastered the skill of "extracting" values from the point-slope form correctly, this is what your friend might come up with:

$$y - 2 = -4(x + 3)$$
$$m = -4 \; ; \; (3, 2)$$

Can you see why this is wrong? It is easy enough to check that it's wrong: looking back at the original description of the line that was used to create the equation, the prompt said it passed through the point $(-3, 2)$. And this is not what is specified above. What happened here? Did I make a mistake writing the equation of the line?

The answer is simple. We first wrote the point-slope form as follows:

$$y - (2) = -4(x - (-3))$$

This version contains the x and y values of the point's coordinate explicitly written (inside the respective parentheses set):

$$y - (2) = -4(x - (-3))$$

point's y value   point's x value

In this version of the equation, most people would correctly express the coordinate as $(-3, 2)$. So what happens when we *simplify* this first draft of the linear equation in point-slope form? Well, the key word here is "*...simplify...*": when the equation is simplified, we end up with a "cleaned-up" version of the equation, where we use the sign-table to consider what "subtracting a negative" is equivalent to, and what "subtracting a positive" is equal to:

$$y - (2) = -4(x - (-3))$$

subtract a positive...   subtract a negative...

...which leads to:

$$y - 2 = -4(x + 3)$$

And this is where the confusion typically begins. Someone that has not yet mastered the skill of *correctly* reading (extracting) the slope and coordinate values of a *simplified* point-slope form of a linear equation will not stop to analyze whether the x and y values of the point's coordinate are being added or subtracted in the equation; this plays a major role in correctly specifying the x and y value's sign.

$$y - 2 = -4(x + 3)$$

the y value is being *subtracted...*   the x value is being *added...*

therefore, $y = 2$ therefore, $x = -3$

Thus, the correct point is $(-3, 2)$.

563

Why is it crucial to notice whether the x or the y value is being added or subtracted? Because if the value is being subtracted, then it means that the value is positive, because when you "*plug in*" a positive coordinate value into the point-slope form, it will end up being **subtracted**. Remember the general equation of the point-slope form:

$$y \underset{\text{\_}}{-} y1 = m(x \underset{\text{\_}}{-} x1)$$

the y_1 value   the x_1 value
is being   is being
subtracted...   **subtracted...**

Thus, since the equation subtracts the x_1 and the y_1 values **by default**, if the coordinate's value is positive, when you simplify the equation the value will end up being **subtracted**. However, if the value is negative, then when you simplify the equation (subtracting a negative), the value will show up as **adding**.

⚓
When "extracting" the slope and the coordinate values of an equation that is in point-slope form, make sure to check if the x_1 and the y_1 values are being added or subtracted (respectively). If the value is being added, extract a negative coordinate value; if it is being subtracted, extract a positive coordinate value. The four possible combinations are:

$y - y1 = m(x - x1) \;\; \dashrightarrow \;\; (\,x_1\,,\,y_1\,)$
$y + y1 = m(x - x1) \;\; \dashrightarrow \;\; (\,x_1\,,\,\text{-}y_1\,)$
$y - y1 = m(x + x1) \;\; \dashrightarrow \;\; (\,\text{-}x_1\,,\,y_1\,)$
$y + y1 = m(x + x1) \;\; \dashrightarrow \;\; (\,\text{-}x_1\,,\,\text{-}y_1\,)$

Try to extract by yourself the correct slope and coordinate values of the following point-slope form linear equations. Then, check your answers to see if you have mastered this key aspect of correctly reading the information contained in an equation that is in point-slope form.

a) $y - 1 = 3(x - 7)$

b) $y - \dfrac{2}{3} = -4(x - 5)$

c) $y + 9 = 6(x + 8)$

d) $y + 4 = -(x - 1)$

e) $y = -\dfrac{3}{4}(x + 5)$ *Careful with this prompt!

f) $y + \dfrac{1}{3} = \dfrac{-3}{7}(x + \dfrac{3}{4})$

g) $y + \sqrt{5} = 3(x + \pi)$

h) $y + \dfrac{3}{2} = -\dfrac{2}{3}(x - 1)$

i) $y - (-4) = -(\dfrac{-4}{5})(x - (\dfrac{-1}{3}))$ *Careful: this equation has not yet been simplified!

Most of these are relatively straight-forward. Just be careful with problems **(e)** and **(i)** ; stay alert as you attempt to solve them and you should be able to handle them as well.

Ready to check your work? Here are the correct slope and coordinate values.

a) $y - 1 = 3(x - 7)$

$$m = 3 \; ; \; x_1 = 7 \; ; \; y_1 = 1$$

Thus, the correct point is $(\,7\,,1\,)$

564

b) $y - \frac{2}{3} = -4(x - 5)$

$m = -4$; $x_1 = 5$; $y_1 = \frac{2}{3}$

Thus, the correct point is $\left(5, \frac{2}{3}\right)$

c) $y + 9 = 6(x + 8)$

$m = 6$; $x_1 = -8$; $y_1 = -9$

Thus, the correct point is $(-8, -9)$

d) $y + 4 = -(x - 1)$

$m = -1$; $x_1 = 1$; $y_1 = -4$

Thus, the correct point is $(1, -4)$

e) $y = -\frac{3}{4}(x + 5)$

$m = -\frac{3}{4}$; $x_1 = -5$; $y_1 = 0$

Thus, the correct point is $(-5, 0)$

The fact that you don't see anything adding or subtracting on the **y** variable on the left side of the equation should allow you to determine that the value of y_1 is equal to **0** ; to prove this, observe what happens when I use the slope and the coordinate values that I specify as the answer and plug them in to the point-slope formula. We should end up with the exact same equation that the prompt specifies, correct? Let's see:

$m = -\frac{3}{4}$; $x_1 = -5$; $y_1 = 0$

$y - (0) = -\frac{3}{4}(x - (-5))$

simplifying the equation we obtain:

$y = -\frac{3}{4}(x + 5)$

exactly what we needed!

f) $y + \frac{1}{3} = \frac{-3}{7}(x + \frac{3}{4})$

$m = -\frac{3}{7}$; $x_1 = -\frac{3}{4}$; $y_1 = -\frac{1}{3}$

Thus, the correct point is $\left(-\frac{3}{4}, -\frac{1}{3}\right)$

g) $y + \sqrt{5} = 3(x + \pi)$

$m = 3$; $x_1 = -\pi$; $y_1 = -\sqrt{5}$

Thus, the correct point is $(-\pi, -\sqrt{5})$

h) $y + \frac{3}{2} = -\frac{2}{3}(x - 1)$

$m = -\frac{2}{3}$; $x_1 = 1$; $y_1 = -\frac{3}{2}$

Thus, the correct point is $\left(1, -\frac{3}{2}\right)$

i) $y - (-4) = -(\frac{-4}{5})(x - (\frac{-1}{3}))$

Simplifies to: $y + 4 = \frac{4}{5}(x + \frac{1}{3})$

$$m = \frac{4}{5} \;;\; x_1 = -\frac{1}{3} \;;\; y_1 = -4$$

Thus, the correct point is $\left(-\frac{1}{3}, -4\right)$

Please note that on this last prompt, you could have extracted the x_1 and the y_1 values directly, without the need to simplify the equation, because it is explicitly showing you their values:

$$y - (-4) = -\left(\frac{-4}{5}\right)\left(x - \left(\frac{-1}{3}\right)\right)$$

Because the parentheses set is being **subtracted**, the y_1 value is precisely **−4** ...

Because the parentheses set is being ***subtracted***, the x_1 value is precisely $-\frac{1}{3}$ ...

However, the slope value would need to be simplified before we can "***extract***" it, or read it from the equation, because the way the equation is originally written in the prompt, the slope consists of "a negative, negative four fifths". Therefore, the slope is equal to "positive four fifths".

And this concludes another chapter of the book: I have to say that this was, by far, the longest chapter yet.

So if you read through it carefully and tried to solve by yourself all of the exercises I suggested throughout the review you should have been able to master all of the skills that we reviewed in this chapter.

See you on the next one!

19

LINEAR FUNCTIONS

This will be a very brief chapter–compared, that is, to the previous one. It is an extension of the idea of linear equations, and, as you will see shortly, a very straight-forward concept to master, assuming you understood most, if not all of, the concepts we reviewed in **Chapter 18**.

We have already established that linear *equations* (not functions) come in different forms: there are three basic forms that we should be able to identify. They are:

⚓
Forms of linear equations

Point-slope form
y = mx + b

Slope-Intercept form
$y - y1 = m(x - x1)$

General form
Ax + By + C = 0

The first two forms allow us to quickly extract key characteristics of the particular equation (their names specifically state which of these are explicitly shown) while the third one does not. However, a given equation can be stated in all three forms, and they would therefore all be equivalent to each other: they possess the same solution pairs regardless of the form that they may be written in.

So what is a *linear function*? Well, a linear function is simply a linear equation expressed in "*function*" format. If you don't remember what functions are to begin with, just go back to **Chapter 17** and review it again. If you recall, a function is of the following form:

$$f(x) = some \ mathematical \ expression$$

For example:

$$f(x) = x^2 + 1$$

$$g(y) = 4y^3 - y - 7$$

$$h(z) = 2x^4y^3 - 5xy^2 + z$$

As you can see from the examples above, the left side contains the "name" of the function, and it also states the variable that defines it. Functions may be defined in terms of more than one variable, of course, such as $h(z)$ above, even though the left side only specifies one variable: the variable **z**. There are more complex functions that do contain multiple variables on the left side, inside the parentheses; for example:

$$f(x, y) = -3x^7 + 5y^2 - \pi$$

but we are not going to explore them here (there are some important differences that would require exploring concepts beyond the scope of the book). We are only considering functions that specifically contain a single variable on its left side, as in the first three examples. But please note that they may have more than one variable on its right side: that is not an issue.

You should also recall that a function such as

$$f(x) = x^2 - x + 1$$

...would be *read* as "*ef* of *ex* is equal to *ex*-squared minus *ex* plus one"; in other words, the left side is always read as "-*the letter that identifies its function-* **OF** -*the variable contained in the parentheses-* **IS EQUAL TO** -*whatever the mathematical statement that is to the right of the equal symbol happens to be-*". Furthermore, functions are particularly easy to evaluate and to explicitly state what they should be evaluated with to begin with. Using the same function above, if I asked you to do the following:

$$f(3) = ?$$

...you should be able to understand what it's asking you to do: replace the variable that was contained inside the set of parentheses when the function was originally defined and replace it on the right side with the current value that is now inside the same set of parentheses:

$$f(x) = x^2 - x + 1$$

$$f(3) = ?$$

$$f(3) = (3)^2 - (3) + 1$$

As you can see on the example above, $f(3) = ?$ is explicitly asking you to replace all "**exes**" that appear in the mathematical statement to the right of the equal symbol with the number **3** and then to solve it using the established rules of math (order of operations, exponents, etc...). The solution process would thus look as follows:

$$f(x) = x^2 - x + 1$$

$$f(3) = ?$$

$$f(3) = (3)^2 - (3) + 1$$
$$f(3) = (3)(3) - 3 + 1$$
$$f(3) = 9 - 3 + 1$$
$$f(3) = 6 + 1$$
$$f(3) = 7$$

Once the function has been solved for a given input value (in this case the input value used was **3**), you can generate the ordered pair that corresponds to this calculation (the input/output solution pair):

Solution pairs are of the form: **(input , output)**

Function notation correspondence: $(x, f(x))$

Solution pair found: **(3 , 7)**

As you can see above, when you are working with functions, once you found the output value for a given input, you can generate the solution pair it represents using the template

$$(x, f(x))$$

...because **x** represents the input value, while $f(x)$ represents the output value.

So what about linear functions, and what do they have to do with linear equations?

Well, linear functions are simply linear equations written in function form.

The problem is that functions are generally less flexible as far as its form is concerned because typically they have the format

$$f(x) = some \quad mathematical \quad expression$$

Notice that the left side of the equal symbol is simply the name of the function and the variable that defines it. Linear equations, on the other hand, are much more flexible because both sides of the equal symbol contain mathematical statements (numbers and operations).

Compare the following two linear *relationships*: one is in equation form, the other in function form:

$$y = 2x + 3$$

$$f(x) = 2x + 3$$

They both represent the exact same linear relationship, which means that they both have the exact same solution pairs. The linear equation is in slope-intercept form, so we know its slope and intercept point explicitly: **m = 2** and **(0 , 3)** respectively. The function form seems to also communicate the slope and the intercept point, correct? That tells us something about linear functions which we will explore in detail momentarily.

But notice that I can express the same linear equation above as follows (keeping the "*equation*" form):

$$y = 2x + 3 \quad \text{---> } \textbf{slope-intercept}$$

$$y - 5 = 2(x - 1) \quad \text{---> } \textbf{point-slope}$$

$$-2x + y - 3 = 0 \quad \text{---> } \textbf{general}$$

Please note that I selected one particular point-slope form, since an infinite version of them may be created (simply choose a different solution pair each time and you will have created a "different" point-slope form equation). However, they are all equivalent, as I proved to you in **Chapter 18**. In this case, I used the solution pair **(1 , 5)** to generate the equation in said form, but I could have instead used the solution pair **(0 , 3)** or **(-2 , -1)** or **(1.3 , 5.6)**, etc..., which would have resulted in superficially

different-looking versions of the same equation (or relationship).

Well, function notation generally doesn't offer the same flexibility. We could take things a bit far and do the following:

$$f(x) = 2x + 3 \quad \text{---> } \textbf{"slope-intercept"}$$

$$f(x) - 5 = 2(x - 1) \quad \text{---> } \textbf{"point-slope"}$$

$$-2x + f(x) - 3 = 0 \quad \text{---> } \textbf{"general"}$$

Strictly speaking, there is nothing wrong with this. However, in practice, this is not typically done. Function notation is thus a bit less flexible than equations, and generally speaking, the left side remains "***untouched***". So we can say that unless you encounter one of the rarely–*if ever*–used versions above, all linear functions are in slope-intercept form:

$$f(x) = mx + b$$

However, note that it is not uncommon to see the following general form for a **linear polynomial**; remember the first couple of chapters of the book?:

$$f(x) = a_1 x + a_0$$

...where a_1 is the coefficient of the **x** term and a_0 is the independent term. Of course, based on what we know about linear relationships, we know that if this *nomenclature* is used, then the following would be true:

$$a_1 = \textit{slope} \quad \text{---> } m$$

$$a_0 = \textit{y-intercept} = (0 , a_0) \quad \text{---> } b$$

So don't be confused if you encounter a linear function defined using "*a-with subscripts*" instead of **m** and **b**. In the end, it would still be in slope-intercept form, and it would still convey explicitly the same information as its linear-equation counterpart: its *slope* and its *y*-intercept.

So there you have it. A linear function is a linear equation that is expressed using function notation, which means that it will generally be expressed in slope-intercept form.

Look at the following examples of linear functions (and their linear-equation counterparts).

$y = x + 1$ $f(x) = x + 1$

$y = -x + 9$ $f(x) = -x + 9$

$y = -\dfrac{5}{4}x - \dfrac{1}{3}$ $f(x) = -\dfrac{5}{4}x - \dfrac{1}{3}$

$y = x$ $f(x) = x$

$y = 2$ $f(x) = 2$

$y = 0$ $f(x) = 0$

$y - 6 = -(x - 2)$ $f(x) = -x + 8$

Observe how in the example above, in order to represent the same linear equation in function form, the linear equation first had to be solved for the variable y...

But what about the following linear equation?

$$x = 2$$

You should recall that this is the equation of a vertical line; it passes through $(2,0)$, and it has an infinite or UNDEFINED slope. Observe its graph.

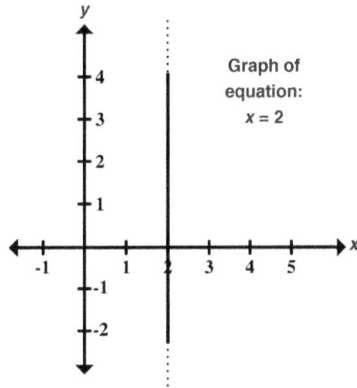

Graph of equation:
$x = 2$

So how can we express this linear equation using function notation?

Clearly, the following would not work

$$f(x) = 2$$

...because this corresponds to a horizontal line, not a vertical line; its "equation-form" counterpart is

$$y = 2$$

as we stated on the list of examples on this page. So what can we do?

Well, you may remember that functions have a very unique limitation: for a given **input** value, only **one output** value may be defined. The problem with the equation

$$x = 2$$

...is that it yields the following input/output pairs:

(2 , –7)
(2 , –3)
(2 , –.82375...)
(2 , 0)
(2 , .333...)
(2 , 4)
(2 , 100)
etc...

In other words, the only allowed input value is **2**, and it is that single input value that provides the *infinite* output **REAL** values that span from negative infinity to positive infinity.

In other words, $x = 2$ is ***NOT A FUNCTION !*** Why? Because it breaks the rule that a given input value may only "point" to a single output value. And for this reason, $x = 2$ cannot be represented using function notation, nor any vertical line, for that matter.

⚓

A Linear Function is of the form:

$$f(x) = mx + b$$

and it is equivalent to the slope-intercept form of a linear equation $y = mx + b$

However, *linear functions* can never be used to represent vertical lines of the form $x = d$ because vertical lines break the function rule that a given input value may only be assigned one output value.

Other than the two differences reviewed so far (that linear functions typically are expressed only in slope-intercept form, and they can never

represent a vertical line), linear functions and linear equations are exactly the same.

All of the principles that we reviewed on the previous chapter regarding the slope and the **y-intercept** point of a linear equation apply to linear functions.

To thoroughly review linear functions, let's explore how to find the x-intercept of a linear function.

For example, if we are asked to define the intercept points of

$$f(x) = 2x + 6$$

... we could easily begin by specifying the **y-intercept**.

$$f(x) = 2x + 6$$

since this linear function is in slope-intercept form, we know its **b** value:

b = 6

therefore, the **y-intercept** is

(0 , 6)

As to its x-intercept, we know that it must occur at **(some value , 0)** ; in other words, we want the solution pair in which the output value is **0**. So, if we "set" the function to zero (make it equal to zero), we should be able to find the input (**x**) value that corresponds to the solution pair, as follows:

$$f(x) = 2x + 6$$

the **output** must be equal to **0** ...

$$0 = 2x + 6$$

...and simply solve for the variable x (which we know corresponds to the input value we are looking for in order to define the **x-intercept**):

$$0 = 2x + 6$$

$$0 - 6 = 2x + 6 - 6$$

$$-6 = 2x$$

$$\frac{-6}{2} = \frac{2x}{2}$$

$$-3 = x \quad \text{or} \quad x = -3$$

And there we have it: if you evaluate the function with   we know its output will be 0 , which corresponds to the following solution pair:

$$(-3 , 0)$$

...in other words, the **x-intercept**.

Whenever you need to express a linear relationship in function form, simply find the linear equation in slope-intercept, and replace y with $f(x)$ or whatever function name you may need or wish to use: $g(x)$ or $h(x)$, etc...

Graphing linear functions works the same way as graphing linear equations. The only difference is that the cartesian plane's vertical axis (in other words, the output axis) must be labeled not as y but as $f(x)$ or whatever function name you may be using ($g(x)$ or $h(x)$, etc...).

Observe the graph of the function we worked with above: $f(x) = 2x + 6$.

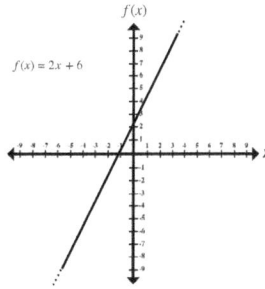

Notice how the vertical axis (the **y-axis** when graphing linear equations) is labeled $f(x)$, clearly indicating the fact that we are using function notation.

This is perfectly aligned with the general function coordinate notation mentioned earlier and in **Chapter 17**:

$$(x , f(x))$$

...in other words, the written-in-stone format of any coordinate: **(input , output)**–well, not in stone, exactly; as I stated earlier in the book, coordinates could very well be defined as having the form **(output , input)** but then it would have to be explicitly and clearly stated in order to avoid any confusions, and it is precisely to avoid such confusions to begin with that it is seldom done.

And this very brief chapter is done! Linear functions are very closely related to linear equations, barring those differences that were detailed in this chapter.

See you on the next one!

20

QUADRATIC EQUATIONS

Welcome to the second-to-last chapter of the book!
This is usually one of the sections of basic algebra
that math students typically struggle with the most,
but I am certain that if you have been able to keep
up with all that has been reviewed in the book so
far, and if you continue to try to solve by yourself
the problems included in this chapter and the next,
that you will realize how simple it is to work with
quadratic equations (and their closely-related

quadratic functions which we will deal with on the next chapter).

So what is a quadratic equation? Well, the *equation* part should be a no-brainer, right? You expect to see something along the lines of

something = something else

...but what does the *quadratic* part mean? Well, remember the chapter on powers (or exponents)? Keep that in mind as you continue reading. Not to bore you with a lesson in language arts, but you should know that the word *quadratic* comes from the Latin word "*quadratus*" which literally means "square", as in the geometric shape that has four equal sides:

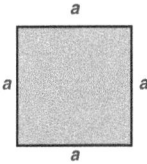

SQUARE: polygon with four sides of equal length

So what does the geometric figure, a square, have anything to do with an equation? As you may recall, I explained early on in the book that to "*square*" a mathematical element means to raise it to the power of **2**, in other words, to *square* it. The reason why mathematicians refer to this

$$(something)^2$$

...as "to *square* something" or "something *squared*" is that historically, our *Homo sapien* species began discovering mathematics through the exploration of geometry. Thus, since in order to find the area of a square you simply multiply its side measurement by itself (which we know means raising its side measurement to the second power, as in

$$\text{area} = side^2 \quad \text{or} \quad \text{area} = a^2$$

...the act of raising something to the second power became known as "squaring". This is also the reason why raising an element to the *third* power also has a specific name: *cubing*... because in order to find the volume of a cube you must multiply its side times itself *three* times, as in

$$side^3 = \text{volume}$$

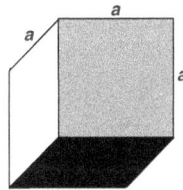

CUBE: a three-dimensional solid bounded by six square faces, with three meeting at every vertex. All of its sides are of equal length

Now, when you look at the name "quadratic equation", I am sure you can envision that it refers to an equation in which a variable is being squared (or being raised to the second power). The difference between a linear equation and a quadratic equation is thus the following:

A **LINEAR EQUATION** must have all of the variables that it is defined by raised to the first power.

A **QUADRATIC EQUATION** must have *only one* of the variables that is defined by raised to the second power.

In general, we define a quadratic equation as follows (using the variables x, y, though these may be exchanged with any other two variables):

Quadratic Equation of Two Variables

$$y = ax^2 + bx + c$$

where $a \neq 0$; b and c any **REAL** number

As you can see, the element that is being squared is the *input* variable; the *output* variable, y, as usual, is by itself on the left side of the equation and *must be* raised to the power of 1, otherwise, the equation would not correspond to a quadratic equation.

Please note that the equation contains a total of three terms on the right side of the equal symbol. The first term, called the *quadratic term*, contains the variable raised to the second power, and has a coefficient labeled a ; the second term is called the *linear term* (can you guess why?); it contains the same input variable, x, raised to the power of 1 ; and finally, the last term, called the *independent* term, consists of any **REAL** number only (in other words, it lacks a variable).

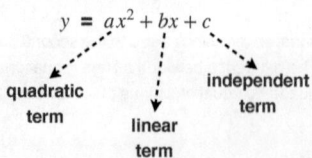

$$y = ax^2 + bx + c$$

quadratic term   linear term   independent term

As with linear equations, we can create solution pairs of the quadratic equation:

(input , output)   or   (x , y)

Furthermore, quadratic equations have a *domain* that is equal to **all REAL numbers**. This is because you can replace the input variable, x, with any **REAL** number whatsoever, and the answer (output) will always be defined. However, the range of a quadratic equation is *NEVER* equal to all **REAL** numbers, as we shall see later on in the chapter.

The first thing I would like you to become familiarized with is being able to detect a quadratic equation just by looking at it. Take a look at the following list of equations, and try to identify those that are quadratic equations.

a) $y = 2x - 4$

b) $y = x^2 + 3x + 1$

c) $y = -4x^2 + 5$

d) $y = 2x^2 - 10x$

e) $y = x^2$

f) $y = 4^2 x + 1$

g) $y = 2x + \pi^2$

h) $x^2 + y = 0$

i) $-2x^2 + x + y - 3 = 0$

j) $y = (7x)^2 - 4$

k) $-y = 3x^2 + 4x + 2$

l) $y = 2x^3 + 6x^2 - 8$

m) $y = (x + 3)(x - 4)$

n) $y = (x)(x)(2) - (3)(x)$

o) $y = -3(x + 1)^2 + 8$

p) $y = x^2 + 4\sqrt{x} + 1$

q) $y^2 = 6x^2 + 3x + 1$

r) $y = \dfrac{4x^2}{5} + \pi x - \dfrac{1}{3}$

s) $y = \dfrac{1}{x^2} + 2x + 3$

t) $\dfrac{1}{y} = 5x^2 - x$

Before you look at the answers below, try to classify them yourself: each equation on the list should either be labeled as a "quadratic equation" or as a "non-quadratic equation". As simple as that.

Ready to check your work?

a) $y = 2x - 4$

NO. This is not a quadratic equation. The x^2 element is missing...

b) $y = x^2 + 3x + 1$

YES. This is a quadratic equation. There is an x^2 element.

c) $y = -4x^2 + 5$

YES. This is a quadratic equation. There is an x^2 element.

d) $y = 2x^2 - 10x$

YES. This is a quadratic equation. There is an x^2 element.

e) $y = x^2$

YES. This is a quadratic equation. There is an x^2 element.

f) $y = 4^2 x + 1$

YES. This is a quadratic equation. There is an x^2 element.

g) $y = 2x + \pi^2$

NO. This is not a quadratic equation. The x^2 element is missing... Do not be confused with the π^2 element: it is NOT the input variable, **x**, squared.

h) $x^2 + y = 0$

YES. This is a quadratic equation. There is an x^2 element. Even though it is on the left side of the equation, it is easy to see that solving for **y** would give us the "familiar-looking" quadratic equation format we have been dealing with so far:

$$x^2 + y = 0$$

$$x^2 + y - x^2 = 0 - x^2$$

$$y = -x^2$$

Do you see it now (if you hadn't seen it before)?

i) $-2x^2 + x + y - 3 = 0$

YES. This is a quadratic equation. There is an x^2 element. Even though it is on the left side of the equation, along with the linear and the independent terms, it is easy to see that solving for **y** would give us the "familiar-looking" quadratic equation format we have been dealing with so far:

$$-2x^2 + x + y - 3 = 0$$

$$-2x^2 + x + y - 3 + 2x^2 - x + 3 = 0 + 2x^2 - x + 3$$

$$y = 2x^2 - x + 3$$

Do you now see why it is a quadratic equation after all (if you hadn't seen it before)?

j) $y = (7x)^2 - 4$

YES. This is a quadratic equation. There is an x^2 element, even though it is "hidden". To explicitly see it, expand the right side of the equation, as follows:

$$y = (7x)^2 - 4$$

$$y = (7x)(7x) - 4$$

$$y = 49x^2 - 4$$

k) $-y = 3x^2 + 4x + 2$

YES. This is a quadratic equation. On this example, even though the output variable, y, has a negative sign attached to it, the equation is still a quadratic. If you want to solve it for y (as opposed to for $-y$, as is in its current state), simply multiply both sides by **–1** and you will see the equation in the same format as the previous quadratic equation examples:

$$-y = 3x^2 + 4x + 2$$

$$(-1)(-y) = (3x^2 + 4x + 2)(-1)$$

$$y = -3x^2 - 4x - 2$$

l) $y = 2x^3 + 6x^2 - 8$

NO. This is not a quadratic equation. Even though there is an x^2 element present in the equation, note that there is an x being cubed (**a.k.a.** raised to the third power): this is therefore NOT a quadratic equation (it is a **cubic equation**, but that is another story).

m) $y = (x + 3)(x - 4)$

YES. This is a quadratic equation. There is an x^2 element, even though it is "hidden". To explicitly see it, expand the right side of the equation, as follows (remember foiling?):

$$y = (x + 3)(x - 4)$$

$$y = x^2 - 4x + 3x - 12$$

$$y = x^2 - x - 12$$

n) $y = (x)(x)(2) - (3)(x)$

YES. This is a quadratic equation. There is an x^2 element, even though it is not in exponent-notation form. To explicitly see it, simplify the terms on the right side using the principles and the rules that we covered earlier in the book:

$$y = (x)(x)(2) - (3)(x)$$

$$y = 2x^2 - 3x$$

o) $y = -3(x + 1)^2 + 8$

YES. This is a quadratic equation. There is an x^2 element, even though it is "hidden". To explicitly see it, expand the right side using the principles and the rules that we covered earlier in the book:

$$y = -3(x + 1)^2 + 8$$

$$y = -3(x + 1)(x + 1) + 8$$

$$y = -3(x^2 + x + x + 1) + 8$$

$$y = -3(x^2 + 2x + 1) + 8$$

$$y = -3x^2 - 6x - 3 + 8$$

$$y = -3x^2 - 6x + 5$$

p) $y = x^2 + 4\sqrt{x} + 1$

NO. This is not a quadratic equation. Even though there is an x^2 element in the equation, the element $\sqrt{x}$ goes against the rules that quadratic equations must adhere to. We could rewrite the same expression as follows, which may help you see why it is NOT a quadratic:

$$y = x^2 + 4\sqrt{x} + 1$$

$$y = x^2 + 4(x^{\frac{1}{2}}) + 1$$

Remember that the "**square root** of something" can be rewritten–using fraction power notation– as "something **raised to the half power**". Also remember that all variables must be raised to the power of **1** –except for one input variable which **must be** raised to the power of **2** –if the equation is going to qualify as a quadratic. Therefore, this is NOT a quadratic equation.

q) $y^2 = 6x^2 + 3x + 1$

NO. This is not a quadratic equation. The output variable **y** is raised to the second power (in other words, it is being squared). This is not allowed. If you tried to remove that power from the **y** variable, you would have to take the square root of both sides, as follows:

$$y^2 = 6x^2 + 3x + 1$$

$$\sqrt{y^2} = \sqrt{6x^2 + 3x + 1}$$

$$y = \pm\sqrt{6x^2 + 3x + 1}$$

At this point, we have changed the problem drastically. The entire right side is now being

"**square-rooted**" (in other words, it is being raised to the **half** power), including the **x** variables, and therefore, it does not adhere to the rules regarding the powers of variables in a quadratic equation. Even though there does appear to be an x^2 element, it is contained within the square root symbol. As if this weren't enough, there is also an **x** term that appears under the root symbol.

By the way, remember that we may **not** cancel the square root with the exponent that is equal to **2** that appears on the right side of the equation...

$$y = \pm\sqrt{6x^2 + 3x + 1}$$

...because the square root is acting on all the terms that are within it, and the terms are (as independent terms are by definition), adding or subtracting each other. We are thus unable to simplify this expression any further.

Also note that the symbol ± is used to indicate that both the **negative** version of the square root and the **positive** version of the square root are possible solutions of the equation. More on this later.

r) $y = \dfrac{4x^2}{5} + (\sqrt{5})x - \dfrac{1}{3}$

YES. This is a quadratic equation. There is an x^2 element, even though it is part of a fraction term. Since it is in the numerator, it is perfectly valid. As far as the middle (linear) term is concerned, the square root symbol is only acting on the **REAL** number **5**, which is defined and is equal to **2.23606...** , an **IRRATIONAL** number. The **x** variable is not part of that square root, so its power is **1**, which is allowed. This same equation may be written as follows, which may help you see clearly why it is, in fact, a quadratic equation:

$$y = \frac{4}{5}x^2 + (\sqrt{5})x - \frac{1}{3}$$

raised to the power of **1**

raised to the power of **2**

raised to the power of **1**

Note how the variables adhere to the rules mentioned earlier regarding quadratic equations.

s) $y = \dfrac{1}{x^2} + 2x + 3$

NO. This is not a quadratic equation. The x^2 element is on the denominator, and that would imply the following exponent if the term was rewritten with the variable on the numerator:

$$y = x^{-2} + 2x + 3$$

Thus, we are clearly missing the x^2 term.

t) $\dfrac{1}{y} = 5x^2 - x$

NO. This is not a quadratic equation. Even though the **x-terms** are not breaking any of the rules mentioned earlier regarding quadratic equations, take a close look at the **y** variable. If we expressed the y-term with the variable on the numerator, we would have the following:

$$y^{-1} = 5x^2 - x$$

Similar to what happened on the previous example, if a variable has a power other than **1** (or **2** in the case of the input variable, typically

the variable **x**), then it cannot qualify as a quadratic equation.

You should now have a very good idea of what to look for before calling an equation a **quadratic equation**.

Next, let's explore the typical behavior of a quadratic equation, and then compare it to the typical behavior of a linear equation.

I will use the simplest of all possible quadratic equations:

$$y = x^2$$

Let's plug in some values (**input**) and compute the result (**output**); we can then create solution pairs of the equation. Remember that at least from a purely mathematical point of view, we may use any **REAL** number whatsoever as the input value (in practice, however, remember that entire groups of numbers may be meaningless as far as considering them as input values is concerned; if we are computing the area of a square, for example, we may need to exclude negative numbers because that would imply a "negative distance", which may or may not be acceptable).

$$y = x^2$$

if **x = 0**

$$y = (0)^2 \ \text{--->} \ y = 0 \ \text{--->} \ (\,0\,,\,0\,)$$

if **x = 1**

$$y = (1)^2 \ \text{--->} \ y = 1 \ \text{--->} \ (\,1\,,\,1\,)$$

if **x = 2**

$$y = (2)^2 \ \text{--->} \ y = 4 \ \text{--->} \ (\,2\,,\,4\,)$$

if **x = 3**

$$y = (3)^2 \ \text{--->} \ y = 9 \ \text{--->} \ (\,3\,,\,9\,)$$

580

As you can see from the solution pairs found above, quadratic equations behave very differently than linear equations. Observe what happens if we were to use the same input values on the simplest of all linear equations: $y = x$:

$$y = x$$

if $x = 0$

$y = (0)$ ---> $y = 0$ ---> $(0,0)$

if $x = 1$

$y = (1)$ ---> $y = 1$ ---> $(1,1)$

if $x = 2$

$y = (2)$ ---> $y = 2$ ---> $(2,2)$

if $x = 3$

$y = (3)$ ---> $y = 3$ ---> $(3,3)$

Now, compare the input/output pairs, side by side:

$y = x^2$	$y = x$
$(0,0)$	$(0,0)$
$(1,1)$	$(1,1)$
$(2,4)$	$(2,2)$
$(3,9)$	$(3,3)$

So... both equations seem to start off the same– their first two solution pairs listed above are equal. But then things quickly begin to change. The quadratic equation takes off, its output value increasing dramatically compared to the linear equation's growth, the latter maintaining its same leisurely pace. Having reviewed everything there

is to know about lines in the previous two chapters, this should not surprise us. The linear equation has a defining characteristic called slope (and in the case of $y = x$ it is equal to **1**), which is a rate of change; as you should now know, the slope is constant for a given linear equation: it never changes, regardless of the input value used.

The quadratic, on the other hand, is behaving very differently. It seems to resemble a linear equation at first (at least in this particular example, and focusing on the input values of **0** and **1**). Its rate of change, or slope, would appear to be, like that of the linear equation we are simultaneously working with, equal to **1** as well, looking only at these two input values and the solution pairs that they yield. However, if we were to use the next two solution pairs on the list of $(1,1)$ and $(2,4)$, we would be forced to say that its slope value would appear to be equal to **3** (remember, slope, **m**, is equal to **rate of change**, which in turn is equal to $\frac{change\ in\ y}{change\ in\ x}$).

And if we used the next two solution pairs on the list, $(2,4)$ and $(3,9)$, we would be forced to say that its slope is equal to **5** ... So what's going on here? Well, the answer is that quadratic equations do not have a constant slope. It changes depending on the pair of values that you may happen to use to compute it; and because this is so, at least for the time being, its best to put on hold the idea of computing the **slope value** of a **quadratic equation**. For now, all we need to focus on is the idea that quadratics have a non-constant–or varying–slope.

To help you visually see the idea of a changing slope, observe the graph of both equations together. I will connect the coordinates found for the quadratic equation using straight lines, to keep things simple. But please note that there is no mathematical reason that supports this decision. In fact, it is one of the key ideas that we are going to explore shortly: that of how to connect points that belong to a quadratic equation when sketching its graph. I will use a solid line for the line of the linear equation, and a dashed line for the quadratic equation's sketch.

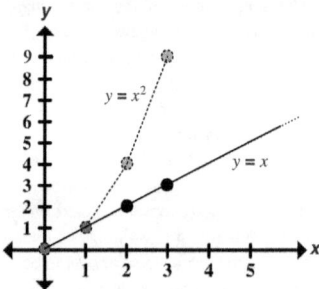

Please note that we are not yet looking at the behavior of the lines for negative input values: we will deal with that later.

See the changing slope of the quadratic equation's graph? If we imagined once again a cyclist riding the graph of the linear and the quadratic equations, it is easy to see that riding the quadratic line will be much more challenging, since its steepness rapidly increases (to the point where it would be humanly impossible to continue to "ride up" the line). Riding ant given linear equation's graph, on the other hand, yields no surprises: its slope remains the same–and thus it would be equally-challenging to ride it– throughout its entire span:

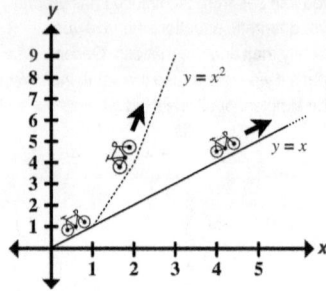

As to why quadratics behave this way, it is easy enough to see. Unlike a linear equation, where the input value is always multiplied by the value of the slope (which is constant for any given linear equation), quadratic equations **square** the input value; this is the source of the changing slope.

So far, we have been able to determine that a quadratic will behave very differently than a linear equation, because the input value will always be squared in one of the terms that form said quadratic (remember that in order for an equation to be called a quadratic, it must have a term with the input variable **squared**, with a coefficient **not equal** to 0).

The next item on the agenda is to explore how to connect the points (solution pairs) that we obtain from a given quadratic equation. It seems a bit arbitrary to decide to join the coordinates using "straight" lines... why not using arcs of a circle instead? Or any other curve for that matter? We do know that since quadratics accept all REAL numbers as input values, that the graph of a quadratic must be continuos; that much is clear.

Well, to find out how to correctly graph a quadratic, lets do the following: first, let's assume that a "straight" line is the correct choice. If it is to be so, it should be able to predict the outputs of any input that corresponds to any point that is part of the drawn line. You may recall that we did something very similar with linear equations: after drawing the graph of a given linear equation (having used only two solution pairs, or points), we "read" other points that were part of the line (identifying the input/output pairings) and tested them on the equation: we found, of course, that the graph did in fact match perfectly the equation's solution pairs set.

Let's try the method with the quadratic we graphed earlier:

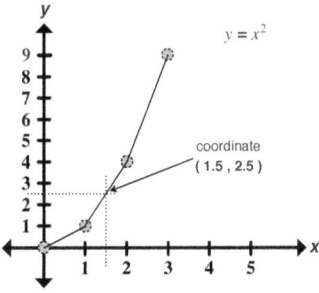

The graph predicts that if we assigned a value of **1.5** to x, that the output, or y value, should be equal to **2.5** ; so is this correct? Let's test this:

$$y = x^2$$

if $x = 1.5$

$y = (1.5)^2$ ---> $y = 2.25$ ---> (**1.5 , 2.25**)

Well, we must discard the idea that we can join the points with "straight" lines. The actual coordinate that should be part of the quadratic equation's graph is (**1.5 , 2.25**) , not (**1.5 , 2.5**) as the *now-we-know-incorrectly-drawn-graph* predicted. If you think about it, it makes perfect sense: we already established that quadratics do not have a constant slope value throughout its span, so why should it have a constant slope between two of its points that we randomly happened to compute?

Let's go back to the drawing board. I am going to plot the five points that we have found so far for this quadratic:

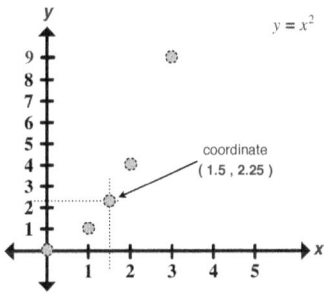

The correct point is lower than the one that the previous graph predicted... It seems like the slope is "continuously" changing, increasing, as we move to the right, not changing in leaps and bounds as the previous graph would imply.

To deal with this, it is perhaps easier to simply find many solution pairs and plot them on the graph to see the actual quadratic's graph shape.

To do this, we can crate a table with input/output values, write them in coordinate format, and plot the points on the **xy-axis**. Computers, of course, are also an option (there are many graphing applications that allow you to specify an equation and it creates many solution pairs, or coordinates, with which it then uses to plot the given equation's graph).

It will be enough for our purposes to find eight additional coordinates. Observe.

$$y = x^2$$

if $x = .25$
$y = (.25)^2$ ---> $y = .0625$ ---> (.25 , .0625)

if $x = .5$
$y = (.5)^2$ ---> $y = .25$ ---> (.5 , .25)

if $x = .75$
$y = (.75)^2$ ---> $y = .5625$ ---> (.75 , .5625)

if $x = 1.25$
$y = (1.25)^2$ ---> $y = 1.5625$ ---> (1.25 , 1.5625)

if $x = 1.75$
$y = (1.75)^2$ ---> $y = 3.0625$ ---> (1.75 , 3.0625)

if $x = 2.25$
$y = (2.25)^2$ ---> $y = 5.0625$ ---> (2.25 , 5.0625)

if $x = 2.5$
$y = (2.5)^2$ ---> $y = 6.25$ ---> (2.5 , 6.25)

if $x = 2.75$
$y = (2.75)^2$ ---> $y = 7.5625$ ---> (2.75 , 7.5625)

Please note that in practice, locating these points on the **x-y plane** can be problematic because the output values contain up to four decimals. If graphing by hand, we would need to choose a scale big enough to allow us to identify a height (output or **y** value) with a precision that is in the ten-thousandths range. As you may recall from

Chapter 13, this would force us to draw a very large **x-y plane**, to the point where it would quickly become impractical. However, our goal is to discover the "shape" of the given quadratic's graph, so we can use the same scale that we used before. I will approximate each point's location.

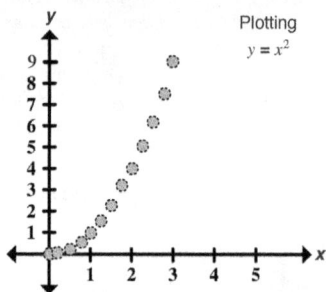

Plotting $y = x^2$

Plotting these thirteen points gives us a very clear indication (if partial) of the shape of this quadratic equation. We don't yet know the shape of this graph for negative number inputs, but we can now draw a **curve** passing through the plotted points, and know that it does correspond to this equation's shape:

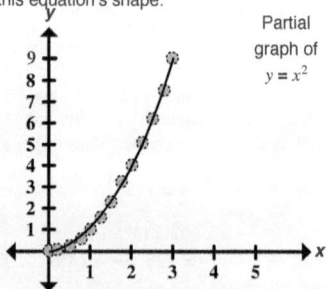

Partial graph of $y = x^2$

Now, if we remove the points and continue drawing the curve to the right, increasing the line's slope as we do so, we will have correctly drawn part of this equation's graph:

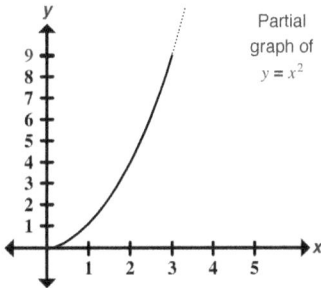

Partial
graph of
$y = x^2$

The graph of this quadratic equation reveals something important about quadratics in general: the output increases (or decreases, as the case may be, as we shall see shortly) in ever-increasing factors, because as the input values increase, the resulting squares of said inputs yield larger and larger outputs. Using the concept of the slope of a line, we can say that there is a section of a parabola's curve where the slope is increasing indefinitely (or decreasing indefinitely as well, as the case may be) as we move further to the right on the **x-axis**:

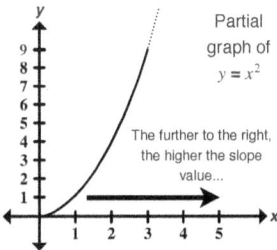

Partial
graph of
$y = x^2$

The further to the right, the higher the slope value...

So what happens if we use negative input numbers? Another very revealing aspect of quadratics and their graphs becomes apparent. Before I compute a few coordinates, can you guess what will happen? Observe.

$$y = x^2$$

if $x = -1$
$$y = (-1)^2 \quad \text{---> } \quad y = 1 \quad \text{---> } \quad (-1, 1)$$

if $x = -2$
$$y = (-2)^2 \quad \text{---> } \quad y = 4 \quad \text{---> } \quad (-2, 4)$$

if $x = -3$
$$y = (-3)^2 \quad \text{---> } \quad y = 9 \quad \text{---> } \quad (-3, 9)$$

Something very interesting happens when we use negative numbers as input values on this quadratic equation: the outputs are always positive! In retrospect, you should be able to quickly see why this happens; in fact, you could have predicted it. Because the input variable x is being squared in the equation we are working with, when you replace the x with a negative number, the computed (squared) value will always be positive, This stems from the basic mathematical principle that when you multiply two negative **REAL** numbers, the answer is always positive. For example:

$$y = (-1)^2$$
$$\text{---> } \quad y = (-1)(-1)$$
$$\text{---> } \quad y = +1$$

Solution pair: $(-1, 1)$

Before analyzing this further, observe the graph of this quadratic equation which includes these three additional points that we have found.

I am connecting the three dots using the same idea of a continuous changing slope, which, not surprisingly, yields a *symmetric* graph:

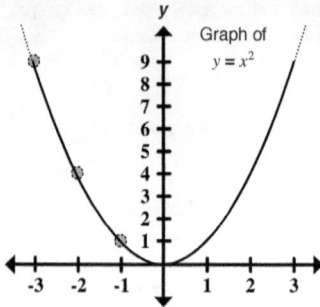

Graph of $y = x^2$

It is to be expected that for this particular quadratic equation, using a *positive*, **REAL**-numbered input will yield the same output as using its *negative counterpart*, as the following table illustrates:

positive x (input, output)	negative x (input, output)
(1 , 1)	(−1 , 1)
(2 , 4)	(−2 , 4)
(3 , 9)	(−3 , 9)

See how the output is the same for a given **REAL** number, regardless of whether it is positive or negative (with a notable exception, which we will explore in a moment)? It is thus easy to see that just as these six points are related, all other points that appear to the right of the *y-axis* have a "to-the-left-of-the-*y-axis*" counterpart.

Thus, the symmetry of the parabola. Observe the following diagram illustrating this idea:

Symmetry of a parabola

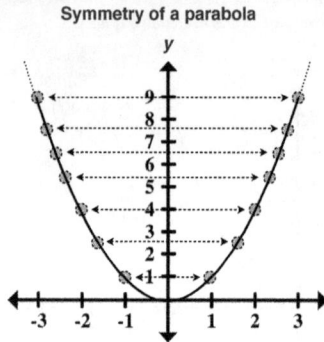

As you can see from the above graph, if you take any point that is on the curve (or line) of this specific quadratic, either to the left of the *y-axis* or to its right, there is a "partner" point that has the same *height* (output value) for an input value that differs only in its sign: if positive, negative; if negative, positive. Again, this applies to this specific quadratic equation: $y = x^2$. Other quadratics behave in a similar manner (as we shall see shortly), with the possibility that the input values of the symmetry pairs will differ not only in sign, but in their value as well. This will be very clear in the next quadratic we will work with shortly.

Once again, the source of this symmetry is the even-numbered exponent (the **2**) that is attached to the input variable (in other words, the fact that the input variable is being squared).

It is important to identify the *axis of symmetry*. The axis of symmetry refers to the type of symmetry that a graph may posses. Observe the following three examples of pairs of shapes:

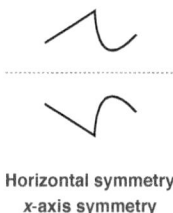

Horizontal symmetry
x-axis symmetry

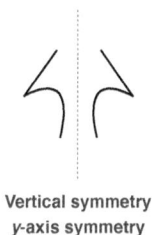

Vertical symmetry
y-axis symmetry

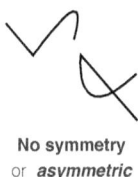

No symmetry
or *asymmetric*

These three pairs of objects clearly illustrate the idea of *x-axis* symmetry, *y-axis* symmetry, and no symmetry (or asymmetry). All quadratic equations yield graphs that correspond to parabolas, and they all present vertical symmetry (or *y-axis* symmetry). The reason we must clarify this is because a parabola may be drawn so that it has horizontal symmetry (simply rotate the parabola drawn earlier 90 degrees to the left or to the right, like a sideways "U" ---> "c" ; this latter rotated version now has *x-axis* symmetry); however, *quadratic equations* never produce such a curve. Their parabola graphs are always *vertically-symmetric*, "U" shaped, pointing either "upwards" as the graph of $y = x^2$, or downwards, as an upcoming example will illustrate.

It is because of this fact that when describing the graph of a parabola, we can talk about "the axis of symmetry". Yes, we know it is going to be vertical, but its location may differ from equation to equation, as we shall see shortly. We can also talk about a very special point that is part of the solution set of all quadratic equations: the *vertex*. Looking at the graph of $y = x^2$, we can see that all points that form the curve of the graph of this particular quadratic equation come in pairs (these pairs will have the same height or output value; the input value of each pair will differ only in the sign: one is positive, while the other is negative). Except for one very important point. Can you see which one it is? Hint: it appears on the very first table we created to determine input/output pairs of values for this equation.

Did you find it? It is the point **(0 , 0)**. So what is so special about this point (*for this particular quadratic*)? Well, it is special because it is the only point that does not have a "partner". Never mind the fact that **0** does not have a sign *per se*, as we established at the very beginning of the book (in other words, **0** isn't positive or negative; it lacks this *characteristic* that the rest of the **REAL** numbers do posses); what's important here is that no other point that is part of the curve–which means that no other solution pair– has the same output as that point, given a different input value...

You may think that the reason for this is that because the input value of this vertex point happens to be **0** , and since **0** can't be positive or negative, a "partner" point is therefore non-existent. But this misconception is, pardon the redundancy, incorrect. The reason it is "partnerless" from a symmetry point of view is that all quadratic equations have a solution pair that has a *unique output*: only one *input* will ever produce it. And in the case of *this particular quadratic*, its unique output value is given by the input value of **0** .

Observe the vertex explicitly indicated in the graph below.

in a positive output, since any **REAL** number squared (which means "times itself") will always be greater than **0** since regardless of how close to zero a number may be, when you square it, the answer will always be greater than **0** (only **0** times **0** is equal to **0**), regardless of whether you consider negative input numbers or positive input numbers. It is only when you use **0** as the input value that the output is equal to **0** ; therefore, **0** is the lowest possible output that this equation is able to provide. And what I mean by "lowest" is the point that belongs to the equation's curve that has a **y** value that is closest to negative infinity, as discussed previously in earlier chapters of the book (and therefore, the "highest" point, if such a point exists for a given curve, will be the point that has a **y** value that is closes to positive infinity). The following graph illustrates this aspect of the "lowest" and "highest" points on a given graph.

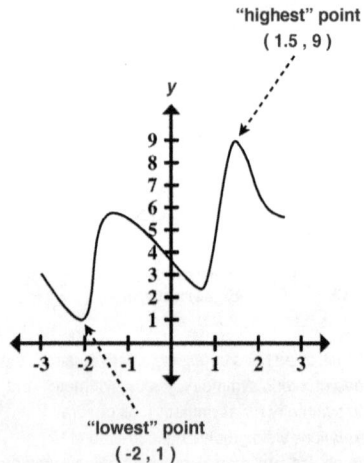

Vertex

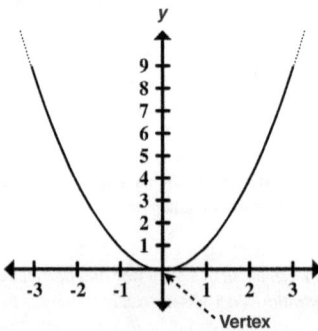

"highest" point
(1.5 , 9)

"lowest" point
(-2 , 1)

This equation can thus be said to have a *lowest* possible output value: **0** . Think about it. Any input value that you use other than **0** will result

As you can see on the graph above, the curve has a lowest point and a highest point: its lowest point corresponds to the **y** value of **1** , and its highest point corresponds to the **y** value of **9** . No other point on this curve achieves that lowest or highest output value.

Okay. Back to the vertex and quadratic equations. It is precisely along the **vertex** point that the **vertical line of symmetry** will always be found for any given quadratic equation. Always. And as we already reviewed in previous chapters, a vertical line's equation is of the form

$$x = d$$

where **d** = any **REAL** number

...and for this particular quadratic equation, since the symmetry line is vertical and must pass through the point **(0 , 0)** , we can say unequivocally that the axis of symmetry is given by the linear equation

$$x = 0$$

...as the following graph illustrates:

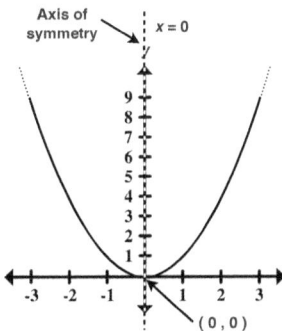

Is a curve symmetric?

A practical way of thinking about symmetry is to imagine placing a mirror along the axis of symmetry: if the image reflected on the mirror (reflecting side) matches the actual shape that is on the opposite side of the reflecting part of the mirror, then the object is said to be symmetric along the axis in which the mirror has been placed.

The curve is symmetric along the dotted vertical axis, because the reflected image on the mirror matches the actual part of the curve that is (in this case) to the right of the mirror (on the opposite side of the reflecting side).

The *range* of this quadratic equation

Remember how I said earlier that quadratic equations have a domain that spans from negative infinity to positive infinity (in other words, all **REAL** numbers), but that the range never does? Well, now you can see why this happens. In the case of this specific quadratic equation,

$y = x^2$, the output values will never be negative. The reason is quite simple... we can only use the following **REAL** number types as input values:

- **Positive numbers**

- **Negative numbers**

- **Zero**

...and given this equation, the outcome (output) of replacing *x* with any one of these three number types can only be:

- **Positive**

or

- **Zero**

The only way to obtain **0** as the output is to use **0** as the input value. As far as the positive output values are concerned, it is easy to see that we can obtain any positive **REAL** output value imaginable... positive infinity being its upper limit (remember that positive infinity is an *idea*, not an actual number: you can never "reach" positive infinity: it simply represents the idea of ever-increasing positive **REAL** numbers). This is why when we draw the graph of this equation, we can use a continuous curve (strictly speaking, a line: remember that a line does not necessarily imply "straightness"). Mathematically speaking, this means that the *range* of this quadratic equation is from **0** (including it) to positive infinity. Remember that range specifies the output values that a given equation is able to yield (or, to state this differently, the range is the set of all *outputs* of a given equation).

And so we have the following key characteristics of this specific quadratic equation we have been working with (a profile list, if you will):

Quadratic equation

$$y = x^2$$

Vertex @ (0 , 0)
Graph type: **parabola**
Graph orientation: ***points upwards***
Lowest output value: **0**
Highest output value: **positive infinity**
Axis of symmetry @ ***x = 0***

Let's explore the behavior of another quadratic equation. It will illustrate the common characteristics that all quadratic equations share, as well as the way in which they may differ.

$$y = -x^2 + 4x - 5$$

It my seem very complicated, but remember that all we need to do is replace *x* with several input values, find the output values they produce, create the coordinates and plot them on the ***x-y plane***. The input values were chosen carefully; at this point in the review, however, I am not going to explain the criteria I used to select them; I will review that later on.

$$y = -x^2 + 4x - 5$$

if *x* = −1
$$y = -(-1)^2 + 4(-1) - 5$$
$$y = -10 \dashrightarrow \textbf{(−1 , −10)}$$

if *x* = 0
$$y = -(0)^2 + 4(0) - 5$$
$$y = -5 \dashrightarrow \textbf{(0 , −5)}$$

if *x* = 1
$$y = -(1)^2 + 4(1) - 5$$
$$y = -2 \dashrightarrow \textbf{(1 , −2)}$$

if $x = 2$

$$y = -(2)^2 + 4(2) - 5$$
$$y = -1 \;\text{--->}\; (\,2\,,-1\,)$$

if $x = 3$

$$y = -(3)^2 + 4(3) - 5$$
$$y = -2 \;\text{--->}\; (\,3\,,-2\,)$$

if $x = 4$

$$y = -(4)^2 + 4(4) - 5$$
$$y = -5 \;\text{--->}\; (\,4\,,-5\,)$$

if $x = 5$

$$y = -(5)^2 + 4(5) - 5$$
$$y = -10 \;\text{--->}\; (\,5\,,-10\,)$$

Please note that I am not showing all the steps necessary to solve each equation for a given input value; the assumption is that you are able to apply the rules and principles that we have covered so far in the book to find the output values by yourself. Any issues, concerns, doubts, and/or mishaps, simply go back to those chapters of the book where these skills were reviewed. Just remember to follow the order of operations, a key part of finding the correct output values.

After plugging in seven input values, we have as many coordinates (points) that are part of this quadratic's graph. To see them clearly, they are:

$(-1,-10)$
$(0,-5)$
$(1,-2)$
$(2,-1)$
$(3,-2)$
$(4,-5)$
$(5,-10)$

Just by looking at this list of coordinates you should be able to see the symmetry and identify the vertex. If you don't, the graph will make it obvious. Observe.

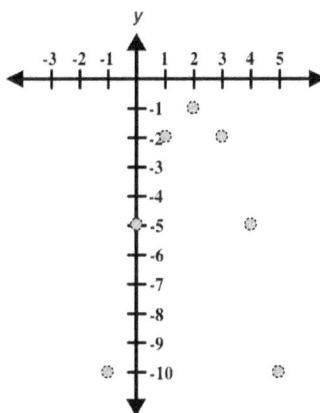

Can you see the parabola that can be perfectly drawn passing through the plotted points? Can you see where the vertex of this particular quadratic equation is located? What about the axis of symmetry? Could you specify its equation?

There are two additional aspects of quadratics that demand our attention: the location or locations, *if any*, where the quadratic's graph crosses the **x-axis**, and the location (singular, and all quadratics have it) where the quadratic's graph crosses the **y** axis. If you look back to the previous equation's graph ($y = x^2$) you should be able to see that it crosses the **x-axis** only once, at the point (**0 , 0**), and that it crosses the **y** axis at the same point, (**0 , 0**). This quadratic, however, never crosses the *x-axis*; this is a critical piece of knowledge about a given quadratic's graph that tells us a lot about the behavior and the type of equation that it belongs to, as we shall see later on. Its *y-axis* crossing occurs at (0 , –5). As stated, all quadratics have a *y-axis* crossing. Every single one of them.

Were you able to specify the other key components of this quadratic we are currently working with? Let's explore its graph, and then I'll show you the correct answers so you can check your work.

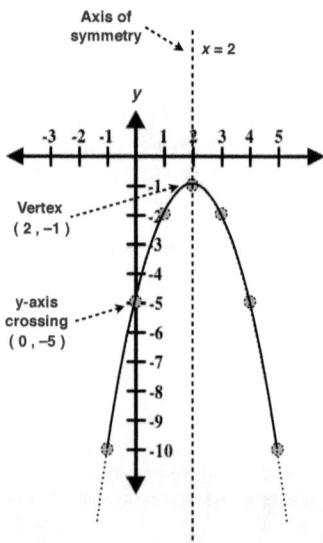

Axis of symmetry

$x = 2$

y

-3 -2 -1   1   2   3   4   5

Vertex
(2 , –1)

y-axis
crossing
(0 , –5)

-1
-2
-3
-4
-5
-6
-7
-8
-9
-10

The two graphs that we have now reviewed allow us to explore the common characteristics that every graph of a quadratic equation must share, *without exception*.

1) All graphs of quadratic equations are *parabolas*.

2) All parabolas of quadratic equations will either "*point up*" (as in the letter "U") or "*point down*" (as in "∩"), but never sideways; this, in turn, means that all graphs of quadratic equations have the remaining characteristics on this list.

3) A vertex, which can be expressed, in general, as the coordinate (*d* , *e*).

4) A *vertical* axis of symmetry; the linear equation that defines it is of the form $x = d$. The *d* value corresponds to the same *d* value of the *x-coordinate* from the vertex (see previous point).

5) A *y-intercept*. All quadratic equations have a single *y-intercept* point; no more, no less.

6) One (and *one* only) of the following:

    a) Two *x-intercepts*
    b) One *x-intercept*
    c) Zero *x-intercepts*

Reading quadratic equations

It is important to know that it is possible to specify all of the characteristics of a given quadratic equation without having to actually graph it. Just like we "read" linear equations (in their various forms: the point-slope form and the slope-intercept from) and reviewed how to be able to specify the slope and either the *y-intercept* or a general point from a given linear equation, so too can quadratic equations be "read". In order to do this, we are going to review three basic forms of quadratic equations.

Quadratic Equation in Standard Form

$$y = ax^2 + bx + c$$
where $a \neq 0$

Quadratic Equation in Vertex Form

$$y = a(x - h)^2 + k$$
where $a \neq 0$

Quadratic Equation in Intercept Form

$$y = a(x - m)(x - n)$$
where $a \neq 0$

It is an interesting exercise to expand and simplify the second and third versions listed above; you will find that they end up having the same form as the standard form (although with different letters representing the constant values; in other words, the coefficients of the **x-squared** term and of the linear term, and the independent term itself.

I will review how to be able to determine each of the defining characteristics given each form mentioned above.

Determining the parabola's orientation

In order to know if the quadratic equation's graph (parabola) points up or down, all you have to do is consider its **a** value, regardless of the form it may be in. Then, use your common sense. Allow me to explain.

Let's consider once more the simples of quadratics, $y = x^2$. I hope you can specify this

quadratic's a value. If not, I'll give you a hint: it is the coefficient of its **x-squared** term. To see why, compare it to the standard form:

$$y = x^2$$
$$y = ax^2 + bx + c$$

The dotted line clearly shows you where to look for it. So, as we have seen many times throughout the book, if you don't see the number that is multiplying the variable, its because it is equal to **1**, and mathematicians agreed that it is not necessary to show it. So, for this equation, **a** = 1. As a bonus, do you know its **b** and **c** values? I hope you said **0** for both. Whenever the term is missing, it means is coefficient is **0** (as is the case of the **bx** term above); and as to the independent term, if it is missing it means its value, **c**, is equal to **0** in that particular equation. Observe:

$$y = x^2$$
$$y = 1x^2 + 0x + 0$$
$$y = ax^2 + bx + c$$

See how the equation $y = 1x^2 + 0x + 0$ simplifies to $y = x^2$? Being able to specify the **a**, **b**, and **c** values of a given quadratic equation (if it is in standard form) is a very important skill when trying to specify its key characteristics.

Okay, so back to the equation $y = x^2$. We know its **a** value is equal to **1**, as in "**positive one**". So why does this a value tell us if the parabola points up or down? Simple. If the **a** value is positive, then it means that the greater the input values we use, the greater the output will be. And by greater, I mean values that are more and more to the right of **0** in the number line:

The greater the number, the more to the right of 0 it is in the number line...

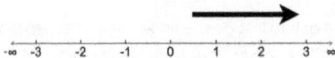

But why does this happen? Well, think about it. The number that you choose as the input value will replace the **x** in the equation. And since the **x** is being squared, this means you will end up multiplying the input number times itself. It therefore follows that the greater **x** is (in other words, the more we approach positive infinity), the greater the output **y** will be. Observe some examples that perfectly illustrate this effect.

As the input grows...

$x = 0$ --> $y = x^2$ --> $y = (0)^2$ --> $y = 0$

$x = 5$ --> $y = x^2$ --> $y = (5)^2$ --> $y = 25$

$x = 15$ --> $y = x^2$ --> $y = (15)^2$ --> $y = 225$

...the output grows as well.

And something interesting happens when we analyze the behavior of this quadratic equation when we use input values that are progressively more and more to the left of **0** on the number line (in other words, negative input values, smaller and smaller, approaching negative

infinity). Remember what happens when you square a negative number? The resulting computation is a positive number, correct? And this means that after the dust settles, using progressively smaller input values results in obtaining progressively greater output values as well! Observe:

As the input decreases...

$x = 0$ --> $y = x^2$ --> $y = (0)^2$ --> $y = 0$

$x = -5$ --> $y = x^2$ --> $y = (-5)^2$ --> $y = 25$

$x = -15$ --> $y = x^2$ --> $y = (-15)^2$ --> $y = 225$

...the output increases!

This, by the way, is the true source of the symmetry of the graph of all quadratic equations.

So what happens when **a** is negative, as in the following equation where its **a** value is specifically equal to **−1** ?

$$y = -x^2$$

Just keep in mind the order of operations as we compute this equation's outputs... When we plug in a positive number, observe the effect that the negative **a** value has on the equation's output.

As the input grows...

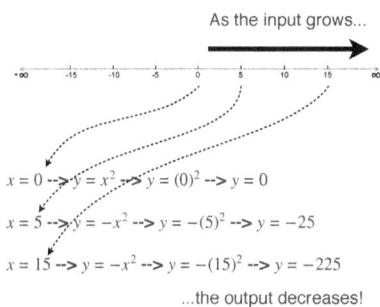

$x = 0 \dashrightarrow y = x^2 \dashrightarrow y = (0)^2 \dashrightarrow y = 0$

$x = 5 \dashrightarrow y = -x^2 \dashrightarrow y = -(5)^2 \dashrightarrow y = -25$

$x = 15 \dashrightarrow y = -x^2 \dashrightarrow y = -(15)^2 \dashrightarrow y = -225$

...the output decreases!

Why does this happen? The answer is simple. As the input value increases (moves farther to the right of **0** on the number line), its square value will, in fact be positive and ever greater; however, the **a** value is a **factor** of that result, meaning it will multiply it **after** computing the square. So if the a value is negative, then this ever-greater square turns automatically into a negative number, progressively farther and farther to the left of **0** towards negative infinity. The same thing will happen as we use negative numbers that are progressively more to the left of **0** on the number line. Observe.

As the input decreases...

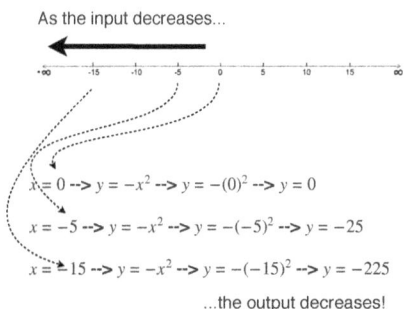

$x = 0 \dashrightarrow y = -x^2 \dashrightarrow y = -(0)^2 \dashrightarrow y = 0$

$x = -5 \dashrightarrow y = -x^2 \dashrightarrow y = -(-5)^2 \dashrightarrow y = -25$

$x = -15 \dashrightarrow y = -x^2 \dashrightarrow y = -(-15)^2 \dashrightarrow y = -225$

...the output decreases!

The inputs are squared, yes, and so the square will be progressively greater in value; however, the factor **a** will turn the positive square result into a negative number. Therefore, the output becomes progressively smaller, found progressively more to the left of **0** on the number line.

This idea of analyzing an equation in terms of what happens to its output as its input is progressively greater or smaller (progressively more to the right or to the left of **0** in the number line, respectively) is important. If you are still unsure what this analysis is all about, the following exercise will help you understand it better.

Let's look at the graphs of these two quadratic equations and perform the same analysis; the graph provides a powerful visual representation of this powerful mathematical concept.

$$y = x^2$$

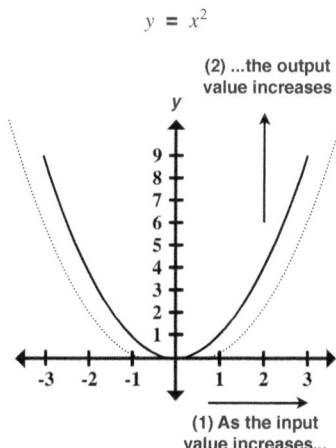

(2) ...the output value increases

(1) As the input value increases...

$$y = x^2$$

(4) ...the output value increases

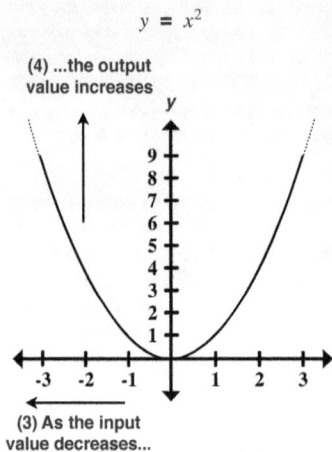

(3) As the input value decreases...

$$y = -x^2$$

(3) As the input value decreases...

(4) ...the output value decreases

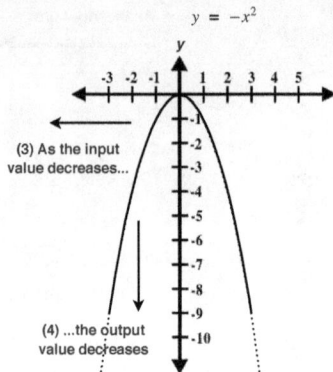

The two graphs above clearly allows us to see that as the input value increases, the output value increases as well. And as we analyzed previously, this is the case of all quadratics that have a positive **a** value.

$$y = -x^2$$

(1) As the input value increases...

(2) ...the output value decreases

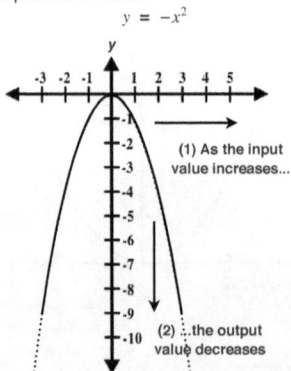

The two graphs above clearly allows us to see that as the input value increases, the output value decreases. And as we analyzed previously, this is the case of all quadratics that have a negative **a** value.

It is therefore very easy to know not only the shape of the graph of a quadratic equation (it s always a parabola), but also to determine if it points up or down simply by analyzing the equation itself, without having to plot any points at all.

Try to determine by yourself the orientation of the parabolas of the following quadratic equations just by looking at the given equations. If they are not in one of the forms defined previously, it is necessary to rewrite them using the rules and principles that we have reviewed throughout the book so that they are in one of the forms (standard, vertex, or intercept form).

a) $y = 3x^2 - 4x$

b) $y = x^2 + 3x + 1$

c) $y = -4x^2 + 5$

d) $y = 2x^2 - 10x$

e) $y = -.5x^2$

f) $y = 4x^2 - 8$

g) $y = (x - 4)^2 - 3$

h) $y = -5(x - 4)^2 - 3$

i) $-2x^2 + x + y - 3 = 0$

j) $y = (-7x)^2 - 4$

k) $-y = 3x^2 + 4x + 2$

l) $y = -(x + 1)^2 + 5$

m) $y = (x + 3)(x - 4)$

n) $y = -2(x + 3)^2 + 1$

o) $y = -4(x + 8)^2$

p) $y = (-2x + 1)^2 + 5$

q) $y = -(-4x - 3)^2 - 8$

Most of these have an *a* value that is very easy to determine. However, some are quite challenging, because they can be very deceiving. You have to carefully analyze whether each of these matches the general form or not (save for the value of the coefficients, or constants, as defined in the corresponding form): if not, you must rewrite them before being able to extract the *a* value.

Ready for the answers? Observe.

a) $y = 3x^2 - 4x$

The quadratic is in standard form. Since *a* = **3**, its graph is a parabola that points up (**3** is positive).

b) $y = x^2 + 3x + 1$

The quadratic is in standard form. Since *a* = **1**, its graph is a parabola that points up (**1** is positive).

c) $y = -4x^2 + 5$

The quadratic is in standard form. Since *a* = **–4**, its graph is a parabola that points down (**–4** is negative).

d) $y = 2x^2 - 10x$

The quadratic is in standard form. Since *a* = **2**, its graph is a parabola that points up (**2** is positive).

e) $y = -.5x^2$

The quadratic is in standard form. Since *a* = **–.5**, its graph is a parabola that points down (**–.5** is negative).

f) $y = 4x^2 - 8$

The quadratic is in standard form. Since *a* = **4**, its graph is a parabola that points up (**4** is positive).

g) $y = (x - 4)^2 - 3$

The quadratic is in vertex form. Since *a* = **1**, its graph is a parabola that points up (**1** is positive).

h) $y = -5(x - 4)^2 - 3$

The quadratic is in vertex form. Since *a* = **–5**, its graph is a parabola that points down (**–5** is negative).

i) $-2x^2 + x + y - 3 = 0$

The quadratic is not in any of the forms defined earlier. We need to solve for **y** before we can specify its **a** value:

$$-2x^2 + x + y - 3 = 0$$
$$-2x^2 + x + y - 3 + 2x^2 - x + 3 = 0 + 2x^2 - x + 3$$
$$y = 2x^2 - x + 3$$

Now that it's in standard form, we can state that **a = 2** ; therefore, its graph is a parabola that points up (**2** is positive).

j) $y = (-7x)^2 - 4$

The quadratic is not in standard form (the term that contains the **input variable squared** must be simplified).

$$y = (-7x)^2 - 4$$
$$y = (-7x)(-7x) - 4$$
$$y = 49x^2 - 4$$

Now that it's in standard form, we can state that since **a = 49** , its graph is a parabola that points up (**49** is positive).

k) $-y = 3x^2 + 4x + 2$

The quadratic is not in standard form (notice the negative sign attached to the output variable).

$$-y = 3x^2 + 4x + 2$$
$$(-1)(-y) = (3x^2 + 4x + 2)(-1)$$
$$y = -3x^2 - 4x - 2$$

Now that it's in standard form, we can state that since **a = -3** , its graph is a parabola that points down (**-3** is negative).

l) $y = -(x + 1)^2 + 5$

The quadratic is in vertex form. Since **a = -1** , its graph is a parabola that points down (**-1** is negative).

m) $y = (x + 3)(x - 4)$

The quadratic is in intercept form. Since **a = 1** , its graph is a parabola that points up (**1** is positive). Remember that the a value in the intercept form is the number (factor) that multiplies the parentheses set; in this case, since there isn't a number explicitly written, it means that it's equal to **1** :

$$y = a(x - m)(x - n)$$
$$y = 1(x + 3)(x - 4)$$

n) $y = -2(x + 3)^2 + 1$

The quadratic is in vertex form. Since **a = -2** , its graph is a parabola that points down (**-2** is negative).

o) $y = -4(x + 8)^2$

The quadratic is in vertex form. Since **a = -4** , its graph is a parabola that points down (**-4** is negative).

p) $y = (-2x + 1)^2 + 5$

This can be a very deceiving equation: it seems to be in vertex form, but a closer inspection reveals a negative sign that is never part of the vertex form: the input variable, **x** , has a negative sign "attached" to it. Furthermore, the x variable inside the parentheses contains a coefficient that is not equal to **1** , so we must take it out of the parentheses (factor it out). To get rid of those two elements, they must be *factored out of the inside of the parentheses set*, as follows:

$$y = (-2x + 1)^2 + 5$$

$$y = [(-2)(x - \frac{1}{2})]^2 + 5$$

if you don't remember how to factor out an element (or elements) from a collection of terms, refer back to **Chapter 11** where we reviewed this process in detail.

Please note that when we factor out the negative sign and the **2** from the terms inside the parentheses set, we must still consider them to be within the "grips" or scope of the exponent, thus the bracket set that encompasses both the factored out elements, and the set of parentheses that contains the two terms...

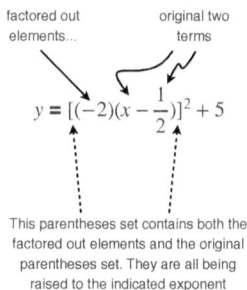

factored out   original two
elements...   terms

$$y = [(-2)(x - \frac{1}{2})]^2 + 5$$

This parentheses set contains both the factored out elements and the original parentheses set. They are all being raised to the indicated exponent

Now that the negative sign and the *x*-term's coefficient have been **factored out** of the original two terms that were inside the original parentheses set, and taking into account the fact that the resultant **−2** is itself contained inside the parentheses set that is being squared, and thus

is also being squared, we may continue simplifying as follows:

$$y = [(-2)(x - \frac{1}{2})]^2 + 5$$

$$y = (-2)^2(x - \frac{1}{2})^2 + 5$$

Why are we allowed to do this? Because remember that if you have **elements** inside a set of parentheses that are being raised to a given power (exponent), and only if there is a continuous multiplication link between those elements, you may expand the expression by assigning each element it own given exponent. This rule can be illustrated as follows:

$$(xyz)^n = x^n y^n z^n$$

This includes "elements" that may consist of sets of parentheses regardless of what they may contain inside, as the following examples illustrate:

$$(3(y - 4)(w + z))^n = 3^n(y - 4)^n(w + z)^n$$

$$((x + 1)(y - 2))^3 = (x + 1)^3(y - 2)^3$$

$$(-5w(x - 2))^{\frac{1}{3}} = (-5)^{\frac{1}{3}}w^{\frac{1}{3}}(x - 2)^{\frac{1}{3}}$$

We can go back to the problem we were in the process of solving.

Once we have the equation with the input variable inside the parentheses set in positive form and with a coefficient of **1**, we can finish simplifying the equation in order to extract the correct *a* value:

$$y = (-2)^2(x - \frac{1}{2})^2 + 5$$

$$y = 4(x - \frac{1}{2})^2 + 5$$

The process is now complete. The quadratic is in vertex form. Since **a = 4**, its graph is a parabola that points up (**4** is positive).

Observe the correct **h** value of this quadratic, based on the general vertex form that we reviewed earlier:

$$y = a(x - h)^2 + k$$

It is easy to see that $h = \frac{1}{2}$. But if you refer back to the original equation, $y = (-2x + 1)^2 + 5$, you would have incorrectly **extracted** an **h** value of **−1**. See why it's important to express the quadratic in true vertex form before extracting any of the desired values (**a** , **h** or **k**)? If this is not done, you will incorrectly define the quadratic equation's defining values.

We encountered a similar situation when we reviewed linear equations: always make sure that the equation you are working with is truly in whatever form you need it to be, with the input variable and the output variable free of any extra elements (like a negative sign, a coefficient other than **1** if this is what is required, etc...), before attempting to "read" or "extract" its key values.

q) $y^2 = -(-4x - 3)^2 - 8$

Once again, as in the previous example, we must rewrite the equation so that it is in true vertex form. If you look closely at the inside of the set of parentheses, you will notice that the input variable, **x**, has a negative sign and a coefficient other than **1** (or, we can simply say that its

coefficient is **−4** instead of **1**). To fix this, we need to factor out the **−4** . Observe the process.

$$y = -(-4x - 3)^2 - 8$$

$$y = -[(-4)(x + \frac{3}{4})]^2 - 8$$

$$y = -(-4)^2(x + \frac{3}{4})^2 - 8$$

$$y = -(16)(x + \frac{3}{4})^2 - 8$$

$$y = -16(x + \frac{3}{4})^2 - 8$$

Now that the quadratic is in true vertex form we can extract its **a** value: **−16** . Since **a** is negative, we know that this quadratic's graph is, of course, a parabola, and that it points down.

Graphing quadratic equations

It's important to be able to graph a quadratic equation regardless of the form that is used to express it. As stated earlier, a quadratic equation can be expressed in any one of three forms (in all forms, recall that **a ≠ 0**):

Quadratic Equation in Standard Form

$$y = ax^2 + bx + c$$

Quadratic Equation in Vertex Form

$$y = a(x - h)^2 + k$$

Quadratic Equation in Intercept Form

$$y = a(x - m)(x - n)$$

If it is not in one of these three forms, you can always rewrite it so that it is in one of the three forms defined above. It is easiest to convert any quadratic equation—regardless of the form it is originally expressed as—into the standard form; the other two forms will vary in difficulty, although generally speaking, the next easiest form is the vertex form, while the intercept form can be the most challenging (again, generally speaking; some quadratic equations can be expressed in intercept form in the blink of an eye).

I am sure it will not come as a surprise to you that each of the three forms is naturally suited to provide you with certain key characteristics of the quadratic equation—and its graph; we encountered something very similar to this when we reviewed linear equations; remember the point-slope form and the slope-intercept form, and what each form explicitly communicates about the equation and/or its graph?

So let's now explore this in detail.

Quadratic Equation in Vertex Form

$$y = a(x - h)^2 + k$$

The quadratic equation in vertex form tells us, as we already stated, whether its graph, which is a parabola, points up or down. In other words, in terms of its input and corresponding output pairs, it tells us whether the output increases (a parabola that points up) or decreases (a parabola that points down) the farther away we move from its vertex (or from the axis of symmetry). This is determined, as we stated previously, by the sign of a: if it is positive, the parabola points up, if it is negative, the parabola points down.

It should not surprise you to know, based on the name of this form, that it also tells us the exact location of the parabola's *vertex*; in other words,

it gives us the lowest point on the graph (if the parabola points up) or the highest point on the graph (if the parabola points down). The vertex is located at (h, k). The following diagram summarizes this for you.

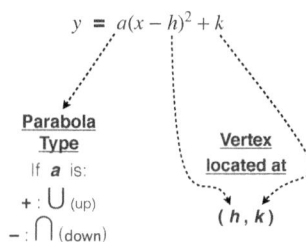

$$y = a(x - h)^2 + k$$

Parabola Type
If a is:
$+ :$ ∪ (up)
$- :$ ∩ (down)

Vertex located at
(h, k)

Do you remember when we reviewed how to extract the point from the *point-slope formula* of a linear equation that is written in such form? Well, similarly, extracting the vertex from a quadratic that is in vertex form requires us to be extra careful as well. The problem is how the vertex form, in its general form, expects to have the equation written in the first place:

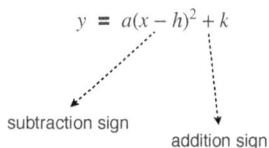

$$y = a(x - h)^2 + k$$

subtraction sign
addition sign

In other words, before extracting the h and k values from a quadratic equation that is written in vertex form, make sure that you consider whether the h and the k values are being subtracted and added respectively. If they are, you can take them at face value (examples to come). If they aren't, you must take that into account, because it will have an impact on the sign of the h and the k value, as we will see shortly.

For example, consider the following quadratic equation:

$$y = 5(x-3)^2 + 4$$

a is + : $\cup$
$a = 5$

Vertex located at

$(\,3\,,\,4\,)$

Notice that the a value is positive, so we know that the parabola will point up. As to the vertex, because the **3** is being subtracted, as the general form expects, we know that its h value is positive, equal to **3** . Moving on to the k value, because the **4** is being added, as the general form expects, we know that the k value is positive, and equal to **4** . Therefore, the x and y values of the coordinate that specifies the vertex of this equation is **3** and **4** respectively, which is why the vertex coordinate is $(\,3\,,\,4\,)$.

Now, observe this next example.

$$y = 2(x+6)^2 - 4$$

If you think that the vertex is located at $(\,6\,,\,4\,)$, then I'm sorry to burst your bubble, but that's incorrect.

To understand why it's incorrect, simply compare the general form and this quadratic:

$$y = a(x-h)^2 + k$$

$$y = 2(x+6)^2 - 4$$

On the general form, the expectation is to have the h value subtracting, while the k value adding; however, that's not what we are seeing:

$$y = a(x-h)^2 + k$$

$$y = 2(x+6)^2 - 4$$

On this specific equation, the **6** is being ***added***, while the **4** is being **subtracted**.

So what can we do to the equation so that we can extract the correct values that serve to specify the coordinate of the vertex?

Well, we can change the expression so that we see a subtracting h value and an adding k value by using the following property of signs:

$$a + b = a - (-b)$$

and

$$a - b = a + (-b)$$

The first property allows us to change an expression so that the second term appears subtracting instead of adding, while the second property allows us to change an expression so that the second term appears adding instead of subtracting.

Observe how to use these two properties on the original equation:

$$y = 2(x+6)^2 - 4$$

$$y = 2(x-(-6))^2 + (-4)$$

Now we are able to extract the correct values for the coordinate of the vertex.

$$y = a(x - h)^2 + k$$

$$y = 2(x - (-6))^2 + (-4)$$

a is + : ∪

$a = 2$

Vertex
located at

$(-6, -4)$

You could memorize the fact that whenever a quadratic equation is in vertex form, if the h value appears adding then it needs to be extracted as a negative number, and that if the k value appears subtracting, then it needs to be extracted as a negative number as well. However, I do not recommend memorizing hundreds of little details like this one, because pretty soon you are going insane trying to cope with all of them. So rather than memorizing details such as these, I recommend you master the art of reading general formulas and how to compare a given expression or equation to its corresponding general form as we did above.

Let's work on two more quadratic equations together, starting with $y = -(x + 5)^2 + 3$.

$$y = -(x + 5)^2 + 3$$

a is − : ∩

$a = -1$

Vertex
located at

$(-5, 3)$

Observe that the parabola will point down because the a value is negative ($a = -1$).

As to the vertex, because the 5 is being added, we know that this equation's h value is equal to **negative five**, not positive five. The other value of the vertex coordinate, k, is positive three because the 3 is being added, as the general form expects.

I circled the **adding five** to help you see why the extracted value should be **negative five**; the adding three does correspond to a **positive three** k value because it's what the general form expects.

If you want to see how the equation can be rewritten so that you may explicitly observe the correct h and k values, here it is:

$$y = -(x + 5)^2 + 3$$

$$y = -(x - (-5))^2 + 3$$

$$(h, k)$$

Let's work on one final quadratic equation together.

$$y = -3(-2x + 4)^2 - 5$$

We cannot extract any of the values associated with the vertex form from this quadratic equation until we rewrite it so that it is in the expected format. The independent variable inside the set of parentheses has a coefficient other than 1 (it's equal to −2 to be exact). This is going to have an impact on the **h** value inside the set of parentheses–do know, even before we begin to work on transforming this equation, that **h** is **not** equal to **−4** , as you will see shortly–, as well as on its **a** value, so we need to work on transforming the equation so that the independent variable has a coefficient of **1** before being able to determine their correct value. The only value that will remain intact throughout the rewriting process will be the **k** value; you should thus be able to see that because the **5** is being subtracted, this equation's **k** value is equal to **−5** .

Let's transform the equation.

$$y = -3(-2x + 4)^2 - 5$$

$$y = -3[(-2)x + (-2)(-2)]^2 - 5$$

$$y = -3[(-2)(x + (-2))]^2 - 5$$

$$y = -3[(-2)(x - 2)]^2 - 5$$

Observe that to factor out the **negative two** from the inside of the set of parentheses, I rewrote the adding **4** as an adding **(−2)(−2)** ; this way, that second term had a **−2** to provide for this process. That left us with an adding **−2** which eventually simplifies to a subtracting **2** . Observe that as on an earlier example, we must contain the factor and the original set of parentheses within a set of brackets that is being squared.

$$y = -3[(-2)(x - 2)]^2 - 5$$

$$y = -3(-2)^2(x - 2)^2 - 5$$

$$y = -3(4)(x - 2)^2 - 5$$

$$y = -12(x - 2)^2 - 5$$

This last version of the equation above expresses the quadratic equation in true vertex form. Now, we can compare it to the general form and extract the correct **a**, **h** and **k** values.

$$y = a(x - h)^2 + k$$

$$y = -12(x - 2)^2 - 5$$

a is − : ∩

a = −12

Vertex
located at

(2 , −5)

As you can see above, the **a** value and the **h** value could not be determined from the original equation that we started out with; only after transforming the quadratic equation and rewriting it so that it was in true vertex form were we able to determine them. The **k** value, on the other hand (and as promised), remained unaffected by this transformation.

Knowing the vertex of the parabola and whether it points up or down based on the sign of the a value allows us to quickly sketch the quadratic's graph. We can be as precise as we need to by obtaining several coordinates from the equation (using some well-chosen input values), or simply provide an idea of how the graph would look like.

Let's revisit the four equations we worked on above: I am going to do a quick sketch of each equations graph using the vertex and the parabola's orientation, and then I am going to graph each equation more precisely by finding out some well-chosen input values.

$$y = 5(x - 3)^2 + 4$$

$a = 5$; parabola *points up*

vertex @ $(3, 4)$

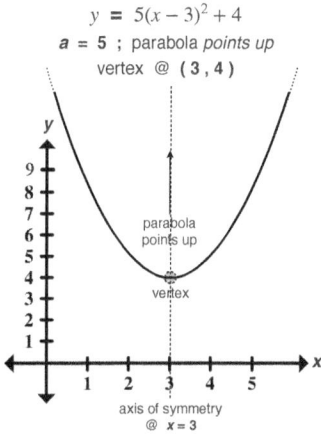

Remember that this is just a sketch... All we know for sure at this point is that the parabola points up and that its vertex is at $(3, 4)$. We could not actually use it to "read" solution pairs from this graph (except for the vertex, of course!).

Let's move on to the next example.

$$y = 2(x + 6)^2 - 4$$

$a = 2$; parabola *points up*

vertex @ $(-6, -4)$

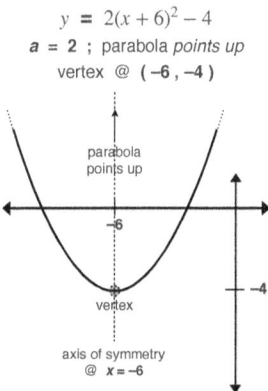

Notice how this last quadratic was sketched using a different style: instead of indicating the tick marks precisely along both axes, simply indicate the specific x and the y values of the vertex on the corresponding axis, and then sketch the graph. This is a perfectly valid sketch, but do know that the information it provides is limited to the location of the vertex and the parabola's orientation. Nothing else can be "read" from this sketch.

$$y = -(x + 5)^2 + 3$$

$a = -1$; parabola *points down*

vertex @ $(-5, 3)$

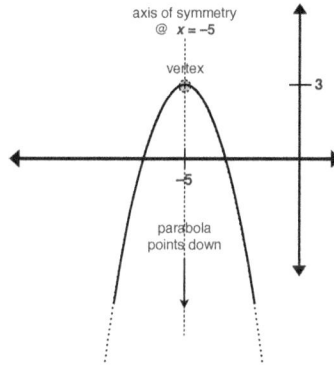

This quadratic points down, as determined by the sign of the a value (a negative value implies a downward-pointing parabola). As we move away either to the left or to the right of the axis of symmetry (the vertical line passing through the vertex), successively farther away input values will yield smaller and smaller output values (in other words, it can be said that the equation's output *tends towards negative infinity*).

The last quadratic of this set now follows. Just remember that we had to rewrite it before being able to extract its *a*, *h*, and *k* values (the original version of the equation does not explicitly show the *a* and *h* true values).

$$y = -3(-2x + 4)^2 - 5$$

a = **-12** ; parabola *points down*
vertex @ (2 , -5)

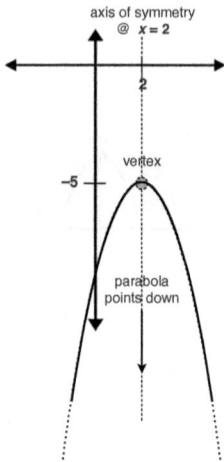

axis of symmetry
@ *x* = 2

vertex

-5

parabola
points down

As you can see from the examples we worked on above, sketching the graph of a quadratic that is in vertex form is relatively simple. Just be careful when you attempt to extract its *a*, *h* and *k* values: the quadratic must be in true vertex form if you want to explicitly see them; otherwise, remember to take into account whether the

values are being added or subtracted, and if the independent variable has a coefficient of 1 or not (this last condition is trickier to take into account without actually rewriting the quadratic, so be careful).

The next item on the agenda is being able to graph the quadratic equation that is in vertex form with more precision. The four previous sketches are good enough as "draft" versions of the quadratics' graphs, each communicating three things precisely: the location of the vertex, the location of the axis of symmetry, and whether the parabola points up or down. You may wonder why I am not including a fourth item: the shape of the quadratic's graph (parabola)... Well, the reason is obvious: all quadratic equations have a graph in the shape of a parabola, so that one is always a given.

So how do we know the precise graph of a quadratic equation that is expressed in vertex form? Well, although there are many different ways to do this, I will review a very simple method that will work every time.

First, remember that all quadratics' graphs are parabolas. Furthermore, keep in mind that they all have a vertex (lowest or highest point, and the location of the (vertical) axis of symmetry). And so the idea is to exploit these afore-mentioned characteristics that all parabolas share, and simply plot two additional points to the left and two to the right of the vertex. With five points plotted on the graph, you can do a very precise sketch of the graph of the quadratic equation. As always, you want to use values that are easy to compute: this means using **INTEGERS** that are close to the vertex.

Of course, depending on the application, and what the graph will be used to begin with, it may be necessary to plot more points, or to simply graph it using a computer or a graphing calculator.

Let's put this method into action.

$$y = 5(x - 3)^2 + 4$$
$$a = 5 \; ; \text{ parabola } \textit{points up}$$
$$\text{vertex @ } (\,3\,,\,4\,)$$

Because the vertex has an **x** value of **3** , which means that the axis of symmetry is at **x = 3** , I am going to find the solution pairs using the input numbers of **1** and **2** (for the two values that are to the left of the vertex), and the numbers **4** and **5** (for the two values that are to the right of the vertex). Using the input number **3** would be a waste of time because we know that its output would be equal to **4** (the vertex point!). Remember that a **solution pair**, a.k.a. **a coordinate**, is formed by an input value (the **x** value) and its corresponding output value (the **y** value). Therefore, the vertex is a solution pair of the equation.

$$x = 1$$
$$y = 5(1 - 3)^2 + 4$$
$$y = 5(-2)^2 + 4$$
$$y = 24$$
Coordinate ---> (1 , 24)

$$x = 2$$
$$y = 5(2 - 3)^2 + 4$$
$$y = 5(-1)^2 + 4$$
$$y = 9$$
Coordinate ---> (2 , 9)

We now have the exact location of two points that are to the left of the vertex. We can now find two that are to its right. Before plugging in the input values, I want you to think about the fact that parabolas are perfectly symmetric: this means, as I mentioned earlier in the chapter) that all points must have a "partner" of equal output, but with a different input value, except, of course, for the vertex.

But why do I mention this now? Because if you think about it, we don't have to actually compute the outputs of the two points that are to the right of vertex, after having found the outputs of the two points that are to its left... The left-right input values chosen are matching pairs! Thus, they should have equal outputs. In other words, the input value of **1** has an output value equal to the input value of **5** since both are equally far apart from the vertex (or from the line of symmetry). The same thing happens with the input values of **2** and **4** . Observe.

$$x = 4$$
$$y = 5(4 - 3)^2 + 4$$
$$y = 5(1)^2 + 4$$
$$y = 9$$
Coordinate ---> (4 , 9)

$$x = 5$$
$$y = 5(5 - 3)^2 + 4$$
$$y = 5(2)^2 + 4$$
$$y = 24$$
Coordinate ---> (5 , 24)

As you can see above, the input values of **1** and **5** have the same output value (**24**), while the input values of **2** and **4** also have the same output value of **9** .

Now that we have these four additional points, a precise graph of the quadratic equation may be sketched.

But what about the graph's scale? After all, although the input values are very manageable (spanning from **1** to **5**), the output values are a bit too far apart (from **4** to **24**). Remember that you typically want to include the origin in your graph as well, which means that the **y-axis** must span from **0** to **24** at a minimum.

Fitting **24** tick marks in one axis can make the graph quite big, so another option would be to use ***five-unit*** tick marks, as I will do below. The disadvantage is that we may lose precision if the points are not drawn correctly.

Before I draw the actual graph, I want you to observe the symmetric coordinate pairs on the **x-y axis** below.

See how points that are symmetry pairs of the parabola have different input values but the same output value? It makes perfect sense because these symmetry pair coordinates have the same height, and height is determined by the output value.

The source of this symmetry, by the way, is the set of parentheses that is being squared, which is part of the quadratic equation in vertex form. If the inside of the parentheses computes to a positive two or a negative two, for instance, squaring it will yield ***positive four*** in both cases, so the output will end up being the same since the rest of the equation is made up of constant values.

Now, all we have to do is draw a parabola that passes through the points that have already been plotted. Observe.

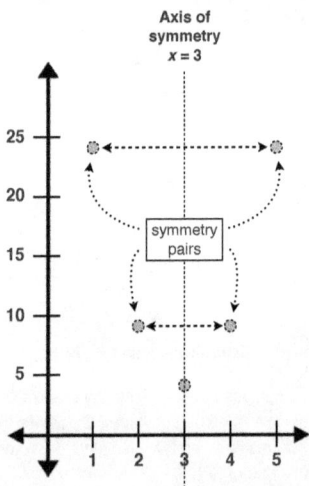

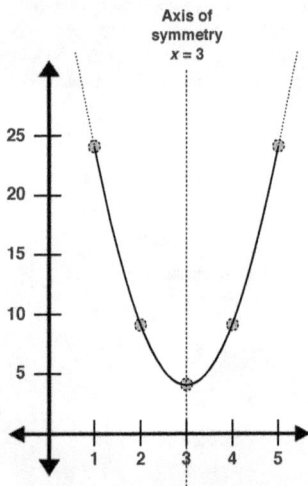

This graph is now much more precise. It provides useful information regarding the quadratic equation's behavior, beyond the vertex and whether it points up or down. Now, we can see how open or closed it is, which is an indication of how fast or how slow its output values increase or decrease the farther away we move from the vertex.

Compare the more precise version of the graph drawn above with the original sketch that I presented to you earlier.

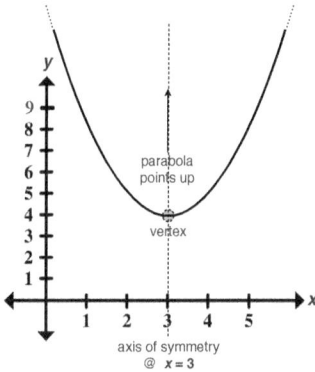

If you try to read the output value that the sketch above predicts for an input value of 2 (or for an input value of 4 since that would be its symmetry "*partner*"), for example, you should estimate it to be equal to about 5 , which is very far from the actual output of 9 . To be fair, remember that it was simply a sketch; the intention was not to provide a graph that could be used to predict an output based on a given input.

Let's graph the rest of the quadratics precisely.

$$y = 2(x + 6)^2 - 4$$
$a = 2$; parabola *points up*
vertex @ (-6 , -4)

Guess which input values I'm going to use? All we need is a couple of points to the left and to the right of the vertex in order to provide a precise graph of the equation. If you are thinking about using the input values of -8 , -7 , -5 , and -4 , then you are absolutely correct. Furthermore, using the idea of symmetry pairs discussed earlier, it will suffice to compute the output for an input of -8 or -4 (these are the *x* values of the coordinates of one symmetry pair) and for -7 or -5 , the inputs of the other two symmetry pairs.

$$x = -4$$
$$y = 2((-4) + 6)^2 - 4$$
$$y = 2(2)^2 - 4$$
$$y = 4$$
Coordinate ---> (-4 , 4)
which also implies ---> (-8 , 4)

$$x = -7$$
$$y = 2((-7) + 6)^2 - 4$$
$$y = 2(-1)^2 - 4$$
$$y = -2$$
Coordinate ---> (-7 , -2)
which also implies ---> (-5 , -2)

We now have five points that we can use to precisely graph this quadratic equation, counting, of course, the vertex.

This time, the *x*-axis must span from -8 to 0 (remember that I will typically include the origin in all graphs to provide an adequate frame of reference) and the **y-axis** must span from **–4** to **4** .

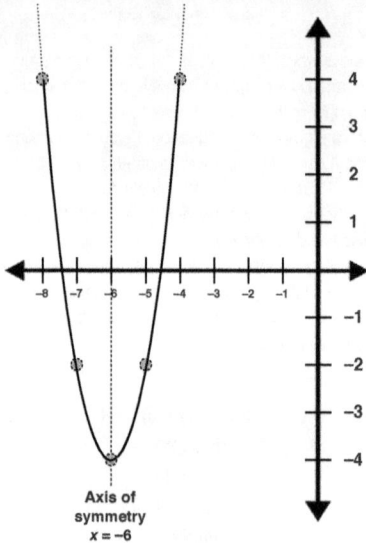

Axis of
symmetry
$x = -6$

You should now attempt to sketch the remaining two quadratic equations using the principles I used on the previous two graphs. Use the idea of symmetry, and choose an adequate scale for the **x-axis** and the **y-axis** that suits each of the graphs you are going to create.

Ready to check your work?

$$y = -(x + 5)^2 + 3$$
a = –1 ; parabola *points down*
vertex @ **(–5 , 3)**

Suggested input values: **–7** , **–6** , **–4** and **–3** .
Symmetry pairs: **–7** and **–3** ; **–6** and **–4** .

$$x = -7$$
$$y = -((-7) + 5)^2 + 3$$
$$y = -(-2)^2 + 3$$
$$y = -1$$
Coordinate ---> (–7 , –1)
which also implies ---> (–3 , –1)

$$x = -6$$
$$y = -((-6) + 5)^2 + 3$$
$$y = -(-1)^2 + 3$$
$$y = 2$$
Coordinate ---> (–6 , 2)
which also implies ---> (–4 , 2)

With these five coordinates, we can now draw a much more precise graph. The **x-axis** should thus span from **0** to **–7** , while the **y-axis** should span from **–1** to **3** (I typically include an additional tick-mark on each end, so I will include the tick-marks for **–2** and **4** on the **y-axis**).

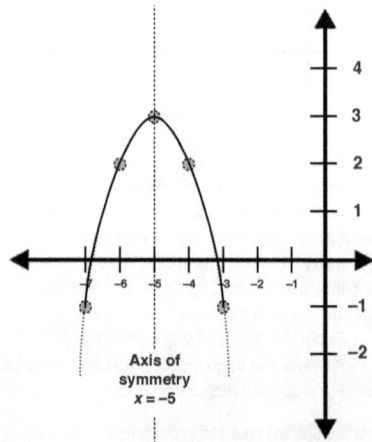

Axis of
symmetry
$x = -5$

Was your graph similar to the one I drew above? If not, make sure yours adheres to the principles and guidelines that we have been reviewing throughout he book regarding graphs. If, after reviewing this last graph you think you need to revisit the graph you made of the last quadratic equation, do so now. You will master these skills a lot faster and a lot better by doing these exercises on your own every chance you get.

Ready for the last graph? Check your work.

$$y = -3(-2x + 4)^2 - 5$$

a = -12 ; parabola *points down*
vertex @ **(2 , -5)**

Suggested input values: **0** , **1** , **3** and **4** .
Symmetry pairs: **0** and **4** ; **1** and **2** .

$$x = 0$$
$$y = -3(-2(0) + 4)^2 - 5$$
$$y = -3(0 + 4)^2 - 5$$
$$y = -53$$

Coordinate ---> (0 , -53)
which also implies ---> (4 , -53)

$$x = 1$$
$$y = -3(-2(1) + 4)^2 - 5$$
$$y = -3(-2 + 4)^2 - 5$$
$$y = -17$$

Coordinate ---> (1 , -17)
which also implies ---> (3 , -17)

With these five coordinates, we can now draw a much more precise graph. The **x-axis** should thus span from **0** to **4** , while the **y-axis** should span from **-53** to **0** . I am sure you can appreciate the fact that we must be careful with the **y-axis** scale, since we need to accommodate a very wide range of values. Assigning a one-unit

spacing for the tick-marks would be very impractical. I will opt to use ten-unit spacings for this graph on the **y-axis**; the **x-axis**, on the other hand, will have one-unit spacings since its span is much narrower and it is thus practical to do so.

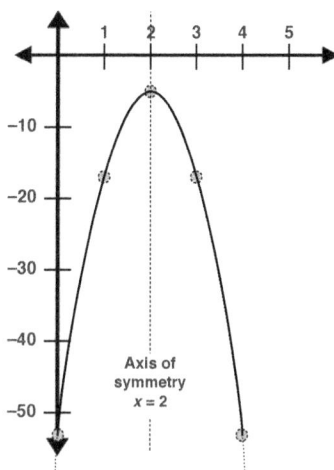

By now you should be able to precisely graph any quadratic equation that is in vertex form; this requires being able to extract its vertex, know beforehand if the parabola points up or down, and using the principle of symmetry pairs, find at least four additional points so that together with the vertex, the parabola can be drawn with a high-degree of precision. It also helps to draw a dotted line representing the axis of symmetry, although please remember that it is not an actual part of the equation's graph. In order to drill this last principle, and to underline the fact that the "thick grey points" are not actual elements of the graphs, allow me to present to you the graph of all four of the equations that we worked on in their **pure state**:

$$y = -(x + 5)^2 + 3$$

$$y = 5(x - 3)^2 + 4$$

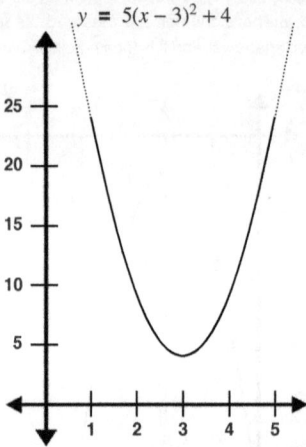

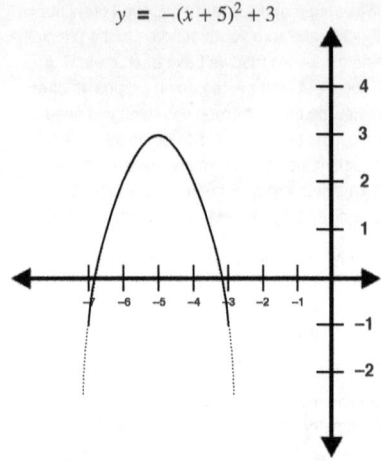

$$y = 2(x + 6)^2 - 4$$

$$y = -3(-2x + 4)^2 - 5$$

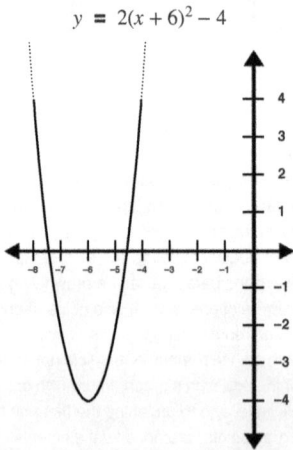

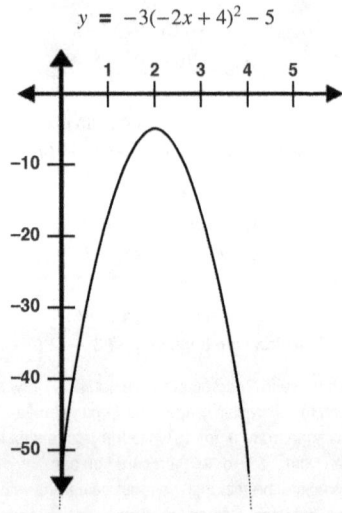

After working with these vertex form examples, you are ready to explore the mathematical behavior of a quadratic equation that the vertex form **encodes**. To do this, we are going to use the most basic quadratic equation there is, a quadratic in its purest form: $y = x^2$.

This quadratic has the vertex at the origin, $(0,0)$, and its behavior is solely determined by the "squaring the input" characteristic of all quadratics, since the value of **a** is equal to **1**. The parabola, of course, points up, which means that the output grows as the input value moves farther away from **0** both to its left and to its right. Its graph is, as we have already reviewed, the following.

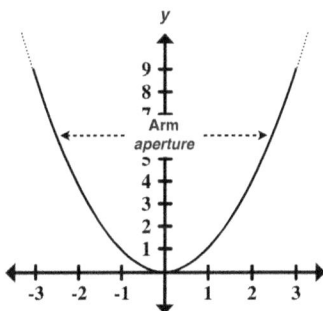

This so-called "parent graph" of all quadratics, which is in its simplest form possible, has four important characteristics:

1) Its "arm-aperture", meaning the wideness of its arms, as the graph above illustrates. This is determined by the value of **a**. A greater absolute value of **a** implies a much narrower aperture (since the output increases (if it points up) or decreases (if it points down) much faster for larger absolute values of **a**).

2) The orientation of the parabola: an indication of whether it points up or down. In this case, the parabola points up, as determined by the sign of the value of **a**.

3) The location of the vertex: this is either the lowest point of the graph (as is the case here, located at $(0,0)$) or the highest point of the graph (this only happens when the parabola points down). This is given by the **h** and **k** values.

4) The number of **x-axis** crossings. As you may recall, all quadratics will either have **0**, **1**, or **2 x-axis** crossings. In the case of this quadratic, it only has **one x-axis** crossing. This is determined by a combination of two factors: the location of the vertex and whether the parabola points up or down.

Thus, we may state the following: the vertex form of a quadratic equation encodes the following behavior, with respect to the parent function:

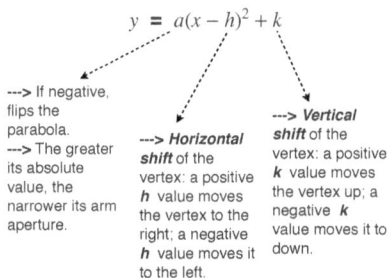

$$y = a(x - h)^2 + k$$

---> If negative, flips the parabola.
---> The greater its absolute value, the narrower its arm aperture.

---> **Horizontal shift** of the vertex: a positive **h** value moves the vertex to the right; a negative **h** value moves it to the left.

---> **Vertical shift** of the vertex: a positive **k** value moves the vertex up; a negative **k** value moves it to down.

The following diagram illustrates this precisely.

Left diagram labels:

Shift up and left
$\ll h \ ; \ \gg k$
$y = (x - 3)^2 - 4$

Shift up and right
$\gg h \ ; \ \gg k$
$y = (x - 3)^2 + 4$

vertex:
(–3 , 4)

vertex:
(3 , 4)

Parent graph
$y = (x - 0)^2 + 0$
vertex:
(0 , 0)

vertex:
(–3 , –4)

vertex:
(3 , –4)

Shift down and left
$\ll h \ ; \ \ll k$
$y = (x + 3)^2 - 4$

Shift down and right
$\gg h \ ; \ \ll k$
$y = (x + 3)^2 + 4$

Right diagram labels:

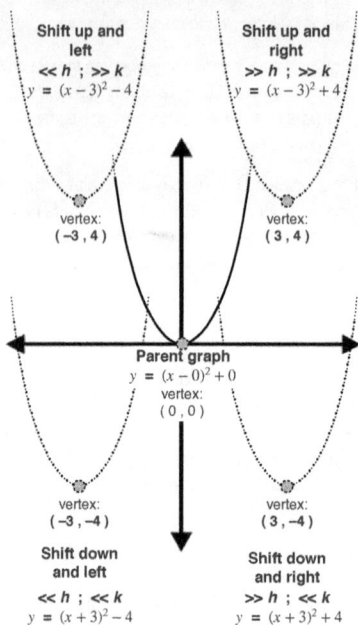

(and its graph) may be represented in vertex form.

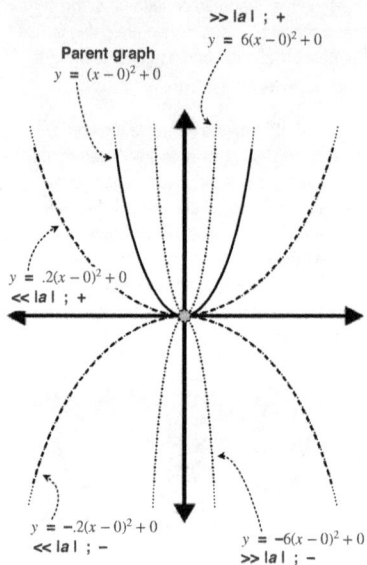

$\gg |a| \ ; \ +$
$y = 6(x - 0)^2 + 0$

Parent graph
$y = (x - 0)^2 + 0$

$y = .2(x - 0)^2 + 0$
$\ll |a| \ ; \ +$

$y = -.2(x - 0)^2 + 0$
$\ll |a| \ ; \ -$

$y = -6(x - 0)^2 + 0$
$\gg |a| \ ; \ -$

As you can see on the diagram above, changing the *h* or *k* values (indicted with $\gg$ and $\ll$ for increase or decrease, respectively) will move the vertex point left/right and up/down respectively. I left the *a* value the same in all cases, equal to **1** .

Observe how the values of *h* and *k* impact the final equation's set up.

Now, let me illustrate how changing the *a* value transforms the parent function's orientation and arm-aperture. Note that by combining all these element transformations, any quadratic equation

Note how the parabola is transformed by the change of its *a* value. We had already seen that the sign determines whether the parabola points up or down; but now, in addition to this, we see the impact of making the absolute value of *a* larger (farther to the right of **0** in the number line) or smaller (closer to **0**). On the diagram above, an *a* value of **.2** makes the parabola much more open (its arms are spread wider apart), while an *a* value of **6** makes it much narrower. In terms of its input/output solution pairs, this means that as the *a* value increases, the output will increase more rapidly as well; while as the *a* value gets smaller and smaller (closer to **0**), its output increases much more slowly.

This idea of "transforming" the parent quadratic equation ($y = x^2$, or $y = (x - 0)^2 + 0$ in explicit vertex form) by changing the **a** , **h** , and **k** values allows you to know even before you do any calculations whatsoever, the kind of graph that you should expect (and thus the type of behavior that the equation exhibits).

Let's move on to the next quadratic form.

Quadratic Equation in Intercept Form

$$y = a(x - m)(x - n)$$
where **a ≠ 0**

The quadratic equation in intercept form encodes three important aspects of the quadratic equation's behavior and its graph. The **a** value will yield the same information that we discussed in the previous form, while the **m** and **n** values correspond to the **x-intercepts**: the point or points where the graph crosses the **x-axis**, which corresponds to the input values that yield an output value of **0** . This stems from the fact that the **x-intercept** always occurs where the graph touches or crosses the **x-axis**, and all points that are **ON** the **x-axis** have a height of **0** (in other words, their **y** value is equal to **0**).

The intercept form can be difficult to work with because in order to cover all possible quadratic equations, we are forced to extend the number system to include imaginary numbers as well as far as the m and n variables are concerned. Why this is so will become obvious shortly.

As far as graphing a quadratic that is in intercept form is concerned, we can do a sketch of the graph with this information; however, in order to draw a precise graph, it will be necessary to find the vertex.

First, let's summarize what this form encodes.

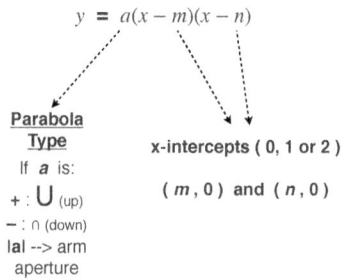

$$y = a(x - m)(x - n)$$

Parabola Type
If **a** is:

$+ : \cup$ (up)

$- : \cap$ (down)

|a| --> arm aperture

x-intercepts (0, 1 or 2)

$(m , 0)$ and $(n , 0)$

As with the other form, remember that the subtraction signs that are inside the sets of parentheses are crucial in determining the values of **m** and **n** ; the form expects a **subtraction sign**: if it is there, the value of **m** or **n** , whichever is being analyzed, is positive, while if it is adding, then it means that the value is negative.

Let's dive right in and explore some examples. We will begin with the "parent" quadratic equation, this time written in intercept form.

$$y = (x - 0)(x - 0)$$

It should be easy for you to distribute (or FOIL) the parentheses so that you amy verify that this is, in fact, the same equation as $y = x^2$. Try it on your own.

Ready to check your work? Observe.

$$y = (x - 0)(x - 0)$$

$$y = (x)(x) - (x)(0) - (0)(x) - (0)(0)$$

$$y = x^2 - 0 - 0 - 0$$

$$y = x^2$$

Analyzing this equation's intercept form we are able to extract the following data:

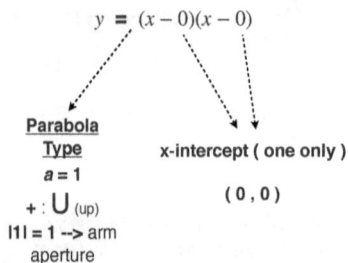

$$y = (x - 0)(x - 0)$$

Parabola Type

$a = 1$

$+ : \cup$ (up)

$|1| = 1$ --> arm aperture

x-intercept (one only)

(0 , 0)

Please note that since both *m* and *n* are equal to each other, this particular quadratic has only one **x-intercept** (we already know this, correct? We are simply testing the intercept form and the data it is supposed to provide...). The *a* value is equal to **1** since in the absence of a number, we know that's what's implied. And thus the parabola points up and its aperture is equal to **1** which corresponds to the quadratic's "parent" equation aperture.

It is worth mentioning that if the quadratic equation has only one x-intercept, it means that the point also corresponds to its vertex. Think about it: since a quadratic has two arms, it only works if it is so. It is a very particular case of the intercept form.

To graph this quadratic equation we would simply locate the point **(0 , 0)** and knowing that it is the vertex *and* the **x-intercept**, sketch a parabola that points up using that coordinate as the parabola's lowest point. If we wanted to be more precise, we would have to use two input values– say, to the left of the vertex–and using the principle of symmetry pairs, locate their two corresponding symmetry partners–those to the right of the vertex. In this case, I would choose the input values of **–2** , **–1** , **1** and **2** .

First, observe the sketch:

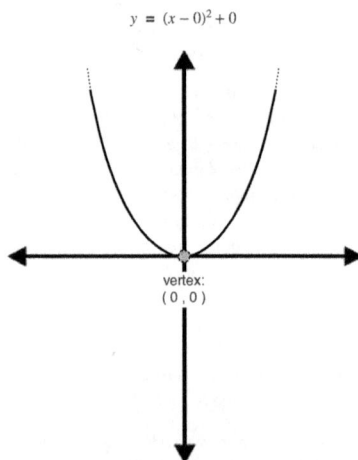

$$y = (x - 0)^2 + 0$$

vertex: (0 , 0)

As to a more precise version of the graph, let's find the input/output solution pairs.

$$x = -2$$
$$y = ((-2) - 0)((-2) - 0)$$
$$y = (-2)(-2)$$

$$y = 4$$

Coordinate ---> (-2 , 4)

which also implies ---> (2 , 4)

$$x = -1$$
$$y = ((-1) - 0)((-1) - 0)$$
$$y = (-1)(-1)$$
$$y = 1$$

Coordinate ---> (-1 , 1)

which also implies ---> (1 , 1)

With these five coordinates, we can now draw a more precise graph.

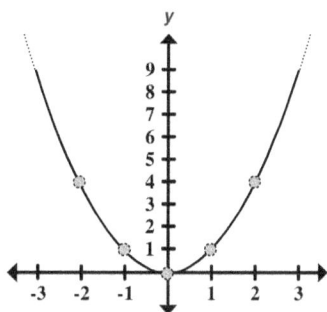

Let's try another quadratic equation in Intercept form.

$$y = (x - 2)(x - 4)$$

You should be able to see right away that this parabola points up, and that its arm aperture is the same as the parent quadratic equation, since its **|a| = 1** . Furthermore, it has two **x-intercepts**: **(0 , 2)** and **(0 , 4)** , based on the equation's **m** and **n** values. Remember that the general intercept form contains a subtraction operation and then the **m** and **n** values respectively; and

because this equation contains a subtracting **2** and a subtracting **4** (the **m** and **n** values respectively), they are both positive. Observe.

$$y = a(x - m)(x - n)$$

$$y = (x - 2)(x - 4)$$

m is positive;   **n** is positive;
m = 2   **n = 4**

Remember that this is one of the most confusing aspects of extracting data from specific equation forms... it is crucial to compare the equation you are working on with the general form, and consider whether the general form expects to see a subtracting value or an adding value.

So now that we know the parabola's orientation (points up), and its two **x-intercepts**, we are able to sketch the graph.

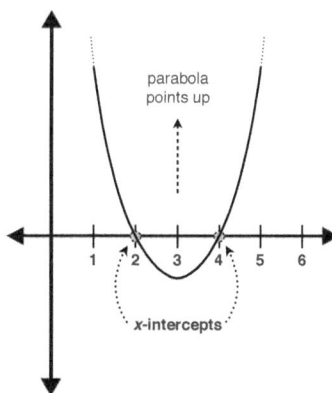

parabola
points up

x-intercepts

As you can see on this last sketch, I am not specifying the *y-axis* scale; the reason for this is simple: since this is only a sketch, and I did not find any other points besides the *x-intercepts*, to include a *y-axis* scale would be presumptuous since I would not be able to accurately portray this equation's actual arm aperture, and thus its true solution pairs. If we want to include a *y-axis* scale we would need to come up with more points as we have done in previous examples.

Let's find them.

But before we do so let's reason something together regarding any quadratic equation's parabola and its *x-intercepts*.

If a parabola has two *x-intercepts*, and we know their location (because we are in possession of the quadratic's intercept form, for example), we can easily determine the *x* value of the vertex. How? Well, think about it. A parabola is perfectly symmetric. All points have a corresponding symmetry pair, except for the vertex which is unique (it is, if you will, its own symmetric partner). So if we know the location of any two points that are part of the parabola, regardless of their height, but both of which have *the same height*, then we can be certain that the vertex has an *x* value that lies in the middle of those two given points, horizontally-speaking, that is! Another way of thinking about this is that the vertical axis of symmetry will always lie *halfway* between any two given points, as long as the two given points have an equal height value (which is equivalent to their *y* value, or to their output value).

In this case, since the two points are (2 , 0) and (4 , 0) , we can be sure that the vertex (and the axis of symmetry) lies halfway between 2

and 4 . All we have to do, then, is add the two *x* values and divide by two, as follows:

$$x \text{ value of the vertex } = \frac{2+4}{2} = 3$$

Now that we know that the vertex must be of the form

$$(3 , ?)$$

...we can easily find the corresponding *y* value, or output value, or its height. Simply use 3 as the input value on the equation, and compute it.

$$\text{If } x = 3 :$$
$$y = ((3) - 2)((3) - 4)$$
$$y = (1)(-1)$$
$$y = -1$$

We now know where the vertex is located.

$$\text{vertex @ } (3 , -1)$$

When working with quadratic equations in intercept form, it will always be possible to apply the above technique to find the precise location of the vertex. Why? Because by definition, the intercept form provides the exact location of two points, both with the same output value or height (that of 0). Rather than memorizing this, it should come naturally to you if you are truly comprehending the basic principles of equations, input/output pairs, what the points represent of a given equation, and the particular behavior of quadratics–and therefore parabolas–in general.

We could attempt to draw a precise graph with these three coordinates (vertex and two *x-intercepts*); however, it helps to have a fourth and its corresponding fifth symmetry partner. To find the additional two points, I will use the following input value (can you guess which one I will use?): **1** . It is typically very easy to evaluate an equation with this input value, and it is thus the logical choice. It will, of course, yield the output value as well of its symmetry point: the coordinate **(5 , ?)** .

$$x = 1$$
$$y = ((1) - 2)((1) - 4)$$
$$y = (-1)(-3)$$
$$y = 3$$

Coordinate ---> (1 , 3)
which also implies ---> (5 , 3)

With these five coordinates, we may now draw a very precise graph of this quadratic equation.

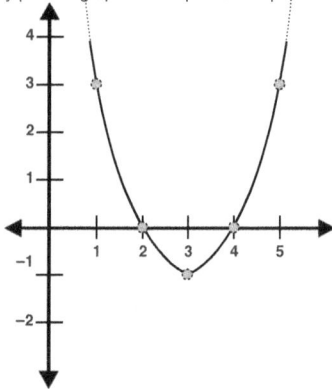

Transforming a quadratic in intercept form is not as simple as you would imagine. For example, if we changed the equation that we graphed above to, say, $y = (x - 1)(x - 5)$, the vertex will still have an **x** value of **3** ; however, its

corresponding **y** value will not remain the same. Observe.

$$x = 3$$
$$y = ((3) - 1)((3) - 5)$$
$$y = (2)(-2)$$
$$y = -4$$

Coordinate ---> (3 , –4)

So what happened here? Well, although this transformed equation's **a** value is the same as the previous equation (**a = 1**), and although its vertex is also located at **x = 3** , the graph has been shifted down and to the right, so the **y** value of the vertex has changed. Observe.

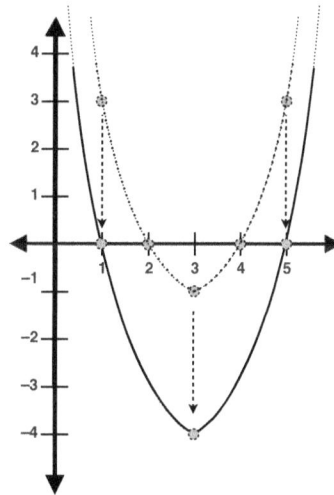

As you can see in the graph above, changing the **m** and **n** values to **1** and **5** respectively essentially caused a downward shift of the graph; besides this shift, nothing else changed since the

a value remained equal to **1** (the same as that of the the equation before transforming it). In other words, because the a value is the same, the arm aperture remained unchanged; combine this with the fact that the *x*-intercepts were altered, and you can now see why the transformation resulted in a downward shift of the graph.

It is worth mentioning that the intercept form is not generally used to create transformations, but rather to have direct access to the ***zeros of a quadratic***.

Zeros of a quadratic

The so-called "zeros" of a quadratic equation correspond to the value or values (a maximum of two values may exist for a given quadratic equation, with a minimum of one) that produce an output value of **0**. As we reviewed earlier, the parent quadratic equation, $y = x^2$, has only one *x*-intercept, and therefore, only one x value that produces a y value of **0** (another way of saying this is that there is only one ***input value*** that produces an ***output value*** of **0**): the input value in this case is, of course, **0**, which is why its *x*-intercept is **(0 , 0)**. Thus it is said that the parent quadratic equation $y = x^2$ has only one zero: (**x = 0**).

The second equation we worked with in intercept form, $y = (x-2)(x-4)$, has two zeros: **x = 2** and **x = 4**. Determining the zeros of a quadratic equation that is in intercept form is extremely simple. All we have to do is make use of the zero product property, which states that whenever we have the product of two entities, if we need said product to be equal to **0**, then the only way to make it happen is to ensure that either the first entity is equal to **0** or that the second entity is equal to **0**. Observe.

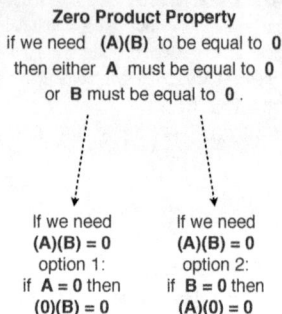

Zero Product Property
if we need **(A)(B)** to be equal to **0**
then either **A** must be equal to **0**
or **B** must be equal to **0** .

If we need	If we need
(A)(B) = 0	**(A)(B) = 0**
option 1:	option 2:
if **A = 0** then	if **B = 0** then
(0)(B) = 0	**(A)(0) = 0**

Observe how we can use this property with the equation $y = (x-2)(x-4)$ in order to find its zeros (yes, we already know what they are, but there is an important lesson in the process).

$$y = (x-2)(x-4)$$

Since our goal is find the input values that will yield an output value of 0 , we may ***set the equation to zero*** as follows:

$$0 = (x-2)(x-4)$$

Now, knowing the zero product property, we can make use of it by thinking along the following lines: "To ensure that the product of these two entities is equal to zero (one is the parentheses set with the x and the subtracting **2**, while the other entity is the other set of parentheses with the x and the subtracting **4**), the only way to make that happen is to either ensure that the first entity be equal to zero or that the second parentheses be equal to zero. This means that we can take the first entity and set it to zero, then solve for the variable x, and then take the second entity and set it to zero and solve it for the variable x as well. Observe.

$$0 = (x - 2)(x - 4)$$

$$(x - 2) = 0 \qquad (x - 4) = 0$$
$$x - 2 = 0 \qquad x - 4 = 0$$
$$x - 2 + 2 = 0 + 2 \qquad x - 4 + 4 = 0 + 4$$
$$x = 2 \qquad x = 4$$

...which leads to the two **x-intercepts** that we had already defined previously:

(2 , 0) and **(4 , 0)**

What is crucial is that you understand why the zero product property is necessary, and how it is applied to this quadratic when trying to determine the zeros of the equation.

The zeros may be either listed as in "the zeros of the quadratic are **2** and **4**", or expressed in terms of the variable **x** as in "the zeros of the quadratic are **x = 2** and **x = 4**". You should be able to construct the coordinate (input/output pair) based on this information, since by definition, a "zero" is an input value that when plugged in to a given equation, produces an output of **0** , so a zero will always have an output (**y**) value of **0** . This why we may also state the zeros along the lines of "...the zeros of the equation are **(2 , 0)** and **(4 , 0)** ".

We will be using this idea of the zeros of a quadratic for the rest of the chapter, so rather than assign some equations for you to work on, I will move on instead. You will have many opportunities to practice this skill by yourself later on.

Let's work on another quadratic equation.

$$y = -(x + 1)(x - 5)$$

You should be able to see right away that this parabola points down, and that its arm aperture is the same as the parent quadratic equation, since its |a| = 1 . Furthermore, it has two **x-intercepts**: **(0 , –1)** and **(0 , 5)** , based on the equation's **m** and **n** values. Remember that the general intercept form contains a subtraction operation and then the **m** and **n** values respectively; and because this equation contains an **adding** 1 and a subtracting **5** (the **m** and **n** values respectively), the former is negative while the latter positive. Observe.

$$y = a(x - m)(x - n)$$

$$y = -(x + 1)(x - 5)$$

m is negative; \qquad **n** is positive;
m = –1 \qquad **n = 5**

If in doubt that the m is in fact negative, observe what happens if we do it in reverse: given an **m** and an **n** value of **–1** and **5** respectively, and an **a** value of **–1** , what quadratic equation do they specify? Using the general intercept form, and replacing **a** , **m** and **n** with these values, watch the equation that they define.

$$y = a(x - m)(x - n)$$

$$y = -1(x - (-1))(x - (5))$$

$$y = -(x + 1)(x - 5)$$

See how the equations are equal?

This proves that the values we extracted were, in fact, correct.

In addition to the x-intercepts we defined, we can also easily figure out the location of the axis of symmetry, which can help us correctly sketch the graph. Simply add the two x values of the coordinates (remember that this can only work if both points have the same output) and dividing that by two (note that it is the same method used on the previous example to find the x value of the vertex):

$$\text{axis of symmetry @ } \frac{-1+5}{2} = 2$$

We may now sketch the graph of this quadratic equation:

$$y = -(x+1)(x-5)$$

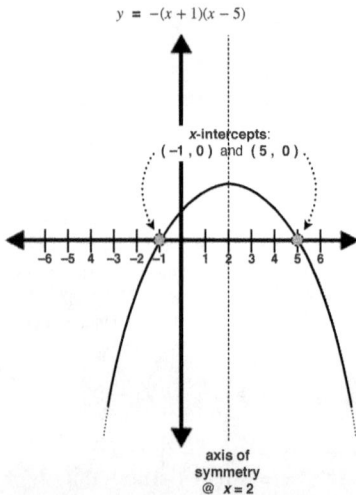

x-intercepts:
$(-1, 0)$ and $(5, 0)$

axis of
symmetry
@ $x = 2$

As I've pointed out before, remember that this is only a sketch; this is why I did not specify a

y-axis scale. In order to make a precise graph of this quadratic equation, we would need to precisely determine the location of the vertex (we know its x value but not its y value), and find two other coordinates (only one need be computed; the other one would correspond to its symmetry partner). Try to do this by yourself; after you are done, check your work (scale, points found, and the drawing of the graph itself) to see if you are mastering these concepts and skills. If you need to refer to the previous example, do so. Just don't look at my answer below until you are finished or you have made an honest effort to graph it yourself.

Ready to check your work?

First, I will find the location of the vertex. Knowing that the axis of symmetry is at $x = 2$ tells us that we need to find the output when the input is 2 in order to find the precise location of the vertex.

$$\text{If } x = 2$$
$$y = -((2)+1)((2)-5)$$
$$y = -(3)(-3)$$
$$y = 9$$

We now know where the vertex is located.

$$\text{vertex @ } (2, 9)$$

Next, we want to find a fourth and fifth coordinate (at a minimum) in order to sketch the graph with sufficient precision. Given the coordinates that we already have (and specifically focusing on their x values: -1, 2 and 5, I will use the input value of -2 which will also provide the output value of the solution pair that has an x value of 6 based on the fact that they are symmetry partners.

If $x = -2$

$$y = -((-2) + 1)((-2) - 5)$$
$$y = -(-1)(-7)$$
$$y = -7$$

Coordinate ---> (-2 , -7)
which also implies ---> (6 , -7)

The graph can now be sketched precisely.

$$y = -(x + 1)(x - 5)$$

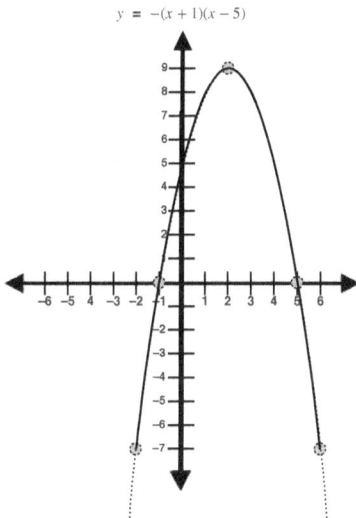

To prove how precise this graph actually is, let's put it to the test. Say we wanted to know the output value when the input value used is 0. Before computing this solution pair using the quadratic's equation, let's read the answer from the graph.

Remember that to do this, you must locate the **x = 0** position on the **x-axis** (since that's the input value we are interested in using), and then locate the point on the graph (by looking above or below that **x-axis** location) and then specifying that point's height: this will be the output value predicted by the graph.

$$y = -(x + 1)(x - 5)$$

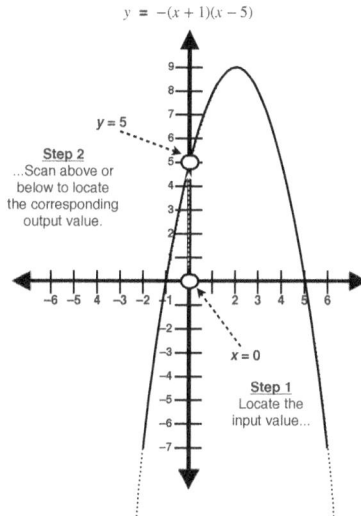

As you can see from the diagram above, using the two step process of reading a graph allows us to know the solution pair predicted by this graph we created, which is supposed to be precise. If it is precise, the solution pair predicted should be reasonably close to the actual value. The coordinate found is **(0 , 5)**, which is another way of saying that if the input value is **0** (the **x** value), that its corresponding output value is **5** (the **y** value).

To unequivocally find the correct output value given an input value, we can use the quadratic equation itself. So let's put the graph to the test.

$$\text{If } x = 0$$

$$y = -((0) + 1)((0) - 5)$$
$$y = -(1)(-5)$$
$$y = 5$$

Coordinate ---> (0 , 5)

We now have proof that the graph is reasonably well drawn, with sufficient precision that we can use it to read solution pairs of the equation.

Do keep in mind that the graph of an equation is supposed to be precise; it's supposed to represent all of the solution pairs of the equation that it embodies. A graph created by a computer or by a graphing calculator will always be precise enough to guarantee this (up to a certain level of precision, of course: a perfect graph that includes all points, or solution pairs, including **IRRATIONAL** number inputs, and **IRRATIONAL** number outputs, is in practice impossible). It is when we draw graphs by hand, especially non-linear graphs (of linear equations), where the graphs may not be precise enough to provide reasonable solution pair readings. But if you follow the techniques reviewed here for quadratic equations, your graph should be reasonably well drawn, as this last graph proved to be.

Let's explore two more examples involving a quadratic equation in intercept form. These will be somewhat more challenging than the previous two equations, but you should still be able to understand how to work them out. I will explain all of the steps involved.

$$y = -2(2x - 1)(2x + 3)$$

Does this equation look intimidating? It's actually quite manageable. First, as always, let's try to specify its **a** value and its **x-intercepts**.

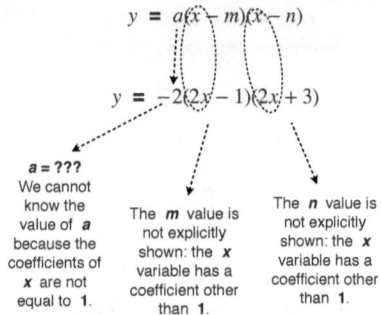

$$y = a(x - m)(x - n)$$

$$y = -2(2x - 1)(2x + 3)$$

a = ???
We cannot know the value of **a** because the coefficients of **x** are not equal to **1**.

The **m** value is not explicitly shown: the **x** variable has a coefficient other than **1**.

The **n** value is not explicitly shown: the **x** variable has a coefficient other than **1**.

As you can see in the diagram above, we cannot extract any of the values that we normally extract from the intercept form because the **x** variables have a coefficient different than **1** . In order to know its precise value, we need to extract the coefficients and rewrite the equation so that the **x** variables do have a coefficient of **1** .

$$y = -2(2x - 1)(2x + 3)$$
$$y = -2(2)(x - \frac{1}{2})(2)(x + \frac{3}{2})$$
$$y = -2(2)(2)(x - \frac{1}{2})(x + \frac{3}{2})$$
$$y = -8(x - \frac{1}{2})(x + \frac{3}{2})$$

Now, we may extract not only the **a** value, but the **m** , and **n** values as well.

Observe.

$$y = a(x - m)(x - n)$$

$$y = -8(x - \frac{1}{2})(x + \frac{3}{2})$$

$$a = -8$$

$$m = \frac{1}{2} \qquad n = -\frac{3}{2}$$

Based on the *a* value of **–8** we know that the parabola points down, and that its arms are very close together. As to the *m* and *n* values, we could have found them by using the original equation. Remember that we cannot know the actual value of *m* or *n* from the original equation because inside their respective set of parentheses, the *x* variable has a coefficient that is not equal to **1** (observe the circle elements on the earlier diagram that point this out explicitly); the general intercept form clearly expects the *x* variables to have a coefficient equal to **1**.

So what could be done other than extracting the coefficients directly? Well, we could use the zero product property, just as we used it before. The difference here–or rather, whenever the *x* variable has a coefficient that is **NOT** equal to one–, is that we have to set up the equations and solve them before being able to unequivocally determine the location of the *x*-intercepts.

Observe this technique in action. It is a very powerful method of finding the *hidden* *x*-intercepts and thus very commonly used.

$$0 = -2(2x - 1)(2x + 3)$$

$(2x - 1) = 0$	$(2x + 3) = 0$
$2x - 1 = 0$	$2x + 3 = 0$
$2x - 1 + 1 = 0 + 1$	$2x + 3 - 3 = 0 - 3$
$2x = 1$	$2x = -3$
$\dfrac{2x}{2} = \dfrac{1}{2}$	$\dfrac{2x}{2} = \dfrac{-3}{2}$
$x = \dfrac{1}{2}$	$x = -\dfrac{3}{2}$

...which leads to the following two *x*-intercepts:

$$(\frac{1}{2}, 0) \text{ and } (-\frac{3}{2}, 0)$$

You can now clearly appreciate why we couldn't simply extract the *m* and *n* values directly from the original quadratic equation in intercept form; using this other method, it was necessary to take the inside of each set of parentheses, set it to zero, and solve for *x* based on the zero product property. The **–2** factor that appears at the left of the first parentheses set is irrelevant because it is a factor (it multiplies the rest of the expression); therefore, if we can make the inside of either one of the sets of parentheses be equal to zero, the entire product will be equal to zero. You could, if you prefer, "send" that element to the left side, dividing **0** (by dividing both sides by **–2**); recall that **0** divided by anything other than **0** is equal to **0** : that would thus get rid of the element altogether. Observe.

$$0 = -2(2x - 1)(2x + 3)$$

$$\frac{0}{-2} = \frac{-2(2x - 1)(2x + 3)}{-2}$$

$$0 = (2x - 1)(2x + 3)$$

The lesson here is that whenever one or both of the sets of parentheses contain the variable x with a coefficient other than **1**, it is necessary to apply the zero product property directly.

The graph is now easy to sketch. Using the two **x-intercepts** found, and knowing that the parabola points down, we proceed as before.

Note that the axis of symmetry is located halfway between these two points, since they have the same output and are thus symmetry pairs:

axis of symmetry @ $\dfrac{.5 + (-1.5)}{2} = -.5$

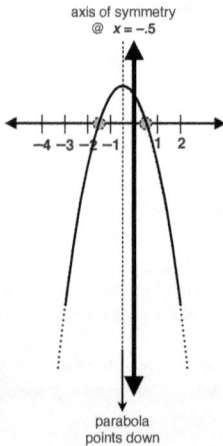

axis of symmetry
@ x = -.5

parabola
points down

To draw a precise graph, we would need to find the exact location of the vertex, and find two additional coordinates. For the vertex, we need to plug in an x value of **-.5** ; as to the other two coordinates, I will use an input value of **-2** (its symmetry partner would be **1**). But rather than use **-2**, I will use **1** instead since the equation is much easier to evaluate with the latter than with the former. Also note that I am using an **INTEGER** input value even though the two coordinates that we obtained (**x-intercepts**) have decimal x values.

For the vertex: if $x = -.5$

$$y = -2(2(-.5) - 1)(2(-.5) + 3)$$
$$y = -2(-2)(2)$$
$$y = 8$$
vertex @ (-.5 , 8)

For the two additional points: if $x = 1$

$$y = -2(2(1) - 1)(2(1) + 3)$$
$$y = -2(1)(5)$$
$$y = -10$$
Coordinate ---> (1 , -10)
which also implies ---> (-2 , -10)

I would like to point out that the original equation, $y = -2(2x - 1)(2x + 3)$, did not allow us to directly extract any of the values that we normally wish to extract from the intercept form. To figure out its a value, we had to remove the coefficients from the x variable not equal to **1**, which led to a version of the equation that also allowed us to extract the m and n values as well. Often, however, we are only interested in extracting the m and n values, so using the second method (setting each parentheses set to zero and solving for x), may make a lot more sense. In addition to this, when working with quadratic equations in general, it is often necessary to "solve them". You will see later in this chapter that using the zero product property as we did here is generally a desirable method that lets us solve a quadratic quickly and easily.

Now, using the vertex and the two additional coordinates that we found, we are ready to draw a precise version of the graph.

$$y = -2(2x - 1)(2x + 3)$$

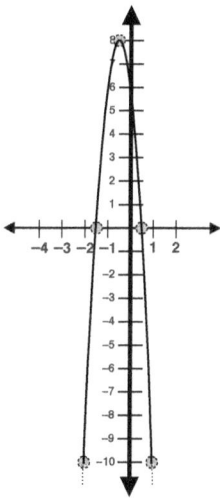

We can now move on to the third form of a quadratic equation, the Standard Form. This is typically a less friendly format because the only value that is explicitly available is the value of a ; anything else requires rewriting the equation or using *formulae*–this means *formulas* in Latin, a word that is also used in the English language—that apply to this specific format.

<u>Quadratic Equation in Standard Form</u>

$$y = ax^2 + bx + c$$
where $\mathbf{a \neq 0}$

The quadratic equation in standard form does not communicate anything beyond its a value, which tells us the orientation of the parabola based on its sign and the parabola's arm aperture. This would pose a major problem if we needed to graph the equation, or if we needed to know the vertex or even the axis of symmetry. Think about it: out of the infinite number of possible input values, how would we know which one corresponds to the vertex and thus to the graph's axis of symmetry? Are we left with trial and error as our only recourse? The answer, of course, is no. We do have options, so let's explore them by using a specific quadratic equation in standard form.

$$y = x^2 + 4x + 3$$

We only know three things about this quadratic equation: first, that it is a quadratic equation; second, that its a value is equal to $\mathbf{1}$ which specify the arm aperture of the equation's graph (parabola); and this leads to our third and last bit of knowledge about this equation: its graph (parabola) points up and thus the equation's output values increase as the input values used stray further and further from the axis of symmetry.

We can't really do much with this information if we wanted to graph the equation. If we randomly begin to compute solution pairs by using any input values we care to use, we would need to test a lot of values and carefully analyze the location of each plotted coordinate until a clear enough picture emerges of the parabola's true shape and form. I am sure you will agree that this is not a very practical method at all.

Instead, we are going to use the rules and principles that we have been using throughout the book to discover the formulae we can use to graph equations that are in standard form, and alternatively, how to "*factor*" the quadratic so that it is in intercept form, or how to "*complete the perfect square*" so that it is in vertex form. Sounds complicated, but it really isn't.

First, let's explore the idea of "factoring" the quadratic that is in standard form so that it is rewritten in intercept form.

The idea is the following: we would like to have the equation that is currently written as

$$y = ax^2 + bx + c$$

and express it instead as

$$y = a(x - m)(x - n)$$

To do this, we need to factor the standard form. It is important to note that all quadratic equations may be rewritten in factored form (intercept form), but some may require the use of **COMPLEX** numbers (remember imaginary numbers from **Chapter 12**?), while other equations end up having an *m* and/or an *n* value that consist of numbers with decimals or even **IRRATIONAL** numbers, which makes it practically impossible for us to come up with those values using the method I am about to explore below. The "factoring" method only works on a very special type of quadratic equation, so do be aware of this technique's limitation.

To attempt to factor a quadratic that is in standard form and that has an *a* value of **1** (this condition is crucial to the method below), all you have to do is the following.

Step 1. Take the original equation in standard form, and make sure that its *a* value is equal to **1**...

$$y = x^2 + 4x + 3$$

Step 2. Rewrite the equation as follows, leaving the blank spaces to be filled in later:

$$y = (x \quad)(x \quad)$$

Step 3. Look at the *b* and *c* values specifically from the original standard form equation and record them:

$$b = 4 \quad ; \quad c = 3$$

Step 4. You need to come up with two numbers, let's call them *d* and *e*, that will satisfy the following conditions:

$$(d)(e) = c$$

$$d + e = b$$

In this case, we need the following to occur:

$$(d)(e) = 3$$

$$d + e = 4$$

I always recommend you start with the product equation, because typically, the values of *d* and *e* are **INTEGERS**, and therefore, there is a very limited number of possibilities involving **INTEGERS** that would satisfy that product condition (since we are basically exploring using all **INTEGER** factors of *c*).

Therefore, since we need to find two numbers that multiplied are equal to 3 , we end up with the following **INTEGER** options:

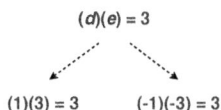

$$(d)(e) = 3$$

$$(1)(3) = 3 \qquad (-1)(-3) = 3$$

In other words, if this factoring method is going to work using **INTEGERS**, then we only have two options: either $d = 1$ and $e = 3$ **OR** $d = -1$ and $e = -3$. The trivial case where we reverse the values in each of those two possibilities (where $d = 3$ and $e = 1$, for example) need not be considered, since they essentially represent the same set up.

We now have two options: either $d = 1$ and $e = 3$ **OR** $d = -1$ and $e = -3$ ... So how do we know if one of these two works? Well, remember that the values also need to satisfy this second condition:

$$d + e = b$$

...which in this case means that

$$d + e = 4$$

Let's test each of the two set of options to see if either one works:

When $d = 1$ and $e = 3$ we find that

$$(1) + (3) = 4$$

is **TRUE!**

We therefore have the values we need to factor the quadratic that is in standard form. Look what happens when we test the other pair of values.

When $d = -1$ and $e = -3$ we find that

$$(-1) + (-3) = 4$$

is **FALSE!**

If we add $(-1) + (-3)$ we get the following correct answer:

$$(-1) + (-3)$$

$$= -1 - 3$$

$$= -4$$

Now that we know that the numbers we need are 1 and 3 , what do we do next?

Step 5. Once you found the correct pair of numbers, "plug" them into the set of parentheses from **Step 2**, adding the number to the x variable if positive, subtracting it if negative.

$$y = (x + 1)(x + 3)$$

We now have the same quadratic equation that was originally written in standard form, but now expressed in intercept form.

There is an optional step, **Step 6**, that verifies that we factored the quadratic equation correctly. You should use this step at your discretion: before you are an expert at this, you may want to always check your factored version by using it.

Step 6. Verify that your factored version is, in fact, the same as the standard version. The idea here is simple. If the factored form version is supposed to be the same as the standard form version, then foiling the factored form should yield the exact same equation that we originally started out with. Let's test the equation we came up with.

$$y = (x + 1)(x + 3)$$

$$y = (x)(x) + (x)(3) + (1)(x) + (1)(3)$$

$$y = x^2 + 3x + x + 3$$

$$y = x^2 + 4x + 3$$

And there you have it! Proof that the factored form we came up with is, in fact, equivalent to the original standard form that we started out with.

Now that the equation is in intercept form, we know its a value ($a = 1$), and we know its two **x-intercepts**: (**0 , –1**) and (**0 , –3**) . The axis of symmetry is thus located at $x = -2$. You could then proceed as we did earlier and sketch this graph precisely. I will leave that as an exercise that you can do on your own; you can always check your graphs by using an online calculator (many colleges and universities, as well as school districts, have links on their websites to free, online graphing calculators).

So there you have it. Factoring quadratics in standard form seems complicated, but its actually rather simple.

There is another way of approaching this technique. It embodies the same principles, but it may be a bit easier for some to use because it is a bot more straight-forward. Observe.

The idea is to write the original equation in standard form, and then below it, the factored form with the blank spaces (like **Step 2** of the previous technique). Then simply ask yourself what two numbers multiply to the independent term and simultaneously, when added, add to the middle term's coefficient (the linear term, remember?). It looks like this:

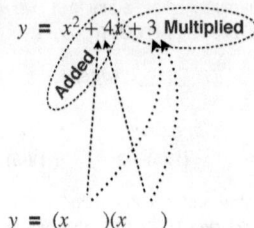

$$y = x^2 + 4x + 3 \quad \text{Multiplied}$$
$$\text{Added}$$

$$y = (x \quad)(x \quad)$$

As you can see from the diagram above, this technique is in essence the same as the previous technique that we reviewed; however, rather than approach it as a step-by-step process, this one simply asks you to think about what two pair of numbers (**INTEGERS**, usually) multiply to the independent term and add to the middle term's coefficient.

A **_CRUCIAL_** point that needs to be emphasized here is that you **_MUST_** take into account whether the independent term is being **_added_**–if so, then you want the product to be equal to a **_positive number_**, but if it is being **_subtracted_**, then you want the product to be equal to a **_negative number_**–, and likewise, whether the middle term is being added (if so, the **_SUM_** must be equal to a **_positive number_**, but if it is being **_subtracted_**, then the SUM must be equal to a **_negative number_**).

You may still want to write out the possible factors of the independent term, and test which pair have a SUM equal to the middle term's coefficient (taking into account the correct sign, as explained in the above paragraph), but that's up to you and to your level of expertise in applying this technique. You will find that in most cases, you can quickly come up with the factored form without having to write anything beyond the factored form itself, performing all your calculations in your mind.

Please remember that this technique ONLY works for quadratic equations in standard form that have an *a* value of 1 ; in other words, for quadratic equations in standard form whose leading term (the *quadratic term*: the term with the variable *x* being squared) is equal to 1 . If this is not the case, there is still hope. We will review shortly how to attempt to factor such quadratic equations in standard form.

For now, let's continue to practice this factoring technique (regardless of the version you wish to use: remember that they are essentially the same).

Let's quickly factor the following quadratic equations in standard form. You should try to solve them on your own: the more we do together, the more you should be able to factor by yourself.

a) $y = x^2 + 5x + 6$
b) $y = x^2 + 8x + 16$
c) $y = x^2 - 4x + 4$
d) $y = x^2 - 3x + 2$
e) $y = x^2 - x - 12$
f) $y = x^2 - 2x + 1$
g) $y = x^2 + 2x + 1$
h) $y = x^2 - 5x + 4$
i) $y = x^2 + x - 6$
j) $y = x^2 + 3x - 4$
k) $y = x^2 - 3x - 10$

Try to solve them by yourself. If you need to see how I solve the first few problems before setting

out to solve the rest of them on your own, do so. But do try to work on as many as you can by yourself, since that is the only way that you will truly learn and master this technique.

I am not going to foil the factored form that I obtain on each of the problems below (to verify that the factored form is, in fact, equivalent to the standard form); I will let you do that so you can practice your foiling skills and so that you may experience first-hand why the factored form is equivalent to the standard form to begin with.

I also suggest that you specify the *x*-intercepts that each factored quadratic specifies. It is a crucial skill that must be mastered.

As to the graphs, you may wish to graph some of these quadratic equations so that you can practice what we reviewed earlier regarding graphing intercept form quadratic equations.

a) $y = x^2 + 5x + 6$

We need two **INTEGERS** that multiply to 6 and whose sum is equal to 5 . Because **(2)(3) = 6** and **2 + 3 = 5** , we know that the factored form of this quadratic equation must be

$$y = (x + 2)(x + 3)$$

x-intercepts @ (–2 , 0) and (–3 , 0)

b) $y = x^2 + 8x + 16$

We need two **INTEGERS** that multiply to 16 and whose sum is equal to 8 . Because **(4)(4) = 16** and **4 + 4 = 8** , we know that the factored form of this quadratic equation must be

$$y = (x + 4)(x + 4)$$

x-intercept @ (–4 , 0)

c) $y = x^2 - 4x + 4$

We need two **INTEGERS** that multiply to **4** and whose sum is equal to **–4** (don't forget to consider whether the middle term and the independent term are being added or subtracted). Because **(–2)(–2) = 4** and **(–2) + (–2) = –4** , we know that the factored form of this quadratic equation must be

$$y = (x - 2)(x - 2)$$

x-intercept @ (2 , 0)

Factoring equations such as this one is typically challenging for those that are attempting to master this technique. The lesson here is this: don't forget to consider the negative version of a factor pair; in other words, when you set out to find two **INTEGERS** that multiply to positive four, remember that although positive two times positive two is an option (which does not work in this example because the sum of the **INTEGERS** has to be negative four, not positive four), so is negative two times negative two because two negatives multiplying yield a positive result. This is the pair of values that actually work because when you find the sum of negative two and negative two, it leads to the following computation:

$$(-2) + (-2)$$

$$= -2 - 2$$

$$= -4$$

...which is the SUM we needed to have. With a bit of practice, you will learn to consider all possible factors, even switching around which number gets the negative sign as in a few of the problems that are coming up shortly.

d) $y = x^2 - 3x + 2$

We need two **INTEGERS** that multiply to **2** and whose sum is equal to **–3** . Because **(–2)(–1) = 2** and **(–2) + (–1) = –3** , we know that the factored form of this quadratic equation must be

$$y = (x - 2)(x - 1)$$

x-intercepts @ (2 , 0) and (1 , 0)

This example is very similar to the previous one we worked with. Remember to consider negative numbers as well when you are trying to find the pair of numbers for the factored form of the equation.

e) $y = x^2 - x - 12$

We need two **INTEGERS** that multiply to **–12** and whose sum is equal to **–1** . Because **(3)(–4) = –12** and **3 + (–4) = –1** , we know that the factored form of this quadratic equation must be

$$y = (x + 3)(x - 4)$$

x-intercepts @ (–3 , 0) and (4 , 0)

f) $y = x^2 - 2x + 1$

We need two **INTEGERS** that multiply to **1** and whose sum is equal to **–2** . Because **(–1)(–1) = 1** and **(–1) + (–1) = –2** , we know that the factored form of this quadratic equation must be

$$y = (x - 1)(x - 1)$$

x-intercept @ (1 , 0)

g) $y = x^2 + 2x + 1$

We need two **INTEGERS** that multiply to **1** and whose sum is equal to **2** . Because **(1)(1) = 1** and **1 + 1 = 2** , we know that the factored form of this quadratic equation must be

$$y = (x + 1)(x + 1)$$

x-intercepts @ (−1 , 0)

h) $y = x^2 - 5x + 4$

We need two **INTEGERS** that multiply to **4** and whose sum is equal to **−5** . Because **(−4)(−1) = 4** and **(−4) + (−1) = −5** , we know that the factored form of this quadratic equation must be

$$y = (x - 4)(x - 1)$$

x-intercepts @ (4 , 0) and (1 , 0)

i) $y = x^2 + x - 6$

We need two **INTEGERS** that multiply to **−6** and whose sum is equal to **1** . Because **(3)(−2) = −6** and **3 + (−2) = 1** , we know that the factored form of this quadratic equation must be

$$y = (x + 3)(x - 2)$$

x-intercepts @ (−3 , 0) and (2 , 0)

j) $y = x^2 + 3x - 4$

We need two **INTEGERS** that multiply to **−4** and whose sum is equal to **3** . Because **(4)(−1) = −4**

and **4 + (−1) = 3** , we know that the factored form of this quadratic equation must be

$$y = (x + 4)(x - 1)$$

x-intercepts @ (−4 , 0) and (1 , 0)

k) $y = x^2 - 3x - 10$

We need two **INTEGERS** that multiply to **−10** and whose sum is equal to **−3** . Because **(−5)(2) = −10** and **(−5) + 2 = −3** , we know that the factored form of this quadratic equation must be

$$y = (x - 5)(x + 2)$$

x-intercepts @ (5 , 0) and (−2 , 0)

So how well did you do on these problems? Once you get into the habit of considering both the negative and the positive options (and their combinations), you will find that factoring quadratic equations that are in standard form is relatively easy to do.

But what happens if the coefficient of the quadratic term is not equal to **1** (in other words, when the value of **a** is not equal to **1**)? This involves using a different technique which can be intimidating, but after doing a few on your own, you will find that it is actually very manageable.

Let's explore it with an example.

$$y = 2x^2 + 7x + 3$$

You can see that the two techniques we reviewed earlier do not apply to this quadratic. The coefficient of the quadratic term is **2** , not **1** as we need it to be. Therefore, we must use this other technique.

Step 1. Find two numbers that multiply to $(a)(c)$ and whose sum is equal to b.

In this case, since $a = 2$, $b = 7$, and $c = 3$, we need to find two numbers that multiply to 6 (because $(2)(3) = 6$) and whose sum is 7.

Clearly, the only two numbers that work are **6** and **1** because

$$(6)(1) = 6$$

$$6 + 1 = 7$$

Now that we know which two numbers satisfy the condition specified earlier, we move on to the next step. It is very different from the previous technique, so do read this carefully.

Step 2. Rewrite the quadratic equation, changing the middle term–also called the linear term, remember?–using the two numbers found, as follows:

$$y = 2x^2 + 7x + 3$$

$$6 \text{ and } 1$$

$$y = 2x^2 + 6x + 1x + 3$$

See how the middle term, **7x**, was rewritten as **6x + 1x**? Of course, we don't actually write **1x** as you well know; I only did it to show you that it is there. The way we would write it would be as follows, without the "**1**" explicitly written:

$$y = 2x^2 + 6x + x + 3$$

Step 3. This is the challenging part of the technique. The idea is to group together the first two terms (the quadratic and one of the linear

terms that we defined in **Step 2**), and to group the last two terms, as follows:

$$y = (2x^2 + 6x) + (x + 3)$$

You have to be very careful with the addition and subtraction symbols and thus the negative signs that may have to be used when doing this. In this example, we have nothing to worry about because all of the terms are positive and adding. We will explore how to cope with a situation like that shortly.

Once grouped, you will attempt to factor out as much as possible from both parentheses sets that you created: the $(2x^2 + 6x)$ and $(x + 3)$.

$$y = (2x)(x + 3) + (x + 3)$$

Note that in this example, the second set of parentheses remained unchanged, since we can't factor anything out from the terms **x** and **3** because they do not have any common elements, unlike the first set where we factored out the **2x** from both terms.

Step 4. You should end up with the same set of parentheses on the two newly created term. Factor it out. Confused? Observe the diagram.

$$y = (2x)(x + 3) + (x + 3)$$

Term 1   Term 2

Addition symbol; it separates both terms

As you can see above, the "element" $(x + 3)$ appears on both terms, so it can be factored out. If this confuses you, remember that the element

$(x + 3)$ can be thought of as a black box, or a variable, or a number, or whatever makes it easier for you to see that it ultimately can be factored out of both terms. To illustrate this, let me replace $(x + 3)$ with **E**:

$$y = (2x)(x + 3) + (x + 3)$$

$$y = (2x)E + E$$

See how replacing $(x + 3)$ with **E** allows you to see it clearly as a repeating element in both terms? We could now factor out the **E** from both terms as follows:

$$y = (2x)E + E$$

$$y = E((2x) + 1)$$

And if we can do that with **E** we can do that with $(x + 3)$ as well. Observe.

$$y = (2x)(x + 3) + (x + 3)$$

$$y = (x + 3)(2x + 1)$$

We now have the equation in intercept form, which was the goal to begin with. If we want to know the **x-intercepts** of the equation, we simply extract the **m** value from the first parentheses set, and set the second parentheses set to zero and solve it for **x** to find **m**.

Remember that this is also called "***finding the zeros of the equation***", and that we can extract the **m** directly because the **x** variable in that set of parentheses has a coefficient equal to **1**.

Observe the process.

$$y = (x + 3)(2x + 1)$$

$(x + 3) = 0$
$x + 3 = 0$
$x + 3 - 3 = 0 - 3$
$x = -3$

$(2x + 1) = 0$
$2x + 1 = 0$
$2x + 1 - 1 = 0 - 1$
$2x = -1$
$\dfrac{2x}{2} = \dfrac{-1}{2}$
$x = \dfrac{-1}{2}$

And so the two **x-intercepts** are $(-3, 0)$ and $(\dfrac{-1}{2}, 0)$. As to the vertex, we can find the sum of these two **x** values and divide it by two, since these two points share the same **y** value and are thus symmetry pairs. We would therefore find the axis of symmetry first, and use this **x** value to find the output in order to determine the precise location of the vertex.

axis of symmetry @ $\dfrac{-3 + (\frac{-1}{2})}{2}$

In equation form: **x** $= \dfrac{-7}{4}$

or, using decimals: **x** $= -1.75$

Now, using this **x** value, we can use it as the input to find its corresponding output in order to locate the vertex.

If $x = \dfrac{-7}{4}$

$$y = ((\frac{-7}{4}) + 3)(2(\frac{-7}{4}) + 1)$$

$$y = (\frac{5}{4})(\frac{-5}{2})$$

$$y = \frac{-25}{8}$$

vertex @ $(\frac{-7}{4}, \frac{-25}{8})$

or, using decimals, @ (–1.75 , –3.125)

We now have the equation in intercept form, and we have found the exact location of its axis of symmetry and the vertex. If we needed to graph it, we would have to find two additional coordinates: I would suggest using the input value of **0** for x because it is very easy to evaluate an equation with this particular input number. Note that its symmetry pair would be (–3.5 , ?) because the distance between the axis of symmetry and this x value of **0** (which is to the right of the axis of symmetry) is **1.75** units long; so, moving to the left of the axis of symmetry **1.75** units, we "land" on the number –3.5 . Observe the diagram.

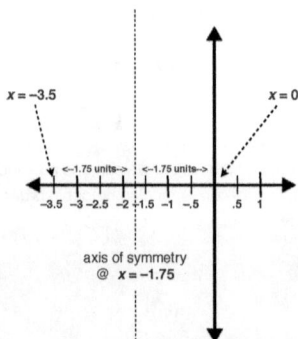

And so to find the y value of these two additional coordinates, I will plug in $x = 0$ into the original equation (the standard form version since it is much easier to evaluate it given this particular input value):

For the two additional points: if $x = $ **0**

$$y = 2x^2 + 7x + 3$$
$$y = 2(0)^2 + 7(0) + 3$$
$$y = 0 + 0 + 3$$
$$y = 3$$

Coordinate ---> (0 , 3)
which also implies ---> (–3.5 , 3)

We now have all we need to draw a precise graph of the equation: a total of five coordinates, which includes the location of the vertex, we now its orientation (it points up), and we know that it is a parabola.

Before we move on to another example, I want you to consider the following situation which you could have encountered in **Step 3** of this process. You may recall that this is where we grouped terms after substituting, in **Step 2**, the middle term (the linear term) with two linear terms using the two numbers found on said step. But what if we had done it as follows instead of how we actually did it earlier? Observe.

Alternate Step 3. Recall that in this step, the idea is to group together the first two terms (the quadratic and one of the linear terms that we defined in **Step 2**), and to group the last two terms. This is what we did originally:

$$y = (2x^2 + x) + (6x + 3)$$

In this alternate version, say we grouped the terms the other way around, as follows:

$$y = (2x^2 + x) + (6x + 3)$$

Once grouped, you will attempt to factor out as much as possible from both parentheses sets that you created: the $(2x^2 + x)$ and $(6x + 3)$.

$$y = (x)(2x + 1) + (3)(2x + 1)$$

Note that in this alternate version, we end up with a completely different (yet equivalent) equation. Compare it to what we originally obtained before:

$$y = (2x)(x + 3) + (x + 3)$$

If you are now worried that we will not end up with the same final factored equation, don't be. Observe what happens when we continue applying the steps to this alternate version.

Alternate Step 4. You should end up with the same set of parentheses on the two newly created term. Factor it out. Confused? Observe the diagram.

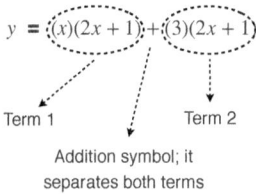

$$y = (x)(2x + 1) + (3)(2x + 1)$$

Term 1   Term 2

Addition symbol; it
separates both terms

As you can see above, this time it is the "element" $(2x + 1)$ that appears on both terms, so it can be factored out. If this confuses you, remember that the element $(2x + 1)$ can be thought of as a black box, or a variable, or a

number, or whatever makes it easier for you to see that it ultimately can be factored out of both terms. To illustrate this, let me replace $(2x + 1)$ with, this time, the variable **R**:

$$y = (x)(2x + 1) + (3)(2x + 1)$$

$$y = (x)R + (3)R$$

See how replacing $(2x + 1)$ with **R** allows you to see it clearly as a repeating element in both terms? We could now factor out the **R** from both terms as follows:

$$y = (x)R + (3)R$$

$$y = R((x) + (3))$$

$$y = R(x + 3)$$

And if we can do that with **R** we can do that with $(x + 3)$ as well. Observe.

$$y = (x)(2x + 1) + (3)(2x + 1)$$

$$y = (2x + 1)(x + 3)$$

We now have the equation in intercept form, which was the goal to begin with. Furthermore, it is the exact same equation we obtained originally.

So what's the lesson here? The lesson is that it does not matter how you group the terms; as long as **Step 1** is done correctly (finding the two numbers that will be used to replace the middle term), everything else will work out automatically.

We can now work on a second example. Let's try to find the factored (intercept form) of the following equation:

$$y = 6x^2 + 5x - 6$$

Because the coefficient of the quadratic term is not equal to **1** (in other words, because its **a** value is not equal to **1**), we must use this other method to attempt to factor this quadratic equation.

Step 1. Find two numbers that multiply to **(a)(c)** and whose sum is equal to **b** (careful how you extract the values of **a** , **b** , and **c** : remember to take into account if the terms they appear in are adding or subtracting, since that will have an impact on whether to consider the values to be positive or negative).

In this case, since $a = 6$, $b = 5$, and $c = -6$, we need to find two numbers that multiply to **−36** (because $(6)(-6) = -36$) and their sum to be equal to **5** .

Something that may help you find the two numbers you are looking for in this step is to explicitly write out all the factors of the product you are looking to obtain (in this case, **−36**):

Factors of **−36** ...

1 , **2** , **3** , **4** , **6** , **9** , **12** , **18** , **36**

We do not need to list the negative factors, since all we need to do is consider a negative number and a positive number so that their product will yield the negative number we are looking for. Keep it simple and always list the positive factors.

Now, since we need the product to be equal to **−36** , we look for pairs of numbers from this list that when multiplied, equal **−36** . Again, simply consider one of numbers positive and the other negative to achieve this. For example, if we start out by looking at **1** and **36** as the pair of

numbers we are looking for, we could think of the following two options:

Option 1

1 and **−36**

Option 2

−1 and **36**

Clearly, the product of the numbers listed in each of the options above do equal **−36** , the product we need. However, remember that the **sum** of the numbers must also be equal to the value of **b** , in this case, **5** . Let's test the two options above to see if one of them works.

Option 1

1 and **−36**

$(1)(-36) = -36$ $(1) + (-36) = -35$

Option 2

−1 and **36**

$(-1)(36) = -36$ $(-1) + (36) = 35$

As you can see above, although the product condition those work out on both cases, the sum condition is not met. Furthermore, the sum is different on each of the two options considered. Why? Because in each of the two options above, the negative sign is attached to a different number, which impacts the result of their sum. This is important to consider because when you are looking for the pair of numbers that do work, when trying to satisfy the sum condition, remember to switch the negative sign between the numbers you are considering before moving on to another pair of numbers.

Moving on, let's consider the following two numbers: 9 and 4 . Clearly, if we assign the negative sign to, say, the 9 , the product of these numbers will be the required **−36** we are looking for:

<u>Option 1</u>

−9 and 4

(−9)(4) = −36

But will the sum condition be satisfied? Let's see:

<u>Option 1</u>

−9 and 4

(−9)(4) = −36   (−9) + (4) = −5

It does not work out. But you see, if you gave up on these two numbers just because this particular negative sign assignment did not work, then you would be giving up on the pair of numbers that does work for this equation! This is why it is important to remember to always explore switching the negative sign around whenever the product is negative (and you thus need one of the numbers to be negative). Observe what happens when we reassign the negative sign to the other number instead in option 2.

<u>Option 2</u>

9 and −4

(9)(−4) = −36   (9) + (−4) = 5

As you can see above, the sum condition is also satisfied!

We may now continue with the process.

Step 2. Rewrite the quadratic equation, changing the middle (linear) term using the two numbers we found on **Step 1**, as follows:

$$y = 6x^2 + 5x - 6$$

9 and −4

$$y = 6x^2 + 9x - 4x - 6$$

Step 3. Remember that this is the challenging part of the technique. The idea is to group together the first two terms (the quadratic and one of the linear terms that we defined in **Step 2**), and to group the last two terms, as follows:

$$y = (6x^2 + 9x) + (-4x - 6)$$

You have to be very careful with the addition and subtraction symbols and thus the negative signs that may have to be used when doing this. In this example, I decided to add the second parentheses set; this meant considering the term 4x as a negative, while maintaining the independent term, 6 , as subtracting. Alternatively, I could have written the same equation as follows:

$$y = (6x^2 + 9x) - (4x + 6)$$

Why is the above equation also equivalent, and correct? Because since we are now subtracting the second parentheses set, then the terms should be positive and adding so that the equality between the versions is maintained. We have been reviewing these skills throughout the book, so I will simply prove to you that the two versions above are, in fact, equivalent, by simplifying them, removing the parentheses as per the Sign Table technique as we have done in earlier chapters of the book. Observe.

Version 1

$$y = (6x^2 + 9x) + (-4x - 6)$$
$$y = 6x^2 + 9x + (-4x) + (-6)$$
$$y = 6x^2 + 9x - 4x - 6$$

Version 2

$$y = (6x^2 + 9x) - (4x + 6)$$
$$y = 6x^2 + 9x - (4x) - (6)$$
$$y = 6x^2 + 9x - 4x - 6$$

As you can see above, both versions are perfectly equivalent. It does not really matter which version you use: if you apply the principles and rules that we have reviewed throughout the book, you will always end up with the same, final, correct answer.

I will solve both versions so that you may witness firsthand that the above statement is true.

Once grouped, you will attempt to factor out as much as possible from both parentheses sets that you created.

Version 1

$$y = (6x^2 + 9x) + (-4x - 6)$$
$$y = (3x)(2x + 3) + (2)(-2x - 3)$$
$$y = (3x)(2x + 3) + (-2)(2x + 3)$$
$$y = (3x)(2x + 3) - (2)(2x + 3)$$

Version 2

$$y = (6x^2 + 9x) - (4x + 6)$$
$$y = (3x)(2x + 3) - (2)(2x + 3)$$

As you can see above, although the first version took a couple more steps to get there, both versions end up being exactly the same. As

promised, they are equivalent, and you never have to worry about which version you may have chosen.

Step 4. Factor out the set of parentheses that is common to both terms.

$$y = (3x)(2x + 3) - (2)(2x + 3)$$
$$y = (2x + 3)((3x) - (2))$$
$$y = (2x + 3)(3x - 2)$$

We now have the equation in intercept form, which was the goal to begin with.

You should be able to do the rest of the process on your own: find the axis of symmetry, the vertex, and the two **x-intercepts**, and graph the equation using these elements.

Just remember that you are going to have to set each of the inside of the parentheses to zero and solve for **x** since you cannot extract the **m** and **n** values directly (both **x** variables have a coefficient that is not equal to **1**).

After you work it, compare it to my answer below.

Ready?

The two **x-intercepts** are...

$$\left(-\frac{3}{2}, 0\right) \text{ and } \left(\frac{2}{3}, 0\right)$$

which may also be written as (in decimal form)...

(−1.5 , 0) and (.666... , 0)

Since these two coordinates are symmetry pairs, the axis of symmetry must lie exactly in the middle. We thus find the sum of the two **x** values of these coordinates and divide by two to find the location of the axis of symmetry:

axis of symmetry @ $\dfrac{-\frac{3}{2} + (\frac{2}{3})}{2}$

$= \dfrac{-\frac{9}{6} + (\frac{4}{6})}{2} = \dfrac{-\frac{5}{6}}{2} = \dfrac{-\frac{5}{6}}{\frac{2}{1}} = \dfrac{-5}{12}$

In equation form @ $x = \dfrac{-5}{12}$

or, using decimals @ $x = -.41666...$

Now we can find the vertex.

If $x = \dfrac{-5}{12}$

$y = (2(\dfrac{-5}{12}) + 3)(3(\dfrac{-5}{12}) - 2)$

$y = (\dfrac{-10}{12} + 3)(\dfrac{-15}{12} - 2)$

$y = (\dfrac{-10}{12} + \dfrac{36}{12})(\dfrac{-15}{12} - \dfrac{24}{12})$

$y = (\dfrac{26}{12})(\dfrac{-39}{12})$

$y = (\dfrac{13}{6})(\dfrac{-13}{4})$

$y = \dfrac{-169}{24}$

vertex @ $(\dfrac{-5}{12}, \dfrac{-169}{24})$

or @ (**−.41666... , −7.041666...**)

Next, to find a fourth and a fifth coordinate, I would use the input value of **0** as last time, and

for the same reasons. The trick here is to correctly specify its symmetry partner. Since the distance between the axis of symmetry and this input value of **0** is, in absolute terms, $\dfrac{5}{12}$ units (to the **right**), all we need to do is move to the **left** of the axis of symmetry that same number of units, from the same spot where the axis of symmetry is located. That leads to the following computation:

Symmetry partner of (**0 , ?**) is located at

$\dfrac{5}{12} + \dfrac{5}{12} = \dfrac{10}{12} = \dfrac{5}{6}$

But we must take into account the fact that the symmetry partner is to the left of the axis of symmetry, so the **x** value must be negative.

Thus, (**0 , ?**) and ($\dfrac{5}{6}$ **, ?**) are symmetry partners.

Another option would have been to start out at the location of the x value of the axis of symmetry, and "walk to the left" $\dfrac{5}{12}$ units, which means in math terms, subtract $\dfrac{5}{12}$ from $\dfrac{-5}{12}$ as follows:

$\dfrac{-5}{12} - \dfrac{5}{12} = \dfrac{-10}{12} = \dfrac{-5}{6}$

...leading, of course, to the same position on the **x-axis**. You can use either one of the two methods discussed above; they are equivalent and lead to the same answer. You do, however, have to make sure you fully grasp the process; to help you master it, observe, once more, its graphic illustration.

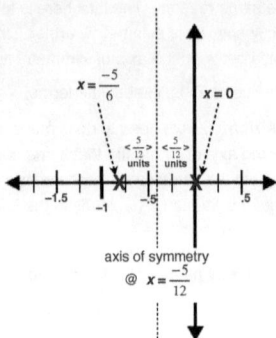

$x = \dfrac{-5}{6}$

$x = 0$

$\xleftarrow{\hspace{0.3cm}} \dfrac{5}{12} \xrightarrow{\hspace{0.3cm}}$ units $\xleftarrow{\hspace{0.3cm}} \dfrac{5}{12} \xrightarrow{\hspace{0.3cm}}$ units

-1.5   -1   -.5   .5

axis of symmetry
@ $x = \dfrac{-5}{12}$

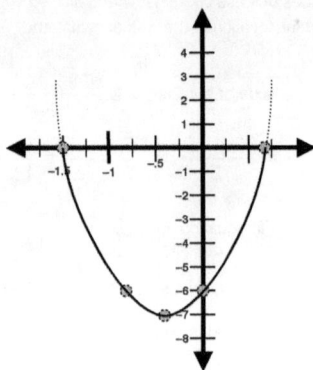

The diagram above offers a visual representation of how to correctly find the symmetry partner of any point that is part of a parabola.

Let's find the output value of these two symmetry pari points.

For the two additional points: if $x = \mathbf{0}$

$$y = 6x^2 + 5x - 6$$
$$y = 6(0)^2 + (0)x - 6$$
$$y = 0 + 0 - 6$$
$$y = -6$$

Coordinate ---> (0 , –6)

which also implies ---> ($\dfrac{-5}{6}$, –6)

We are now ready to sketch the graph.

Using the vertex, the two **x-intercepts** that we found earlier, and the two additional coordinates that we found above, we are ready to draw a precise graph of the equation.

Compare it to yours.

The last two examples have been quite challenging. If you were able to work these problems on your own, you are truly mastering these skills. If you still feel like you need more practice, we will continue to work on more problems, so don't worry. Just remember to always try to work out the problems by yourself before looking at my answers.

After working on these last two examples, you are ready to understand why the following statement is true:

The standard form of a quadratic equation explicitly provides the following information regarding the quadratic and its graph:

1. The value of **a** , which means that we can know right away the orientation of the parabola, which means we know whether it points up or down; or, to state this differently, we are able to know the type of behavior that the equation embodies: its output either increases (tends towards positive infinity) the farther away we

642

move from the axis of symmetry, or its output value decreases (tends towards negative infinity) the farther away we move from the axis of symmetry. This "moving away" applies in both directions: to the left and to the right of the axis of symmetry.

The *a* value also provides us with knowledge regarding the parabola's arm aperture. Remember that as the absolute value of *a* increases, the arm aperture becomes tighter and tighter; and as the absolute value of *a* approaches **0**, the arm aperture becomes wider and wider.

2. The *y*-intercept. Although I have not mentioned this explicitly before, note that on the last two examples we worked on, whenever an input value of **0** was used, we ended up with an output value that is equal to the independent term of the quadratic's equation in standard form. And it makes perfect sense. The other terms have the variable *x* multiplying whatever else may be on the term; so it stands to reason that if we use an input value of **0**, those terms will end up being equal to **0** since anything times **0** is **0**; thus, plug in **0** for *x* (which corresponds to the location of the *y*-intercept point) and the output (*y*) will be equal to the independent term. Observe this on the last example we worked on.

$$y = 6x^2 + 5x - 6$$

Plug in **0** and both of these terms will equal **0**

$$y = 6(0)^2 + 5(0) - 6$$
$$y = 0 + 0 - 6$$

Which is why the standard form gives away explicitly the **y-intercept** point. Although it is not enough to graph the quadratic with, it is still a very useful point to have when graphing the equation.

We can now summarize all that the different quadratic forms explicitly specify, and I will also include a few useful formulae that can be useful.

⚓

Quadratic Equation in Standard Form

$$y = ax^2 + bx + c$$
where **a ≠ 0**
y-intercept @ **(0 , c)**
axis of symmetry @ $x = \dfrac{-b}{2a}$
vertex @ $(\dfrac{-b}{2a}, \dfrac{4ac - b^2}{4a})$

Quadratic Equation in Vertex Form

$$y = a(x - h)^2 + k$$
where **a ≠ 0**
vertex @ **(h , k)**
axis of symmetry @ *x = h*

Quadratic Equation in Intercept Form

$$y = a(x - m)(x - n)$$
where **a ≠ 0**
x-intercepts @ **(m , 0)** and **(n , 0)**
axis of symmetry @ $x = \dfrac{m + n}{2}$

For all forms:
a ---> if + : up ; if − : down
| *a* | = *arm aperture*

You can check the two formulas indicated in the standard form section above using the last two examples we worked on. You should find that they work perfectly.

So what is so important about quadratic equations and finding the x-intercepts anyway?

Well, many real-world problems involve setting up a quadratic equation. It is then necessary to solve it so as to find the value or values that satisfy the conditions of the problem that led to the quadratic equation to begin with. Which means, in turn, that if we want to solve the equation, we MUST set it to zero and then use any one of the methods used earlier to attempt to find its *zeros*, or *solutions*, or *x-intercepts*.

Let's explore this idea with an example.

Say that someone throws a ball straight up in the air, releasing it exactly **2m** (**m** = meters) above the ground. Its initial velocity is **3m/s** (**m/s** = *meters per second*).

How long would it take for the ball to reach the ground?

If this problem seems too complicated, it's actually not. You do have to know—from a course in physics, for instance—that when an object is launched straight up in the air, its motion can be modeled using the formula for a free-falling object, which gives its velocity in terms of how many seconds have passed. The formula is:

$$V = -\frac{at^2}{2}$$

The negative sign is there because in the frame of reference that we are using (which is how it's commonly defined in these type of physics problems), traveling away from Earth is considered a positive velocity, while traveling towards Earth is considered a negative velocity, and clearly, a free falling object is being pulled towards the planet by Earth's gravity, thus the negative sign. FYI, speed differs from velocity in that speed is always considered to be positive: unlike velocity, it lacks a sense of direction.

So, back to our problem. The height of the ball is going to be given by the following sum: the starting height of the ball (that's 2), plus the velocity with which it travels initially (at the moment of release, its velocity is $3t$, positive, since it is directed away from Earth), and minus the pull of Earth's gravity, $-\frac{at^2}{2}$. Since the **a** in this formula stands for acceleration and it is equal to approximately **9.81m/s²**, we can round the entire term to $-5t^2$, because if we round the acceleration to **10m/s²** we obtain

$$-\frac{10t^2}{2} = -5t^2$$

And so the formula we end up with that governs the position (height, or **h**) of the ball, based on the amount of time (**t**) that goes by, is

$$h = 2 + 3t - 5t^2$$

Our goal is to know how many seconds will go by right before the ball lands on the ground. The reason we will have this measurement of time in seconds is because the velocity and the acceleration used in the equation were both defined in terms of seconds, as should be. All formulas must be consistent in terms of the units used. If, for example, you were given the velocity of the ball in meters per hour, then you would have to change the acceleration to meters per hour so that they both have matching units, or change the velocity of the ball to meters per second; the lesson is that all units must match.

So how do we solve this? Well, the key is to know that we want h to be equal to 0. We can thus do as follows:

$$0 = 2 + 3t - 5t^2$$

Now does this look familiar? It is a quadratic equation, and we want to solve it for t (think of this variable t as being the x we have been using; it's in essence equivalent).

But what does this equation represent? Well, it represents the idea that we need the right side to be equal to zero (the zero is what's to the left of the equal sign). And this means that we want to know which input value for t will yield an output equal to 0.

$$0 = 2 + 3t - 5t^2$$

What *input* value must be used so that this *right side* is equal to 0 ?

$$t = ?$$

For example, let's say that we were to guess that it will take the ball 2 seconds to reach the floor (in other words, the height of 0 that was initially requested in the prompt). This implies assigning the value of 2 to the input variable t.

Well, let's plug it in to the equation and see if it satisfies the condition that the output must be equal to 0. Notice the question mark above the equality sign, since we don't know if this value works or not.

If $t = 2$

$$0 \overset{?}{=} 2 + 3(2) - 5(2)^2$$
$$0 \overset{?}{=} 2 + 6 - 20$$
$$0 = -12$$

As you can see, a value of $t = 2$ does not satisfy the condition that the right side be equal to 0. The answer tells us a lot, by the way. The fact that after **2 seconds** the ball is **12 meters** below a height of 0 means that the ball will reach the floor (the height of 0) before the **2** second mark, because by the **2 second** mark the equation tells us that it would be below the floor (**12 meters** below, to be precise), assuming, of course, that there wasn't a floor to prevent it from continuing on its downward journey.

We could try $t = 1.5$, for **1.5 seconds**, or $t = 1$ for **1 second**, and on and on and on (*ad nauseam*) until we found the answer. It is the "trial and error" method, which can work, but is extremely impractical. Instead we can use all the knowledge we have reviewed about quadratic equations, and find the zeros of the equation!

Think about it. The zeros or solutions or x-intercepts of a quadratic equation are always of the form (**?** , **0**), meaning the output they produce is always 0. Isn't that exactly what we are trying to find here? Yes it is.

If the equation was in factored form, we would be able to easily know its zeros. However, the equation we have is in standard form. So, we need to factor it in order to solve it. The alternative is to use what is called the "quadratic formula", which we will explore shortly.

So, lets find the factored form of the quadratic equation.

$$0 = 2 + 3t - 5t^2$$

First, I will flip the sides.

$$2 + 3t - 5t^2 = 0$$

Next, rearrange the terms. Remember that the terms have to be in descending order in terms of the power attached to the variables that they contain. The independent term is always listed at the end.

$$-5t^2 + 3t + 2 = 0$$

Next, find two numbers that multiply to $(-5)(2) = -10$ and whose sum is **3**. Replace the middle (linear) term with two terms using these numbers.

$$-5t^2 + 5t - 2t + 2 = 0$$

Next, group and factor as much as possible from each set of parentheses. Remember to be careful with the subtraction symbols and negative signs.

$$(-5t^2 + 5t) + (-2t + 2) = 0$$

$$(-5t)(t - 1) + (-2)(t - 1) = 0$$

At this point in the proceedings, you should always end up with a set of parentheses that appears on both terms and that contains the

exact same terms; in this case, it is $(t - 1)$. Factor it out of both terms and simplify as much as possible.

$$(-5t)(t - 1) + (-2)(t - 1) = 0$$

$$(t - 1)((-5t) + (-2)) = 0$$

$$(t - 1)(-5t - 2) = 0$$

If the steps are executed correctly, you should end up with the factored form of the quadratic.

Now, since we need the product of the two sets of parentheses to be equal to zero, we can use the zero product property; simply make either the first set of parentheses equal to **0** or the second one equal to **0**.

$$(t - 1)(-5t - 2) = 0$$

$(t - 1) = 0$	$(-5t - 2) = 0$
$t - 1 = 0$	$-5t - 2 = 0$
$t - 1 + 1 = 0 + 1$	$-5t - 2 + 2 = 0 + 2$
$t = 1$	$-5t = 2$
	$\dfrac{-5t}{-5} = \dfrac{2}{-5}$
	$t = \dfrac{-2}{5}$

We now have the two input values that when plugged in to the equation, yield an output value of **0**. In the context of the real-world problem we are working with here, these two time values, after the moment of reales, guarantee a height of **0**. Why? Because we found the **x-intercepts**, and thus the following two coordinates (notice their **y** value):

The two **x-intercepts** are...

$$(1 , 0) \text{ and } (\frac{-2}{5} , 0)$$

which may also be written as (in decimal form)...

$$(1 , 0) \text{ and } (-.4 , 0)$$

So, the first solution, **(1 , 0)**, literally translates to "**1** second after releasing the ball, it reaches a height of **0**", while the second solution, **(–.4 , 0)**, literally translates to "**–.4** seconds after releasing the ball, it reaches a height of **0**". Clearly, the first solution is the solution we want, for what does "**–.4**" seconds after releasing the ball even mean? Is it suggesting we go back in time? Clearly, this second solution does not have any real-world significance in this particular problem which the equation is trying to model, and so we know that the first solution, which does make sense, is the answer to the original prompt.

One second after releasing the ball, it will have gone up a certain amount of distance (hint: up to the vertex point), and then it will have come back down until it reachers the floor (height of **0**), and bounces up again, only to fall back down, etc..., until it eventually comes to rest.

To prove that this is in fact correct, simply plug in the input value of **t = 1** and check to see if the equation's output is, in fact, equal to **0** .

If $t = 1$

$$0 \overset{?}{=} 2 + 3(1) - 5(1)^2$$
$$0 \overset{?}{=} 2 + 3 - 5$$
$$0 \overset{\checkmark}{=} 0$$

As you can see, the answer we found is correct!

The graph of the height equation can also help you understand the prompt. Observe.

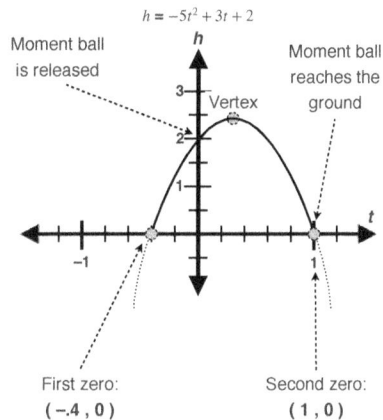

$h = -5t^2 + 3t + 2$

Moment ball is released

Moment ball reaches the ground

Vertex

First zero: **(–.4 , 0)**

Second zero: **(1 , 0)**

The graph clearly illustrates the relationship between the height of the ball and the number of seconds that go by after the moment of release. The ball initially rises (its height increases because at the moment of release, it has a positive velocity, which gravity is not able to immediately revert); then, it reaches a maximum height of   and begins to descend back to Earth, which is why its height begins to decrease to the right of the vertex point. The part of the graph that lies to the left of the **y-axis** does not have any meaning in this particular real-world problem, but it does show you why the equation is providing two zeros (or solutions) to the equation. It is therefore necessary to interpret the results and use the one that makes sense given the context of the problem. In this case, the ball lands on the ground at **t = 1second** , as the graph clearly shows.

The vertex, located at $(\frac{3}{10}, \frac{49}{20})$ or $(.3 , 2.45)$ in decimal form, tells us precisely the highest point reached by the ball (**2.45 meters**), and how many seconds go by until it reaches it (**.3 seconds**). It is at that height that the ball begins its descent back to Earth, due to the planet's gravitational pull.

I am sure that you can now appreciate the power of being able to solve quadratic equations. Setting them to zero always offers a convenient way of solving them, because you can use the factoring technique and then the zero product property to find the value or values you needed.

But what if the quadratic equation's *a*, *b*, and *c* values consist of numbers with decimals, fractions, or irrational numbers? Even many quadratics that do have **INTEGER** coefficients are still humanly impossible to figure out how to factor using the techniques discussed earlier. Do you simply give up?

Fortunately. there are two additional techniques that can be used to solve quadratic equations (that have been set to zero).

The first method is called *Completing the Perfect Square*, while the second method uses the afore-mentioned *Quadratic Formula*.

Completing the Perfect Square

This method allows us to transform a quadratic equation that is in standard form into the vertex form. Always, regardless of the values of the coefficients. It is a bit tricky and involves several steps, but it is always guaranteed to work. And unlike the two techniques we explored earlier that allows us to rewrite (some) standard form equations into the intercept form (by factoring

them) and which are not always easy to apply, completing the perfect square is guaranteed to work every time, and for every quadratic imaginable. No exceptions.

So what's a perfect square anyway? Answering this question is probably the best place to start.

Well, a perfect square occurs whenever the factored form of the quadratic is defined by two sets of parentheses that are exactly the same. Observe the following examples of perfect squares.

$$y = (x + 2)(x + 2)$$

$$y = (x - 1)(x - 1)$$

$$y = (x + 5)(x + 5)$$

$$y = (x + \frac{3}{4})(x + \frac{3}{4})$$

etc...

The advantage of a perfect square is that it can be simplified by squaring the common parentheses set, as follows:

$$y = (x + 2)(x + 2) \dashrightarrow y = (x + 2)^2$$

$$y = (x - 1)(x - 1) \dashrightarrow y = (x - 1)^2$$

$$y = (x + 5)(x + 5) \dashrightarrow y = (x + 5)^2$$

$$y = (x + \frac{3}{4})(x + \frac{3}{4}) \dashrightarrow y = (x + \frac{3}{4})^2$$

It should make perfect sense: since we have two items that are qual to each other multiplying each other, we can use power (exponent) notation to simplify it.

In general, when we are completing the perfect square, we are trying to force the quadratic so that it conforms to a "perfect square" set up. The problem is that most quadratic equations are not perfect squares to begin with. So, we need to get a bit creative.

When completing the perfect square, the goal is to end up with the following:

$$y = (x + m)^2$$

Now let's explore the technique using a quadratic equation as an example.

$$y = x^2 + 4x + 2$$

First of all, it should be clear that this is not a perfect square quadratic equation. Never mind that it is in standard form; there is no way to rewrite the right side as $(x + m)^2$. No m value will achieve it. The only way to end up with the middle term of 4m is to use **2** for m, as in $(x + 2)^2$; however, it will give us the wrong independent term. Observe.

$$y = (x + 2)^2$$

$$y = (x + 2)(x + 2)$$

$$y = (x)(x) + (x)(2) + (2)(x) + (2)(2)$$

$$y = x^2 + 2x + 2x + 4$$

$$y = x^2 + 4x + 4$$

See the problem? We end up with **4** as the independent term's value, where we would need it to be equal to **2** . So what can we do to fix this? Easy. Watch the trick involved.

$$y = x^2 + 4x + 2$$

First, add **2** to both sides of the equation:

$$y + 2 = x^2 + 4x + 2 + 2$$

Next, simplify the right side (**2 + 2 = 4**):

$$y + 2 = x^2 + 4x + 4$$

Now, you have a perfect square in your hands! The right side may now be simplified as follows:

$$y + 2 = (x + 2)(x + 2)$$

$$y + 2 = (x + 2)^2$$

And finally, solve for y :

$$y + 2 - 2 = (x + 2)^2 - 2$$

$$y = (x + 2)^2 - 2$$

And there it is... the original standard form equation transformed into a perfect square (which ends up looking like the vertex form, since we have that extra independent term adding or subtracting the perfect square itself).

Now, we can explore what happens when we expand the general form, which will provide us with a way to complete the perfect square of any quadratic equation. Observe.

$$y = (x + m)^2$$

$$y = (x + m)(x + m)$$

$$y = x^2 + 2mx + m^2$$

The expanded form of a perfect square (last equation above) will be key in what we need to do to any quadratic equation in order to convert it to a perfect square.

According to the equation we obtained,

$$y = x^2 + 2mx + m^2$$

...the middle term must have a coefficient equal to $2m$ while the independent term must be equal to m^2. Whenever this is the case, the standard form is a perfect square and may thus be expressed as

$$y = (x + m)^2$$

...where the inside of the parentheses contains the value m.

We can now establish the following relationship:

For any quadratic equation in standard form

$$y = x^2 + bx + c$$

...if we want to express it as a perfect square

$$y = (x + m)^2$$

...then the following must be true:

$$2m = b$$

which implies

$$m = \frac{b}{2}$$

...and

$$m^2 = c = (\frac{b}{2})^2$$

Let's put this to the test.

Say I give you the equation

$$y = x^2 + 6x + 9$$

...is it a perfect square? If so, can you write it in perfect square form?

First, compare the general standard form with this example, so that we may know the values of **b** and **c**:

$$y = x^2 + bx + c$$

$$y = x^2 + 6x + 9$$

$$b = 6 \qquad c = 9$$

Next, according to the relationships established above, the following must hold:

$$m = \frac{b}{2} \dashrightarrow m = \frac{6}{2} = 3$$

and...

$$m^2 = c = (\frac{b}{2})^2 \dashrightarrow m^2 = 9 \overset{\checkmark}{=} (\frac{6}{2})^2$$

As you can see, the relationships do hold, so we can write the quadratic in factored form–as a perfect square–as follows:

$$y = (x + m)^2 \dashrightarrow y = (x + 3)^2$$

To check that this is true, simply expand the perfect square version. We should end up with the original equation given in standard form.

$$y = (x + 3)^2$$
$$y = (x + 3)(x + 3)$$
$$y = x^2 + 6x + 9$$

Let's try another example.

$$y = x^2 + 10x + 25$$

...is it a perfect square? If so, can you write it in perfect square form?

We can be more efficient this time and simply follow these steps:

1. If the middle term's coefficient (the value of **b**) is equal to **10**, then it means that

$$2m = b \dashrightarrow 2m = 10$$

...which means that (solving for **m**)

$$m = \frac{10}{2} = 5$$

2. Now, since the value of the independent term must be equal to m^2, we must see the following relationship hold, if the equation is a perfect square:

$$m^2 = c = (\frac{b}{2})^2 \dashrightarrow m^2 = 25 \overset{\checkmark}{=} (\frac{10}{2})^2$$

It holds! Now we know that

$$m = 5$$

...so we can rewrite the equation in factored form as:

$$y = (x + m)^2 \dashrightarrow y = (x + 5)^2$$

Let's see this again, but this time, in diagram form.

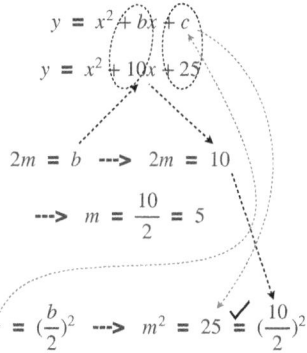

$$y = x^2 + bx + c$$
$$y = x^2 + 10x + 25$$

$$2m = b \dashrightarrow 2m = 10$$

$$\dashrightarrow m = \frac{10}{2} = 5$$

$$m^2 = c = (\frac{b}{2})^2 \dashrightarrow m^2 = 25 \overset{\checkmark}{=} (\frac{10}{2})^2$$

If the diagram seems confusing, stick to the step by step process described earlier.

Let's work on two more examples.

$$y = x^2 + 8x + 16$$

...is it a perfect square? If so, can you write it in perfect square form?

We can be more efficient this time and simply follow these steps:

1. If the middle term's coefficient (the value of **b**) is equal to **8**, then it means that

$$2m = b \dashrightarrow 2m = 8$$

...which means that (solving for **m**)

$$m = \frac{8}{2} = 4$$

2. Now, since the value of the independent term must be equal to m^2, we must see the following relationship hold, if the equation is a perfect square:

$$m^2 = c = (\frac{b}{2})^2 \ \text{---}\!\!> \ m^2 = 16 \ \checkmark{=} \ (\frac{8}{2})^2$$

It holds! Now we know that

$$m = 4$$

...so we can rewrite the equation in factored form as:

$$y = (x + m)^2 \ \text{---}\!\!> \ y = (x + 4)^2$$

On the the last example of this series.

$$y = x^2 - 2x + 1$$

...is it a perfect square? If so, can you write it in perfect square form?

We can be more efficient this time and simply follow these steps:

1. If the middle term's coefficient (the value of **b**) is equal to **−2**, then it means that

$$2m = b \ \text{---}\!\!> \ 2m = -2$$

...which means that (solving for **m**)

$$m = \frac{-2}{2} = -1$$

2. Now, since the value of the independent term must be equal to m^2, we must see the following relationship hold, if the equation is a perfect square:

$$m^2 = c = (\frac{b}{2})^2 \ \text{---}\!\!> \ m^2 = 1 \ \checkmark{=} \ (\frac{-2}{2})^2$$

It holds! Now we know that

$$m = -1$$

...so we can rewrite the equation in factored form as:

$$y = (x + m)^2 \ \text{---}\!\!> \ y = (x + (-1))^2$$
$$\text{---}\!\!> \ y = (x - 1)^2$$

This last exercise served as an example of how to correctly cope with any negative signs or subtraction symbols that may appear in a quadratic equation that is a perfect square and that is written in standard form.

Another method you can use that allows you to determine if a given quadratic equation written in standard form is a perfect square is to check for the following:

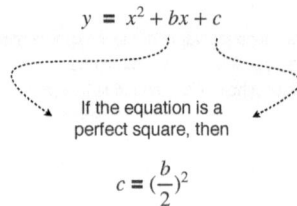

$$y = x^2 + bx + c$$

If the equation is a perfect square, then

$$c = (\frac{b}{2})^2$$

So, we can check the value of **c** whenever we are given a quadratic equation in standard form; if it is equal to the square of half of its **b** value, then we know that it is a perfect square, and we know that we can factor it by writing it as

$$y = (x + \frac{b}{2})^2$$

Observe what happens when we test the previous two examples using this method.

$$y = x^2 + 8x + 16$$

$$b = 8$$

$$c = 16$$

To test it, we proceed as follows:

Since we must have $\quad c = (\frac{b}{2})^2$

$$16 \overset{?}{=} (\frac{8}{2})^2$$

$$16 \overset{\checkmark}{=} 16$$

The equation, as we already determined, is a perfect square. We can thus rewrite it in factored form as

$$y = (x + \frac{b}{2})^2$$

$$y = (x + \frac{8}{2})^2$$

$$y = (x + 4)^2$$

Let's revisit the last equation we worked on, and use this same technique.

$$y = x^2 - 2x + 1$$

$$b = -2$$

$$c = 1$$

To test it, we proceed as follows:

Since we must have $\quad c = (\frac{b}{2})^2$

$$1 \overset{?}{=} (\frac{-2}{2})^2$$

$$1 \overset{\checkmark}{=} 1$$

The equation, as we already determined, is a perfect square. We can thus rewrite it in factored form as

$$y = (x + \frac{b}{2})^2$$

$$y = (x + \frac{-2}{2})^2$$

$$y = (x + (-1))^2$$

$$y = (x - 1)^2$$

So there you have it. To determine if a quadratic equation is a perfect square you can use any of the methods I discussed above. Use the one that best suits your style (I for one prefer this last method).

Now it's your turn. Check to see if the following equations are perfect squares, and if so, rewrite them in factored form.

a) $y = x^2 + 12x + 36$
b) $y = x^2 - 8x + 16$
c) $y = x^2 - 6x + 9$
d) $y = x^2 + x + .25$
e) $y = x^2 + \frac{2}{3}x + \frac{1}{9}$

f) $y = x^2 + 10x + 20$

g) $y = x^2 + 4x + 8$

Remember that you can look back to the examples we worked on together, but don't look at my answers below until you have tried your best to solve all of them.

Ready to check your work?

Observe.

a) $y = x^2 + 12x + 36$

$$b = 12 \quad ; \quad c = 36$$

We need to check if $c = (\frac{b}{2})^2$

$$36 \overset{?}{=} (\frac{12}{2})^2$$

$$36 \overset{?}{=} (6)^2$$

$$36 \overset{\checkmark}{=} 36$$

Knowing that the equation is a perfect square, we can factor it as

$$y = (x + \frac{b}{2})^2$$

...and since $b = 12$ we can write it as:

$$y = (x + 6)^2$$

b) $y = x^2 - 8x + 16$

$$b = -8 \quad ; \quad c = 16$$

We need to check if $c = (\frac{b}{2})^2$

$$16 \overset{?}{=} (\frac{-8}{2})^2$$

$$16 \overset{?}{=} (-4)^2$$

$$16 \overset{\checkmark}{=} 16$$

Knowing that the equation is a perfect square, we can factor it as

$$y = (x + \frac{b}{2})^2$$

...and since $b = -8$ we can write it as:

$$y = (x + (-4))^2$$

$$y = (x - 4)^2$$

c) $y = x^2 - 6x + 9$

I will now take a different approach. Simply observe that the square of half of the value of **b** is equal to **c** (half of **–6** is **–3** , and **–3** squared is equal to **9**). Therefore, the equation is a perfect square. It can thus be factored using half of the value of **b** , as follows:

$$y = (x + (-3))^2$$

$$y = (x - 3)^2$$

d) $y = x^2 + x + .25$

This prompt seems difficult, but it's actually quite simple. Just follow the same steps we have been using, and let the math do all the work. Let's go back to using the same process used on the first two exercises of this set.

Observe.

$$y = x^2 + x + .25$$

$$b = 1 \quad ; \quad c = .25$$

We need to check if $c = (\dfrac{b}{2})^2$

$$.25 \overset{?}{=} (\dfrac{1}{2})^2$$

$$.25 \overset{?}{=} (.5)^2$$

$$.25 \overset{\checkmark}{=} .25$$

Knowing that the equation is a perfect square, we can factor it as

$$y = (x + \dfrac{b}{2})^2$$

...and since $b = 1$ we can write it as:

$$y = (x + \dfrac{1}{2})^2$$

$$y = (x + .5)^2$$

Regardless of the fact that **c** has a decimal value, we were nonetheless able to apply the same steps and determine that the equation is a perfect square. Rewriting it in factored form was thus simple to do.

e) $y = x^2 + \dfrac{2}{3}x + \dfrac{1}{9}$

After solving the previous prompt I'm sure you can tackle this one on your own (if you weren't able to do so before looking at my answers).

Observe the process.

$$y = x^2 + \dfrac{2}{3}x + \dfrac{1}{9}$$

$$b = \dfrac{2}{3} \quad ; \quad c = \dfrac{1}{9}$$

We need to check if $c = (\dfrac{b}{2})^2$

$$\dfrac{1}{9} \overset{?}{=} (\dfrac{\frac{2}{3}}{2})^2$$

$$\dfrac{1}{9} \overset{?}{=} (\dfrac{\frac{2}{3}}{\frac{2}{1}})^2$$

$$\dfrac{1}{9} \overset{?}{=} (\dfrac{2}{6})^2$$

$$\dfrac{1}{9} \overset{?}{=} \dfrac{4}{36}$$

$$\dfrac{1}{9} \overset{\checkmark}{=} \dfrac{1}{9}$$

Knowing that the equation is a perfect square, we can factor it as

$$y = (x + \dfrac{b}{2})^2$$

...and since $b = 1$ we can write it as:

$$y = (\dfrac{\frac{2}{3}}{2})^2$$

$$y = (x + \dfrac{2}{6})^2$$

$$y = (x + \dfrac{1}{3})^2$$

It's not difficult to work with these equation types. Even if the coefficients are in fraction form, as in this prompt, or in decimal form, as in the previous prompt, simply follow the steps and use the same relationship formulas we have already established, and you will be able to determine if they are perfect squares, and if so, how to easily factor them.

f) $y = x^2 + 10x + 20$

Observe that in this equation, when you check to see if the square of half of the value of **b** is equal to **c**, you find that it is not the case:

half of 10 is 5

5 squared is equal to 25

and 25 is **not equal** to 20

So c is **not equal** to $(\frac{b}{2})^2$

Therefore, this equation is not a perfect square. We could change it so that it is a perfect square, but the prompt did not require us to do this.

g) $y = x^2 + 4x + 8$

Observe that as in the previous prompt, when you check to see if half of the value of **b** squared is equal to **c**, you find that it is not the case:

half of 4 is 2

2 squared is equal to 4

and 4 is **not equal** to 8

So c is **not equal** to $(\frac{b}{2})^2$

Therefore, this equation is not a perfect square.

Transforming a quadratic equation that is not a perfect square into a perfect square

I briefly showed you how to transform an equation that is not a perfect square into a perfect square format. We are not breaking any math rules when we do this, because we will always end up with an extra element that will compensate for forcing the equation into the mold of a perfect square.

Let's revisit the second to last equation of the set solved earlier.

$$y = x^2 + 10x + 20$$

We already determined that it is not a perfect square, because its **c** value is **not equal** to half the value of **b** squared.

To fix this, we can simply add or subtract whatever is necessary to both sides of the equation (to keep things equal) so that the value of **c** is equal to half its **b** value squared.

In this case, we would like the **c** value to be equal to 25, because the square of half of its **b** value (recall that **b** is equal to 10) is 25:

We need $c = (\frac{b}{2})^2$

$$(\frac{10}{2})^2 = (5)^2 = 25$$

so **c** must be equal to **25**

We already have the number **20** adding in the equation (the independent term); to change it to **25** we must add **5** to it. However, we must add **5** to both sides of the equation in order to keep things equal.

$$y = x^2 + 10x + 20$$

adding
5
to both sides

$$y + 5 = x^2 + 10x + 20 + 5$$

We can now simplify the right side (combine like terms):

$$y + 5 = x^2 + 10x + 25$$

See how we ended up with the extra element on the left side? Compare this to the original equation we started out with.

$$y = x^2 + 10x + 20$$

So yes, we forced it to conform to the perfect square format, but in the process, to maintain the equality between the versions, we ended up with an adding **5** on the left side.

At this point, we know the right side is a perfect square quadratic, because we forced the **c** value to be equal to the square of half the value of **b**. We can therefore rewrite in factored form, using half of **b** (half of **10** in this case, which is **5**), as follows:

$$y + 5 = x^2 + 10x + 25$$

may be written in factored form as:

$$y + 5 = (x + 5)^2$$

As a last step, we could subtract **5** from both sides of the equation in order to isolate the dependent variable **y**.

Observe.

$$y + 5 = (x + 5)^2$$

$$y + 5 - 5 = (x + 5)^2 - 5$$

$$y = (x + 5)^2 - 5$$

And there it is. The same quadratic equation as the original that we started out with, but now in factored form.

I would like to point out that this factored form matches the vertex format we reviewed earlier:

$$y = a(x - h)^2 + k$$

In fact, even a "*pure*" perfect square quadratic equation fits the mold (those that do not need to be forced to fit the perfect square mold):

$$y = (x + \frac{b}{2})^2$$

...simply add **0** to the right side, as follows:

$$y = (x + \frac{b}{2})^2 + 0$$

As a general rule, all "pure" perfect square quadratics have a vertex point that lies somewhere along the **x-axis**, since its **k** value is always **0**. Not so those quadratics that we must force into the perfect square mold (or, to coin a new term, any "impure" perfect square quadratic).

If you think about it, a pure perfect square quadratic will always have the vertex somewhere along the **x-axis** precisely because it must have only one zero (or solution, or **x-intercept**). After all, a perfect quadratic, by definition, is of the type

$$y = (x + m)(x + m)$$

...which means that the two zeros are equal to each other, which means it only has one zero to begin with, not two. This in turn means that its graph must have one single x-axis crossing point, not two, because if it had two, then it would have two distinct zeros or solutions or **x-intercepts**, so it would not be a **pure** perfect square to begin with! If this sounds a bit confusing to you, I suggest you reread it carefully, and think about the following two graphs that serve to illustrate this logic.

$$y = (x + m)^2$$

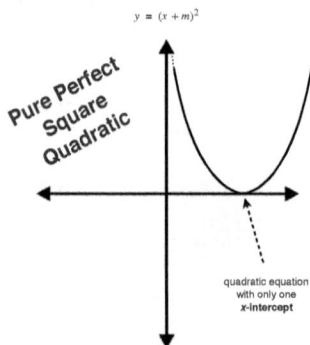

Pure Perfect Square Quadratic

quadratic equation with only one **x-intercept**

$$y = (x - h)^2 + k$$

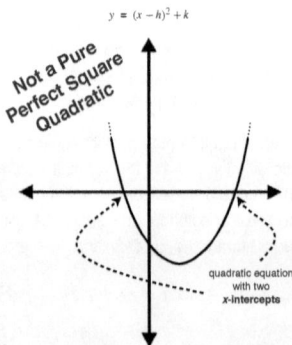

Not a Pure Perfect Square Quadratic

quadratic equation with two **x-intercepts**

Let's work on some additional examples.

Let's revisit the equation

$$y = x^2 + 4x + 8$$

from the previous set of problems (it was the last equation of the set, prompt "**g)**"). As we already established, this is not a "pure" perfect square quadratic because its **c** value is not equal to the square of half its **b** value; the equation's independent term would have to be equal to **4** .

So how do we fix it? Well this time, we need to subtract a number to both sides of the equation, since the independent term is currently equal to **8** , not **4** . We thus need to subtract **4** to both sides of the equation, as follows:

$$y = x^2 + 4x + 8$$

subtracting
4
to both sides

$$y - 4 = x^2 + 4x + 8 - 4$$

We can now simplify the right side (combine like terms):

$$y - 4 = x^2 + 4x + 4$$

See once again how we ended up with the extra element on the left side? Compare this to the original equation we started out with.

$$y = x^2 + 4x + 8$$

So yes, we forced it to conform to the perfect square format, but in the process, to maintain the equality between the versions, we ended up with a subtracting **4** on the left side.

658

At this point, we know the right side is a perfect square quadratic, because we forced the *c* value to be equal to the square of half the value of *b*. We can therefore rewrite in factored form, using half of *b* (half of **4** in this case, which is **2**), as follows:

$$y - 4 = x^2 + 4x + 4$$

may be written in factored form as:

$$y - 4 = (x + 2)^2$$

As a last step, we could add **4** from both sides of the equation in order to isolate the dependent variable *y*.

Observe.

$$y - 4 = (x + 2)^2$$

$$y - 4 + 4 = (x + 2)^2 + 4$$

$$y = (x + 2)^2 + 4$$

And there it is. The same quadratic equation as the original that we started out with, but now in factored form.

So what do you think? Isn't this easy? I certainly think so! But don't worry if you are still a bit unsure about this process of completing the perfect square. We are still going to practice this skill together, and then you will have an opportunity to solve some problems on your own.

Let's try the following equation together.

$$y = x^2 - 14x + 40$$

First, this is not a "pure" perfect square quadratic because its *c* value is not equal to the square of half its *b* value; the equation's independent term would have to be equal to **49** for that to be true.

So how do we fix it? Well, as before, we need to add a number to both sides of the equation, since the independent term is currently equal to **40**, not **49**. We thus need to add **9** to both sides of the equation, as follows:

$$y = x^2 - 14x + 40$$

adding
9
to both sides

$$y + 9 = x^2 - 14x + 40 + 9$$

We can now simplify the right side (combine like terms):

$$y + 9 = x^2 - 14x + 49$$

See once again how we ended up with the extra element on the left side? Compare this to the original equation we started out with.

$$y = x^2 - 14x + 40$$

So yes, we forced it to conform to the perfect square format, but in the process, to maintain the equality between the versions, we ended up with an adding **9** on the left side.

At this point, we know the right side is a perfect square quadratic, because we forced the *c* value to be equal to the square of half the value of *b*. We can therefore rewrite in factored form, using half of *b* (half of **−14** in this case, which is **−7**), as follows:

$$y + 9 = x^2 - 14x + 49$$

may be written in factored form as:

$$y + 9 = (x + (-7))^2$$

$$y + 9 = (x - 7)^2$$

As a last step, we could subtract **9** from both sides of the equation in order to isolate the dependent variable **y**.

Observe.

$$y + 9 = (x - 7)^2$$

$$y + 9 - 9 = (x - 7)^2 - 9$$

$$y = (x - 7)^2 - 9$$

And there it is. The same quadratic equation as the original, but now factored using the complete the perfect square technique.

Have you noticed something about all of the examples we have worked with so far, since introducing the idea of a perfect square equation? If so, it's that all of the equations we have worked with have an **a** value of **1**. In other words, their quadratic or leading term coefficient has always been equal to **1**. So does this mean that if **a** is not **1** that we cannot use this technique? Of course not. Observe the following example.

$$y = -2x^2 + 16x - 10$$

So where do we even begin to force this equation into the perfect square mold? It seems impossible, since its **a** value is not equal to **1**. Well, all we need to do is force the **a** value to be equal to **1**. The rest of the process is going to be exactly the same as what we have already established. Observe.

First, to make the **a** value be equal to **1**, we need to divide both sides of the equation by the current value of **a**, which is **-2**. Why both sides? Because remember that in order to maintain the equality (left-side with right-side), we must be fair and balanced.

So let's do this.

$$y = -2x^2 + 16x - 10$$

dividing by
-2
both sides

$$\frac{y}{-2} = \frac{-2x^2 + 16x - 10}{-2}$$

Seems daunting, doesn't it? I hope not! Don't forget all the skills we have reviewed earlier in the book. Let's simplify the right side, by dividing each term that appears on the numerator by the denominator (**-2**).

$$\frac{y}{-2} = \frac{-2x^2}{-2} + \frac{16x}{-2} - \frac{10}{-2}$$

Now, we can simplify the fractions that appear on the right side.

$$\frac{y}{-2} = x^2 + (-8x) - (-5)$$

...and simplify the expression using the Sign Table:

$$\frac{y}{-2} = x^2 - 8x + 5$$

We now have the quadratic equation, still in standard form, but now with an **a** value of **1**. The dividing **-2** that appears on the left side of the equation will be dealt with later.

Let's now complete the perfect square.

$$\frac{y}{-2} = x^2 - 8x + 5$$

This equation's **b** value is equal to **-8** . Half of this number, then squared, is equal to **16** . So our goal is to have the independent term be equal to **16** . It is currently **5** , so we need to add **11** to both sides of the equation, as follows:

$$\frac{y}{-2} = x^2 - 8x + 5$$

adding
11
to both sides

$$\frac{y}{-2} + 11 = x^2 - 8x + 5 + 11$$

$$\frac{y}{-2} + 11 = x^2 - 8x + 16$$

Now, we know that the right side is a perfect square. The left side does look a bit messy, but we will fix it in a moment.

First, let's factor the right side.

$$\frac{y}{-2} + 11 = (x - 4)^2$$

Now we can isolate the **y** once again. Guess how to achieve this. Do you have an idea? If you are thinking along the lines of "...subtract **11** to both sides of the equation, and then multiply both sides by **-2**...", then you are absolutely correct:

$$\frac{y}{-2} + 11 = (x - 4)^2$$

First, subtract **11** to both sides of the equation to get rid of the adding **11** that appears on the left side...

$$\frac{y}{-2} + 11 - 11 = (x - 4)^2 - 11$$

$$\frac{y}{-2} = (x - 4)^2 - 11$$

Next, multiply both sides by **-2** to get rid of the dividing **-2** that appears on the left side...

$$(\frac{y}{-2})(-2) = [(x - 4)^2 - 11](-2)$$

Be careful with this step. Notice the set of brackets that I introduced on the right side of the equation so that the entire right side is multiplied by the **-2** ? If this is not done, then the equation would be altered to the point where it would not be equal to the original, and thus the technique would be incorrectly applied.

Now, we can simplify both sides. The typical course of action is as follows:

$$(\frac{y}{-2})(-2) = [(x - 4)^2 - 11](-2)$$

$$y = -2(x - 4)^2 - (-22)$$

$$y = -2(x - 4)^2 + 22$$

And there you have it. We started out with an impossible-looking quadratic equation, and after manipulating it a bit, we ended up with a perfectly factored version of itself. Having it in this form is very helpful, because we can easily state its vertex.

What do you think? Is this easy? Let's try another equation.

$$y = -3x^2 - 12x - 15$$

First, to make the a value be equal to 1, we need to divide both sides of the equation by the current value of a, which is -3.

$$y = -3x^2 - 12x - 15$$

dividing by
-3
both sides

$$\frac{y}{-3} = \frac{-3x^2 - 12x - 15}{-3}$$

Let's now simplify the right side, by dividing each term that appears on the numerator by the denominator (-3).

$$\frac{y}{-3} = \frac{-3x^2}{-3} - \frac{12x}{-3} - \frac{15}{-3}$$

Now, we can simplify the fractions that appear on the right side.

$$\frac{y}{-3} = x^2 - (-4x) - (-5)$$

...and simplify the expression using the Sign Table:

$$\frac{y}{-3} = x^2 + 4x + 5$$

We now have the quadratic equation, still in standard form, but now with an a value of 1. The dividing -3 that appears on the left side of the equation will be dealt with later.

Let's now complete the perfect square.

$$\frac{y}{-3} = x^2 + 4x + 5$$

This equation's b value is equal to 4. Half of this number, then squared, is equal to 4. So our

goal is to have the independent term be equal to 4. It is currently 5, so we need to subtract 1 to both sides of the equation, as follows:

$$\frac{y}{-3} = x^2 + 4x + 5$$

subtracting
1
to both sides

$$\frac{y}{-3} - 1 = x^2 + 4x + 5 - 1$$

$$\frac{y}{-3} - 1 = x^2 + 4x + 4$$

Now, we know that the right side is a perfect square. The left side does look a bit messy, but we will fix it in a moment.

First, let's factor the right side.

$$\frac{y}{-3} - 1 = (x + 2)^2$$

Now we can isolate the y once again:

$$\frac{y}{-3} - 1 = (x + 2)^2$$

First, add 1 to both sides of the equation to get rid of the subtracting 1 that appears on the left side...

$$\frac{y}{-3} - 1 + 1 = (x + 2)^2 + 1$$

$$\frac{y}{-3} = (x + 2)^2 + 1$$

Next, multiply both sides by -3 to get rid of the dividing -3 that appears on the left side...

$$(\frac{y}{-3})(-3) = [(x + 2)^2 + 1](-3)$$

662

Once again, be careful with this step. Make sure you shield the entire right side before you include the multiplying element.

Now, we can simplify both sides. As on the last problem, the typical course of action is as follows:

$$(\frac{y}{-3})(-3) = [(x+2)^2 + 1](-3)$$

$$y = -3(x+2)^2 + (-3)$$

$$y = -3(x+2)^2 - 3$$

As a bonus, let's specify this equation's vertex:

vertex @ $(-2, -3)$

We also know that since its **a** value is equal to **–3**, this equation's graph (its parabola) points down. Furthermore, its arms are less spread apart that the parent equation that has an **a** value of **1**, because |–3| = 3, which is greater than **1**. If we wanted to draw a precise graph of this equation, we would want to find two additional coordinates that can guide our drawing. I would use, as usual, an input number of **0** since it is typically very easy to evaluate. Remember that we would want to use the original equation, since the y-intercept point is readily available (its always equal to the independent term, remember?):

$$y = -3x^2 - 12x - 15$$

$$y = -3(0)^2 - 12(0) - 15$$

$$y = -15$$

Coordinate ---> (0 , –15)
which also implies ---> (–4 , –15)

How did I come up with this coordinate's symmetry point? Can you see it?

Simple. Since the vertex is located at $(-2, -3)$, this means that the axis of symmetry is located at $x = -2$; thus it follows that the point we just computed, the **y-intercept** **(0 , –15)**, must have a symmetry partner on the other side of the axis of symmetry, and equally far away from it. Since **(0 , –15)** is **2** units **to the right** of the axis of symmetry, then it follows that its symmetry partner must be **2** units **to the left** of the axis of symmetry, at **(–4 , ?)**; and since they are symmetry partners, they must have equal output values (or heights, or **y** values). This thus leads to the correct symmetry partner point of **(–4 , –15)**. Observe the following diagram which illustrates this reasoning:

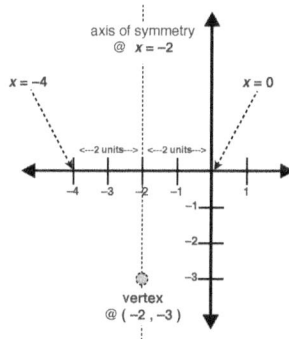

Now that we have three coordinates (including the location of the vertex), we could sketch the graph with reasonable precision. However, to be very precise, we would need at least two more coordinates. Wouldn't it be nice if we knew the zeros of this quadratic (in other words, its **x-intercept** or the solution of the equation if we were to set it to zero)? But how could we do that?

Solving quadratic equations that are in vertex form

To find the zeros or **x-intercepts** of any quadratic equation that is in vertex form, all we need to do is the following:

1) Set the equation to zero.

2) Isolate the parentheses that is being squared.

3) Take the square root of both sides of the equation.

4) Consider the positive and negative version of the solution. I will explore this in detail shortly.

Let's get to work on the last equation.

$$y = -3x^2 - 12x - 15$$

Which we know can be expressed as follows in vertex (factored) form:

$$y = -3(x + 2)^2 - 3$$

If we want to know this equation's **x-intercepts**, we need to set it to zero and solve it:

$$0 = -3(x + 2)^2 - 3$$

So the first step of the process is complete. Let's move on to the second step: isolate the set of parentheses that is being squared (in other words, the $(x + 2)^2$):

$$0 = -3(x + 2)^2 - 3$$

First, add **3** to both sides and simplify to get rid of the subtracting **3** :

$$0 + 3 = -3(x + 2)^2 - 3 + 3$$

$$3 = -3(x + 2)^2$$

Next, divide both sides by **–3** and simplify, to get rid of the multiplying **–3** that is on the right side:

$$\frac{3}{-3} = \frac{-3(x + 2)^2}{-3}$$

$$-1 = (x + 2)^2$$

So we know have the equation with the set of parentheses being squared, isolated. On to the next step.

We now need to take the square root of both sides of the equation. Why? Because our goal is to solve for **x**, and since the **x** variable is currently inside a set of parentheses that is being squared, we must eliminate the squaring element (in other words, the exponent) before we can hope to solve the equation for **x**.

As you may recall from **Chapter 10**, the opposite operation of "raising something to a power" is "to take the root of that something". In other words, exponents and roots are inverse operations of each other, just like multiplication and division are inverses of each other, and addition and subtraction are inverses of each other.

And so we must apply a root to both sides of the equation. But which root? Well, the root that has the same power as the exponent that we want to get rid of, of course. In this case (as in all quadratic equations), we thus need to take the square root of both sides of the equation as follows:

$$-1 = (x + 2)^2$$

$$\sqrt{-1} = \sqrt{(x + 2)^2}$$

But wait a minute! What is wrong with this picture? Do you see something on the left side of the equation that should make us be concerned? I hope you do! The left side has now been transformed into a **COMPLEX** (or imaginary) number! Why? Because the square root of negative one is an imaginary number, as we saw in **Chapter 12**.

Does this mean that we cannot move forward? Well, yes and no. Strictly speaking, we would be forced to say that this quadratic does not have **x-intercepts**. In fact, we could have deduced that a while ago. Notice that the vertex of this equation is located at **(−2 , −3)** , which lies below the **x-axis**. And since this equation's **a** value is equal to **−3** , we know its graph points down. Now think about it. Vertex below the **x-axis**; the parabola points down which would mean its vertex is the parabola's highest point... These two facts taken together would thus imply that the graph never crosses the **x-axis**. Observe:

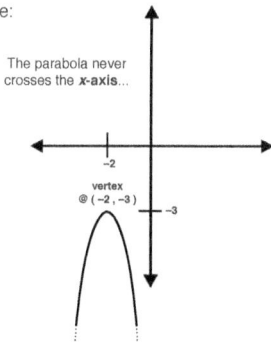

The parabola never crosses the **x-axis**...

vertex
@ (−2 , −3)

−2

−3

Do know that mathematicians took a long time before they came to accept the fact that you can simply shrug off the fact that **COMPLEX** numbers have crept up into our solution, and to just continue to move forward.

Watch what happens when we take this hiccup in stride and simply continue solving the equation.

$$\sqrt{-1} = \sqrt{(x + 2)^2}$$

The right side can now be simplified: we have a square root and an exponent equal to **2** ; they cancel each other out. However, we must now introduce the "plus/minus" element mentioned earlier. Why? Because the expression contains a variable being squared, and we are about to eliminate the power by taking the square root of both sides of the equation. We need a detour for this.

Brief detour. The plus/minus symbol in quadratic equation solutions.

Let's consider this simple quadratic equation case (rather trivial I'd say):

$$x^2 = 16$$

To solve for **x** , we can take the square root of both sides, correct? Let's do it:

$$\sqrt{x^2} = \sqrt{16}$$

Which you would think leads to:

$$x = 4$$

But wait! That's not the only answer that works! Look back at the original expression: $x^2 = 16$. This literally reads "Take some value **x** , square it, and the answer must be **16**... What number **or** numbers may be replaced for *x* that satisfy this condition?...". Well; so **4** works, we know that much:

$$(4)^2 = 16$$
$$(4)(4) = 16$$
$$16 = 16$$

But that's only a partial answer. So what are we missing here? Well, the fact that the number **−4** also works! Watch:

$$(-4)^2 = 16$$
$$(-4)(-4) = 16$$
$$16 = 16$$

See? This means that the equation has two possible correct answers (it shouldn't be a surprise to you anymore: after all, it is a quadratic equation, and you have seen many examples of quadratic equations that yield two answers to a given set up).

And so as a rule, whenever you have an equation with a variable that is being squared, and in the process of manipulating the equation mathematically you get rid of the power by taking the square root of both sides of the equation, you must introduce a "**plus/minus**" sign that will serve to correctly consider all possible answers of the equation. The process looks like this:

$$x^2 = 16$$
$$\sqrt{x^2} = \sqrt{16}$$
$$x = \pm\sqrt{16}$$

You typically introduce the plus/minus symbol when you cancel the exponent with the root. And

how do you deal with the plus/minus sign? Easy. Branch off into two separate paths: one where the plus/minus sign is positive, and the other where the plus/minus sign is negative.

$$x = \pm\sqrt{16}$$

$$x = +\sqrt{16} \qquad x = -\sqrt{16}$$

Of course, the positive version does not need to have the "+" sign explicitly indicated. We would thus branch off the equation as follows, and then continue simplifying/solving:

$$x = \pm\sqrt{16}$$

$$x = \sqrt{16} \qquad x = -\sqrt{16}$$
$$x = 4 \qquad x = -4$$

Typically, a **subindex** is used next to the *x* variable in order to number the solutions, as follows:

$$x_1 = 4 \qquad x_2 = -4$$

In case you are wondering, you must introduce the plus/minus sign to any equation during its solving process whenever an even power is canceled by taking its inverse-powered root. In general we can express this as follows:

$$x^n = d \quad ; \quad \text{if } n \text{ is even and for all}$$
$$\textbf{COMPLEX numbers } d$$

$$\sqrt[n]{x^n} = \sqrt[n]{d}$$
$$x = \pm\sqrt[n]{d}$$

Note that this rule is valid whenever **COMPLEX** numbers are concerned; recall that **COMPLEX** numbers include **REAL** numbers and **IMAGINARY** numbers:

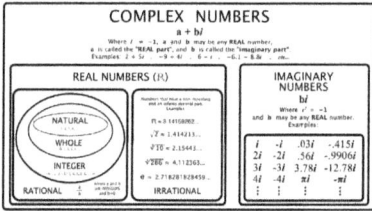

Let's work on a three additional examples.

a) Solve $x^2 = 100$

$$x^2 = 100$$

$$\sqrt{x^2} = \sqrt{100}$$

$$x = \pm\sqrt{100}$$

$$x = \sqrt{100} \qquad x = -\sqrt{100}$$

$$\boldsymbol{x_1} = 10 \qquad \boldsymbol{x_2} = -10$$

b) Solve $x^6 = 64$

$$x^6 = 64$$

$$\sqrt[6]{x^6} = \sqrt[6]{36}$$

$$x = \pm\sqrt[6]{36}$$

$$x = \sqrt[6]{36} \qquad x = -\sqrt[6]{36}$$

$$\boldsymbol{x_1} = 2 \qquad \boldsymbol{x_2} = -2$$

Often, the following mistake is made when solving equations like these: instead of computing the **6th** root of **36** as had to be done in this example, people will compute the **square** root of **36** . This, of course, would be a terrible mistake. Do not forget to consider the power of the root, always. Although you typically do have to work with square roots, be alert: if the root's power is not **2** you must take that into account, as we did here.

On to the last example in this detour.

c) Solve $x^2 = -25$

$$x^2 = -25$$

$$\sqrt{x^2} = \sqrt{-25}$$

$$x = \pm\sqrt{-25}$$

$$x = \sqrt{-25} \qquad x = -\sqrt{-25}$$

$$x = \sqrt{25}\sqrt{-1} \qquad x = -\sqrt{25}\sqrt{-1}$$

$$\boldsymbol{x_1} = 5i \qquad \boldsymbol{x_2} = -5i$$

In this case, we had to introduce the i symbol, since our solutions consist of **IMAGINARY** numbers. Notice how I separated the square root of a negative number into its two components: *the square root of a positive number* (in this case of **25** , which can be easily computed) times *the square root of* **−1** ; it is the latter that i is equal to, hence the switch.

How about a non-example? I just want to make sure you understand how this rule works: that you know which cases it applies to, and to which cases it does not.

d) Solve $x^3 = -1000$

$$x^3 = -1000$$

$$\sqrt[3]{x^3} = \sqrt[3]{-1000}$$

$$x = \sqrt[3]{-1000}$$

$$x = -10$$

In this case, the plus/minus symbol need not be introduced (in fact, it would be wrong to do so). Why? Because we canceled out an *odd* exponent, not an *even* one. Furthermore, the answer consists of a **REAL** number, not an **IMAGINARY** (or **COMPLEX**) answer, because the third root of **–1000** is **–10** . Recall that as a rule, the odd root of a negative number will always be equal to a **REAL** number (negative, but still **REAL**).

End of detour

Let's get back to the problem we were working with before taking this brief but important detour. We were solving the equation

$$y = -3(x + 2)^2 - 3$$

...which we first set to zero (to find its zeros, of course, or solutions, or *x*-intercepts)

$$0 = -3(x + 2)^2 - 3$$

...and ended up with the following equation:

$$-1 = (x + 2)^2$$

...which lead to:

$$\sqrt{-1} = \sqrt{(x + 2)^2}$$

Now, after the detour we took moments ago, when you look at this equation above you should be able to laugh at its simplicity.

Because we are about to cancel an even-numbered exponent that is attached to a variable (maybe not directly, but certainly indirectly, since the set of parentheses that is being squared contains the variable *x*), we need to make sure that we introduce the plus/minus symbol on the left side when the power and the root are canceled out on the right side of the the equation, as follows:

$$\sqrt{-1} = \sqrt{(x + 2)^2}$$

$$\pm\sqrt{-1} = (x + 2)$$

$$\sqrt{-1} = (x + 2) \qquad -\sqrt{-1} = (x + 2)$$

$$i = x + 2 \qquad -i = x + 2$$

Now, we simply solve for *x* by removing unwanted elements from its side:

$$i - 2 = x + 2 - 2 \qquad -i - 2 = x + 2 - 2$$

$$i - 2 = x \qquad -i - 2 = x$$

$$x_1 = i - 2 \qquad x_2 = -i - 2$$

And there you have it. The two "zeros", or solutions, or *x*-intercepts, of the quadratic equation. Now, I want you to recall this equation's graph. It never crosses the **x-axis**:

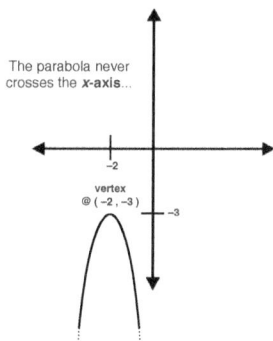

The parabola never crosses the **x-axis**...

vertex
@ (-2 , -3)

So what does a **COMPLEX** answer even mean? Are we supposed to create the following two coordinates, and what would they even mean?

$$(i - 2 , 0) \quad \text{and} \quad (-i - 2 , 0)$$

The two coordinates are valid, but because they are **COMPLEX**-numbered coordinates, they cannot be graphed directly on the **x-y axis** that we have been using all along (we would need to use the **COMPLEX** plane that we reviewed in **Chapter 12** to plot them). As to the meaning of these "**COMPLEX** roots" (the term "roots of a quadratic equation" refers to its solutions or **x-intercepts**), I will just say that they form a very important (and crucial) part of mathematical analysis, with a wide range of real-world applicability. Many physics equations and problems expect and require **COMPLEX** solutions, which in turn serve to model real-world phenomena. Given the scope of the book, I will not review this in detail, but do know that there is

nothing "imaginary" about these **IMAGINARY** roots.

As far as our review is concerned, however, you do need to know how to use these roots to rewrite the original quadratic equation in full factored form (not in vertex form, but rather in intercept form), much like, say $y = x^2 + 5x + 6$, in standard form, can be expressed as $y = (x + 2)(x + 3)$, in *intercept form*.

How do we do this? First, recall that we started out with the following equation:

$$y = -3x^2 - 12x - 15$$

Then, we expressed this equation in vertex form, as follows

$$y = -3(x + 2)^2 - 3$$

Next, to find the zeros, roots, solutions, or **x-intercepts** (recall that they are all equivalent mathematical terms), we had to set **y = 0** and then we set out to solve for **x** :

$$0 = -3(x + 2)^2 - 3$$

$$0 + 3 = -3(x + 2)^2 - 3 + 3$$

$$3 = -3(x + 2)^2$$

$$\frac{3}{-3} = \frac{-3(x + 2)^2}{-3}$$

$$-1 = (x + 2)^2$$

$$\pm\sqrt{-1} = \sqrt{(x + 2)^2}$$

$$\sqrt{-1} = (x + 2) \qquad -\sqrt{-1} = (x + 2)$$

$$i = x + 2 \qquad -i = x + 2$$

$$i - 2 = x + 2 - 2 \qquad -i - 2 = x + 2 - 2$$

$$i - 2 = x \qquad -i - 2 = x$$

$$x_1 = i - 2 \qquad x_2 = -i - 2$$

or, using the preferred **COMPLEX** number format

$$x_1 = -2 + i \qquad x_2 = -2 - i$$

And we thus obtained the two zeros of the equation.

So how can we use these two zeros to express the original equation in intercept form?

Remember the general intercept form is defined as follows:

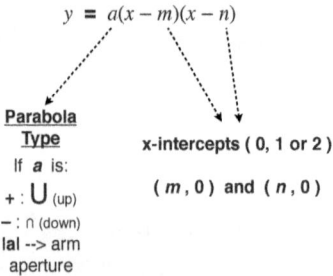

$$y = a(x - m)(x - n)$$

Parabola Type
If **a** is:

+ : U (up)
− : ∩ (down)
|a| --> arm
aperture

x-intercepts (0, 1 or 2)

$$(m, 0) \text{ and } (n, 0)$$

Notice the expected form of the **x-intercepts**: the **m** and **n** values correspond to the **x** values of the coordinates.

Pay particular attention to the **a** value since this is a key component of this form. Our original equation (in standard form) is

$$y = -3x^2 - 12x - 15$$

...and recall that in general, the standard form corresponds to

$$y = ax^2 + bx + c$$

...which means that the **a** value of this equation is **−3** .

Now, knowing the zeros (roots, or **x-intercepts**, or solutions), as well as the **a** value, we can express the equation in intercept form as follows:

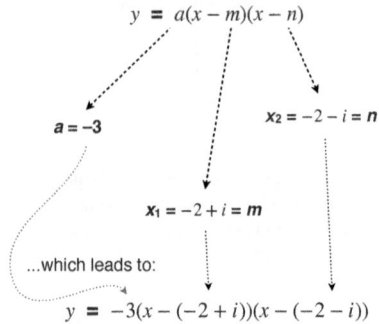

$$y = a(x - m)(x - n)$$

$$a = -3$$

$$x_2 = -2 - i = n$$

$$x_1 = -2 + i = m$$

...which leads to:

$$y = -3(x - (-2 + i))(x - (-2 - i))$$

We may, of course, simplify the equation on the right side (inside the sets of parentheses), using the Sign Table, as follows:

$$y = -3(x + 2 - i)(x + 2 + i)$$

And there it is. The original equation is now in fully factored form (in other words, in intercept form).

To recap, observe the same equation in its three different (yet equivalent) forms:

Standard Form $y = -3x^2 - 12x - 15$

Vertex form $y = -3(x + 2)^2 - 3$

Intercept Form $y = -3(x + 2 - i))(x + 2 + i)$

To prove that the intercept form is, in fact, equivalent to the original standard form equation that we started out with, we can simplify it. The **IMAGINARY** elements that may be cause for alarm will disappear during the distribution (FOIL) process. Observe.

The resulting expansion contains many terms, so I will have to spread it over several lines (note the use of the "...".

$$y = -3(x + 2 - i))(x + 2 + i)$$

$$y = -3[(x)(x) + (x)(2) + (x)(i)$$
$$... + (2)(x) + (2)(2) + (2)(i)$$
$$... - (i)(x) - (i)(2) - (i)(i)]$$

$$y = -3[x^2 + 2x + xi + 2x + 4 + 2i - xi - 2i - i^2]$$

To simplify, combine like terms, and recall that i^2 is equal to **−1** (the terms $+ix$ and $-ix$ cancel each other out, as do the terms $-2i$ and $+2i$):

$$y = -3[x^2 + 2x + xi + 2x + 4 + 2i - xi - 2i - i^2]$$

$$y = -3[x^2 + 4x + 4 - (-1)]$$

$$y = -3[x^2 + 4x + 4 + 1]$$

$$y = -3[x^2 + 4x + 5]$$

$$y = -3x^2 - 12x - 15$$

When you compare it to the original equation we started out with, you find that they are exactly the same, as expected.

We can now establish the following important fact regarding all quadratic equations.

⚓

All quadratic equations have TWO roots (also called solutions, x-intercepts, or zeros). These will consist of either

2 different REAL roots
2 equal REAL roots
2 COMPLEX roots

If the quadratic equation has 2 different REAL roots, it is of the form

$$y = a(x - m)(x - n)$$
where **a ≠ 0**
and its two roots are **m** and **n**

If the quadratic equation has 2 equal REAL roots, it is a PERFECT SQUARE of the form

$$y = a(x - m)^2$$
where **a ≠ 0**
and its root is **m**

If the quadratic equation has 2 COMPLEX roots, it is of the form

$$y = a(x - (d + ei))(x - (d - ei))$$
where **a ≠ 0**
and its two roots are $d + ei$ and $d - ei$

Notice that if the quadratic equation has **COMPLEX** roots, it will always have TWO of them (never one). Furthermore, they are always of the from

$$d + ei \quad \text{and} \quad d - ei$$

...which means they are "*complex conjugates*" of each other. This implies that if you are given one **COMPLEX** root of a quadratic equation, you can automatically deduce its other pair.

Let's find the roots of two more quadratic equations, and use them to express the original equation in intercept (fully factored) form, beginning with the following equation:

$$\frac{y}{2} + 5x - 17 = \frac{x^2}{2}$$

First, solve for **y**:

$$\frac{y}{2} + 5x - 17 - 5x + 17 = \frac{x^2}{2} - 5x + 17$$

$$\frac{y}{2} = \frac{x^2}{2} - 5x + 17$$

$$(2)(\frac{y}{2}) = (2)(\frac{x^2}{2} - 5x + 17)$$

$$y = x^2 - 10x + 34$$

Next, if you try to express it in fully factored form by using the technique we reviewed earlier of finding two **REAL** numbers that multiplied equal **34** and whose sum is **−10**, you are going to find yourself stuck.

The best course of action based on what we have reviewed so far is thus to complete the perfect square, as follows:

$$y = x^2 - 10x + 34$$

We first check to see if **a** is equal to **1**, which it is. Next, we need the independent term to be equal to the square of half of the value of **b**; since **b = −10**, this means the independent term

should be equal to **25**. The independent term is currently equal to **34** so we need to subtract **9** to both sides of the equation to force the right side's independent term to be equal to **25**:

$$y - 9 = x^2 - 10x + 34 - 9$$

$$y - 9 = x^2 - 10x + 25$$

Now, we can factor the right side using knowing that it is a perfect square, as we reviewed earlier:

$$y - 9 = (x - 5)^2$$

Finally, we add **9** to both sides to have the equation solved for **y**:

$$y - 9 + 9 = (x - 5)^2 + 9$$

$$y = (x - 5)^2 + 9$$

At this point, we need to keep in mind that we would like to find the zeros (or roots, or solutions, or x-intercepts) of the equation. Therefore, we need to set **y = 0** and solve for **x**:

$$0 = (x - 5)^2 + 9$$

$$0 - 9 = (x - 5)^2 + 9 - 9$$

$$-9 = (x - 5)^2$$

$$\sqrt{-9} = \sqrt{(x - 5)^2}$$

$$\pm 3i = (x - 5)$$

$$3i = x + 5 \qquad -3i = x + 5$$

$$3i - 5 = x + 5 - 5 \qquad -3i - 5 = x + 5 - 5$$

$$3i - 5 = x \qquad -3i - 5 = x$$

$$x_1 = 3i - 5 \qquad x_2 = -3i - 5$$

or, using the preferred **COMPLEX** number format

$$x_1 = -5 + 3i \qquad x_2 = -5 - 3i$$

Now, we can express the original standard form equation in intercept form, using the above roots and the **a** value of **1** defined earlier:

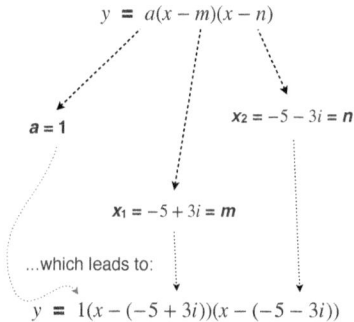

$$y = a(x - m)(x - n)$$

$a = 1$

$x_2 = -5 - 3i = n$

$x_1 = -5 + 3i = m$

...which leads to:

$$y = 1(x - (-5 + 3i))(x - (-5 - 3i))$$

We may, of course, simplify the equation on the right side (inside the sets of parentheses), using the Sign Table, as follows:

$$y = (x + 5 - 3i)(x + 5 + 3i)$$

I will not foil this intercept form equation to prove that it is equivalent to the original standard form equation, but rather will leave it as an exercise for you to do on your own.

Let's work on one final example.

$$-\frac{y}{2} = -x^2 + 6x - 13$$

First, solve for **y**:

$$(-2)(-\frac{y}{2}) = (-2)(-x^2 + 6x - 13)$$

$$y = 2x^2 - 12x + 26$$

Next, if you try to express it in fully factored form by using the technique we reviewed earlier of finding two **REAL** numbers that multiplied equal **26** and whose sum is **−12**, you are going to find yourself stuck.

The best course of action based on what we have reviewed so far is thus to complete the perfect square, as follows:

$$y = 2x^2 - 12x + 26$$

We first check to see if **a** is equal to **1**, which it isn't. We therefore need to define its **a** value for later use, set to zero, and then worry about the leading term's coefficient. Let's first define **a**:

$$a = 2$$

...then set to zero (make **y = 0**). Wondering why we set the equation to zero before we factor it? Well, since the quadratic's equation is not equal to **1**, we must make it be equal to **1**; combine this with the fact that we want to know the zeros of the equation, nothing else, and you have your answer: it makes the process easier to manage.
First set to zero, then divide both sides by **2**, and you will have found an equation that can be easily factored using the perfect square method. Keeping the variable **y** is unnecessary and can be more confusing. Just know that the new equation is equivalent to the original equation as far as its zeros (or solutions, roots, or **x-intercepts**) are concerned, but that some information may be lost in the process (for example, the true value of **a**). So, we set to zero first:

$$0 = 2x^2 - 12x + 26$$

Now, before completing the perfect square, we need to force this quadratic's *a* value to be equal to **1** . Dividing by **2** (both sides) will take care of that:

$$\frac{0}{2} = \frac{2x^2 - 12x + 26}{2}$$

...and then simplify the equation:

$$0 = x^2 - 6x + 13$$

We may now complete the perfect square. Because *b* is equal to **−6** , we need *c* to be equal to **9** ; therefore, we need to subtract **4** from both sides of the equation :

$$0 - 4 = x^2 - 6x + 13 - 4$$

$$-4 = x^2 - 6x + 9$$

Now that the right side of the equation is a perfect square, we may factor it using the method reviewed earlier.

$$-4 = (x - 3)^2$$

Now, we may solve for *x* to find the two zeros (or roots, or solutions, or *x*-intercepts):

$$-4 = (x - 3)^2$$

$$\sqrt{-4} = \sqrt{(x - 3)^2}$$

$$\pm\sqrt{-4} = (x - 3)$$

$$\pm 2i = x - 3$$

$$2i = x - 3 \qquad -2i = x - 3$$

$$2i + 3 = x - 3 + 3 \qquad -2i + 3 = x - 3 + 3$$

$$2i + 3 = x \qquad -2i + 3 = x$$

$$x_1 = 2i + 3 \qquad x_2 = -2i + 3$$

or, using the preferred **COMPLEX** number format

$$x_1 = 3 + 2i \qquad x_2 = 3 - 2i$$

Now, we can express the original standard form equation in intercept form, using the above roots and the *original a* value of **2** defined earlier:

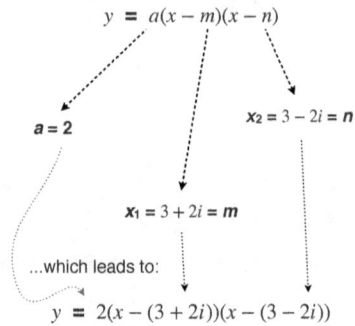

$$y = a(x - m)(x - n)$$

$$a = 2 \qquad x_2 = 3 - 2i = n$$

$$x_1 = 3 + 2i = m$$

...which leads to:

$$y = 2(x - (3 + 2i))(x - (3 - 2i))$$

We may, of course, simplify the equation on the right side (inside the sets of parentheses), using the Sign Table, as follows:

$$y = 2(x - 3 - 2i)(x - 3 + 2i)$$

After working on these three examples you should be able to express any quadratic equation in intercept form (in other words, in fully factored form), regardless of the values of the coefficients *a* , *b* , and *c* .

We can now review another method of solving quadratic equations that just like completing the perfect square, will always work, for any quadratic equation, regardless of the values of *a* , *b* , and *c* ; it requires the use of the so-called "***quadratic formula***". It is, perhaps, easier to apply than any of the other methods, though not necessarily easier to compute (the solving process can be long and tedious).

Rather than just give you the quadratic formula, I will show you how it can be derived from the general form of a quadratic equation written in standard form. I strongly recommend you try to follow along as I show you the steps involved and then try to derive it yourself: it really helps understand and master the ***quadratic formula***.

Please note that even though we are going to use the "complete the perfect square" method, the end result will be a formula that can be used to solve any quadratic equation simply by extracting its *a* , *b* , and *c* values, so don't be thrown off by the steps that are coming up shortly.

We start out with the general form of a quadratic equation written in standard form. It is very easy to express any quadratic equation in standard form, simply by solving for *y* , hence the practicality of the quadratic formula.

$$y = ax^2 + bx + c$$

First, we set the equation to zero, which means replacing *y* with **0** ; we do this because we want to find the roots of the quadratic.

$$0 = ax^2 + bx + c$$

Next, we assume that the quadratic is not a perfect square (if we did make that assumption, then any formula derived having made that assumption would only work for ***perfect square*** quadratic equations, greatly reducing the

practicality of the quadratic formula. Do note that making the assumption that the quadratic equation is not a perfect square will still work for quadratic equations that are perfect squares, for reasons that will become obvious as we move forward). Therefore, we first divide both sides by *a* because we want to ensure that the coefficient of the leading term (a.k.a. the quadratic term or the **x-squared** term, or simply *a*) is equal to **1** . Remember that this is one of the main requirements for completing the perfect square of any quadratic equation.

$$\frac{0}{a} = \frac{ax^2 + bx + c}{a}$$

Next, we simplify.

$$0 = \frac{ax^2}{a} + \frac{bx}{a} + \frac{c}{a}$$

$$0 = x^2 + \frac{b}{a}x + \frac{c}{a}$$

Since we are assuming that the quadratic is not a perfect square, we want to force its independent term to be equal to the square of half of the linear term's coefficient (in the equation above, the square of half of $\frac{b}{a}$).

$$0 = x^2 + \frac{b}{a}x + \frac{c}{a}$$

We want this to be equal to:

$$(\frac{\frac{b}{a}}{2})^2 = (\frac{\frac{b}{a}}{2})^2 = (\frac{b}{2a})^2 = \frac{b^2}{4a^2}$$

The best way to guarantee that the equation's

675

independent value is equal to $\frac{b^2}{4a^2}$ is to first subtract $\frac{c}{a}$ from both sides of the equation so that it is not standing in the way of the value we want to have in its place, the value of $\frac{b^2}{4a^2}$. We thus first subtract $\frac{c}{a}$ from both sides, and then we add $\frac{b^2}{4a^2}$ to both sides, guaranteeing that the right side is a perfect square.

$$0 - \frac{c}{a} = x^2 + \frac{b}{a}x + \frac{c}{a} - \frac{c}{a}$$

$$0 - \frac{c}{a} = x^2 + \frac{b}{a}x$$

$$-\frac{c}{a} = x^2 + \frac{b}{a}x$$

Now, we add $\frac{b^2}{4a^2}$ to both sides...

$$-\frac{c}{a} + \frac{b^2}{4a^2} = x^2 + \frac{b}{a}x + \frac{b^2}{4a^2}$$

At this point, we know for certain that the right side is a perfect square, so we can factor it as reviewed earlier (using half of the linear term's coefficient as the value to use inside the set of parentheses):

$$-\frac{c}{a} + \frac{b^2}{4a^2} = (x + \frac{\frac{b}{a}}{2})^2$$

We can simplify the fraction on the right side, and combine the two fractions that are on the left side of the equation, as follows. First, multiply the first fraction by $\frac{4a}{4a}$ so that both fractions have the

same denominator, as we reviewed in earlier chapters of the book:

$$-\frac{(4a)(c)}{(4a)(a)} + \frac{b^2}{4a^2} = (x + \frac{b}{2a})^2$$

Then simplify and assign the negative sign of the first fraction on the left side to its numerator to help ease the computation...

$$\frac{-4ac}{4a^2} + \frac{b^2}{4a^2} = (x + \frac{b}{2a})^2$$

Next, combine the fractions...

$$\frac{-4ac + b^2}{4a^2} = (x + \frac{b}{2a})^2$$

And then switch the terms of the numerator so that its leading term is positive...

$$\frac{b^2 - 4ac}{4a^2} = (x + \frac{b}{2a})^2$$

At this point, we are ready to solve for x; all we need to do is take the square root of both sides of the equation as we reviewed earlier:

$$\sqrt{\frac{b^2 - 4ac}{4a^2}} = \sqrt{(x + \frac{b}{2a})^2}$$

Remember to introduce the plus/minus sign to the other side of the equation when canceling an even exponent by taking the even root of both sides, as we reviewed earlier:

$$\pm\sqrt{\frac{b^2 - 4ac}{4a^2}} = x + \frac{b}{2a}$$

The quadratic formula is now close at hand. First, let's simplify the left side: recall that the root of a fraction is equal to the root of its numerator divided by the root of its denominator. Therefore, we may simplify the left side as follows...

$$\pm\frac{\sqrt{b^2 - 4ac}}{\sqrt{4a^2}} = x + \frac{b}{2a}$$

We can now simplify the denominator: the square root of $4a^2$ is equal to $2a$ (that's because $(2a)(2a) = 4a^2$):

$$\pm\frac{\sqrt{b^2 - 4ac}}{2a} = x + \frac{b}{2a}$$

Next, we conveniently attach the plus/minus sign to the numerator, as follows:

$$\frac{\pm\sqrt{b^2 - 4ac}}{2a} = x + \frac{b}{2a}$$

We are now ready to solve for x. All we need to do is subtract $\frac{b}{2a}$ from both sides of the equation to isolate the x variable, as follows:

$$\frac{\pm\sqrt{b^2 - 4ac}}{2a} - \frac{b}{2a} = x + \frac{b}{2a} - \frac{b}{2a}$$

$$\frac{\pm\sqrt{b^2 - 4ac}}{2a} - \frac{b}{2a} = x$$

...and then simplify the left side. The two fractions have equal denominators, so we may combine them...

$$\frac{\pm\sqrt{b^2 - 4ac} - b}{2a} = x$$

To avoid making mistakes with the numerator's root (inadvertently including the subtracting b

into the root symbol), the numerator's terms are switched, as follows:

$$\frac{-b \pm \sqrt{b^2 - 4ac}}{2a} = x$$

And finally, we switch the equation's sides, so that the formula, as read from left to right, starts off with " x is equal to...":

$$x = \frac{-b \pm \sqrt{b^2 - 4ac}}{2a}$$

Do note that the plus/minus sign that appears on the numerator before the root can be though of as an "add/subtract" symbol instead. The reason for this is that the numerator has two terms, and by definition terms are either adding or subtracting each other. This means that when we switched the terms of the numerator (starting it off with $-b$), strictly speaking we would have been left with the following (note the dotted circle):

$$x = \frac{-b + (\pm\sqrt{b^2 - 4ac})}{2a}$$

But using the Sign Table, we know that adding a positive term (the root symbol in this case) implies adding it, and that adding a negative term (again, the root symbol) implies subtracting it. Hence, the plus/minus sign becomes an add/subtract symbol instead.

I would also like to point out that because there may be two solutions to the equation, it is common to see the following version of the quadratic formula:

$$x_{1,2} = \frac{-b \pm \sqrt{b^2 - 4ac}}{2a}$$

The subindex "$_{1,2}$" stands for x_1 and x_2: to find these potentially two different values of x that solve the quadratic equation, simply plug in the values of a, b, and c of the quadratic equation that you are trying to solve into the quadratic formula, and branch off your computation into two versions: one where the root symbol is being added (that would yield x_1) and one where the root symbol is being subtracted (that would yield x_2).

In general, this branching off is as follows:

$$x = \frac{-b \pm \sqrt{b^2 - 4ac}}{2a}$$

$$x_1 = \frac{-b + \sqrt{b^2 - 4ac}}{2a}$$

$$x_2 = \frac{-b - \sqrt{b^2 - 4ac}}{2a}$$

Notice that the only difference between the two branches is that in the first case, the root is being added, while in the second case the root is being subtracted.

So now that you know what the quadratic formula is and how to derive it (try to retrace the steps by yourself so that you can derive the quadratic formula all on your own), let's put it to use.

I am going to revisit one of the equations that we solved earlier using the "complete the perfect square" method, but solve it using the quadratic formula instead.

$$y = -3x^2 - 12x - 15$$

Using the complete the perfect square method, we found its two zeros to be:

$$x_1 = -2 + i \qquad x_2 = -2 - i$$

Let's test the quadratic formula. It should provide us with the exact same zeros (or roots, or solutions, or *x*-intercepts).

Step 1. We must extract the equation's a, b, and c values: these will be used to replace the corresponding letters of the quadratic formula. To do this, we must make sure that the quadratic equation is solved for the variable y. Since the given equation is solved for y we may extract its values without having to do any extra work. They are:

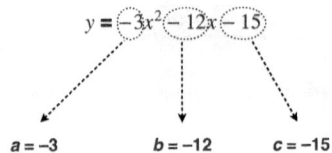

$$y = -3x^2 - 12x - 15$$

$$a = -3 \qquad b = -12 \qquad c = -15$$

Step 2. Replace the a, b, and c values of the quadratic formula with the extracted values from **Step 1**, being careful to use sets of parentheses for each of the values being replaced:

$$x = \frac{-b \pm \sqrt{b^2 - 4ac}}{2a}$$

$$a = -3 \qquad b = -12 \qquad c = -15$$

$$x = \frac{-(-12) \pm \sqrt{(-12)^2 - 4(-3)(-15)}}{2(-3)}$$

Step 3. Solve the quadratic equation. Remember that it is very important to follow the correct order of operations, and to apply all the rules and principles that we have reviewed so far in the book.

$$x = \frac{-(-12) \pm \sqrt{(-12)^2 - 4(-3)(-15)}}{2(-3)}$$

$$x = \frac{12 \pm \sqrt{144 - (-12)(-15)}}{-6}$$

$$x = \frac{12 \pm \sqrt{144 - (180)}}{-6}$$

$$x = \frac{12 \pm \sqrt{-36}}{-6}$$

$$x = \frac{12 \pm 6i}{-6}$$

We can simplify the fraction above, but we must be careful to branch off the two answers that the add/subtract symbol yields: one branch will contain an addition symbol, while the other a subtraction symbol, as follows:

$$x = \frac{12 \pm 6i}{-6}$$

$$x_1 = \frac{12 + 6i}{-6} \qquad x_2 = \frac{12 - 6i}{-6}$$

$$x_1 = \frac{12}{-6} + \frac{6i}{-6} \qquad x_2 = \frac{12}{-6} - \frac{6i}{-6}$$

$$x_1 = -2 + (-i) \qquad x_2 = -2 - (-i)$$

$$x_1 = -2 - i \qquad x_2 = -2 + i$$

When you compare the two roots we found using the quadratic formula with the two roots using the complete the perfect square method, you can see that they are exactly the same:

Complete the perfect square method:

$$x_1 = -2 + i \qquad x_2 = -2 - i$$

Quadratic formula method:

$$x_1 = -2 - i \qquad x_2 = -2 + i$$

Don't be alarmed by the difference in the subindex pairings... ultimately, the subindexes can be switched around since they are just labels used to conveniently differentiate between the different answers that are obtained. Using the complete the perfect square method we obtain two zeros: $-2 + i$ and $-2 - i$; using the quadratic formula we get the same two zeros: $-2 - i$ and $-2 + i$. We can call the first zero "x_1" or "x_2", it is not mathematically relevant. Based on what the first zero is called, then the other zero would have to be called (respectively) either "x_2" or "x_1" to differentiate it from the first one. Thus, both methods yield the same answer.

You may therefore use the quadratic formula to solve any given quadratic equation; simply make sure it is solved for y, extract its a, b, and c values, plug them in to the quadratic formula, and solve and simplify, making sure to branch off into its two possible answers.

Let's revisit another quadratic equation we worked on earlier:

$$y = x^2 + 5x + 6$$

We used the factoring method to find its **x-intercepts** (in other words, we looked for two numbers that multiplied equal **6** and that added equal **5** ; since **2** and **3** satisfy both conditions, we were able to express the equation in fully factored (intercept) form as $y = (x + 3)(x + 2)$, and then, using the zero product property, determined that its two **x-intercepts**/zeros/roots/ solutions/are **−3** and **−2** , which may be expressed as $x_1 = -3$ and $x_1 = -2$).

Let's test the quadratic equation once more. This time, try to use it by yourself before looking at my solution process.

Did you get the same zeros we found earlier? If not, then you must have made a mistake along the way. Make sure you try to fix your mistake before following my solution process.

Ready? Observe.

Step 1. We must extract the equation's **a** , **b** , and **c** values: these will be used to replace the corresponding letters of the quadratic formula. To do this, we must make sure that the quadratic equation is solved for the variable **y** . Since the given equation is solved for y we may extract its values without having to do any extra work. They are:

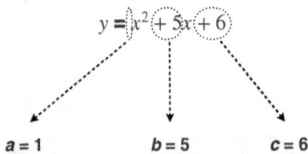

$$y = x^2 + 5x + 6$$

$$a = 1 \qquad b = 5 \qquad c = 6$$

Step 2. Replace the **a** , **b** , and **c** values of the quadratic formula with the extracted values from **Step 1** , being careful to use sets of parentheses for each of the values being replaced:

$$x = \frac{-b \pm \sqrt{b^2 - 4ac}}{2a}$$

$$a = 1 \qquad b = 5 \qquad c = 6$$

$$x = \frac{-(5) \pm \sqrt{(5)^2 - 4(1)(6)}}{2(1)}$$

Step 3. Solve the quadratic equation. Remember that it is very important to follow the correct order of operations, and to apply all the rules and principles that we have reviewed so far in the book.

$$x = \frac{-(5) \pm \sqrt{(5)^2 - 4(1)(6)}}{2(1)}$$

$$x = \frac{-5 \pm \sqrt{25 - (4)(6)}}{2}$$

$$x = \frac{-5 \pm \sqrt{25 - 24}}{2}$$

$$x = \frac{-5 \pm \sqrt{1}}{2}$$

$$x = \frac{-5 \pm 1}{2}$$

We can simplify the fraction above, but we must be careful to branch off the two answers that the add/subtract symbol yields: one branch will contain an addition symbol, while the other a subtraction symbol, as follows:

$$x = \frac{-5 \pm 1}{2}$$

$$X_1 = \frac{-5 + 1}{2} \qquad X_2 = \frac{-5 - 1}{2}$$

$$X_1 = \frac{-4}{2} \qquad X_2 = \frac{-6}{2}$$

$$X_1 = -2 \qquad X_2 = -3$$

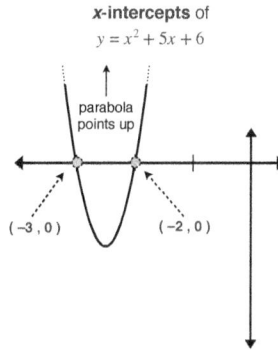

parabola
points up

$(-3 , 0)$ $(-2 , 0)$

As you can see once again, the quadratic formula works perfectly well. We are able to correctly find the two solutions, or zeros, or roots, or **x-intercepts**, of the equation.

I would like to drill once more the fact that the **x-intercepts**, strictly speaking, correspond to a location on the **x-y plane**, and should thus be defined by a coordinate. Therefore, the roots/zeros/solutions found correspond to the **x-intercepts** of:

$$(-2 , 0) \quad \text{and} \quad (-3 , 0)$$

Remember that the output value is **0** in both cases because that's precisely what **y** is set to when solving a quadratic equation. In the graph, these two coordinates correspond to the location where the graph of the equation $y = x^2 + 5x + 6$ crosses the **x-axis**, which must have a **y** value of **0** since any point that lies on said **x-axis** must have such height value.

The graph of this quadratic equation must pass through the **x-intercepts** specified above. Furthermore, we know that it points up because its **a** value is positive. Knowing these facts about the equation and its graph, we can sketch it as follows:

Observe once again how the two zeros (or roots or solutions) of the equation end up being the **x-intercepts** of its graph. This is a very important connection that you need to make; so much so, that it deserves its own summary box.

⚓

The **roots**, **zeros**, or **solutions** of a quadratic equation correspond to the **x-intercepts** of the equation's graph.

$$y = ax^2 + bx + c$$

X_1 X_2

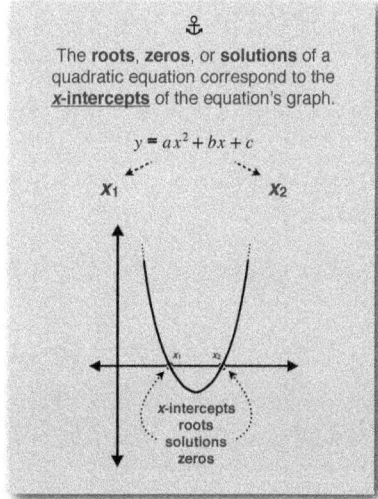

x-intercepts
roots
solutions
zeros

From this correspondence, it follows that if the equation's graph crosses the *x*-axis twice, then the equation has two **REAL** solutions (or roots or zeros); if it only crosses it once (this happens when the parabola's vertex lies exactly on the *x*-axis, as we have seen before) then the equation will have two **REAL** solutions equal to each other (or roots or zeros); finally, if the graph does not cross the *x*-axis at all, then the solutions (or roots or zeros) will be **COMPLEX**, in which case, they will always be conjugates of each other, of the from $d + ei$ and $d - ei$, as we defined earlier as well.

Quadratic equation with two **COMPLEX** solutions

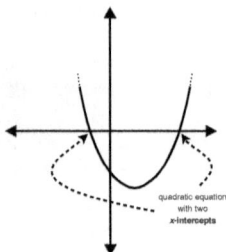

quadratic equation
without any
x-intercepts

Quadratic equation with two **REAL** solutions

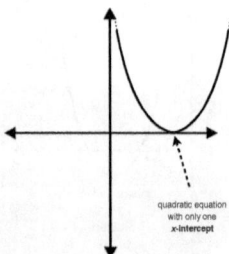

quadratic equation
with two
x-intercepts

Quadratic equation with only one **REAL** solution

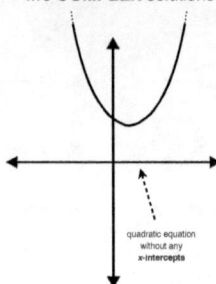

quadratic equation
with only one
x-intercept

Now that you know how to use the quadratic formula you have a practical method for solving any quadratic equation whatsoever. Compare it to the factoring method that we explored earlier in the chapter (finding two numbers that multiplied equal the independent term's value, and that added equal the middle term's (linear term's) coefficient), and you can see that while the factoring method, without the aid of computers, is only (humanly) practical for certain equations, the quadratic formula can always be used to solve any quadratic equation, no computers or calculators needed. The "complete the perfect square" method also works for any equation, so it becomes a matter of preference which method you use on a given prompt.

Before solving a wide range of quadratic equation prompts, we need to explore one last equation that enables you to quickly find the vertex of any quadratic equation written in standard form.

Let's deduce the formula ourselves.

We start out with the general form of a quadratic equation written in standard form. It is very easy to express any quadratic equation in standard form, simply by solving for **y** , hence the practicality of the quadratic formula.

$$y = ax^2 + bx + c$$

Next, we assume that the quadratic is not a perfect square (if we did make that assumption, then any formula derived having made that assumption would only work for **perfect square** quadratic equations, greatly reducing the practicality of the quadratic formula. Do note that making the assumption that the quadratic equation is not a perfect square will still work for quadratic equations that are perfect squares, for reasons that will become obvious as we move forward). Therefore, we first divide both sides by **a** because we want to ensure that the coefficient of the leading term (a.k.a. the quadratic term or the **x-squared** term, or simply **a**) is equal to **1** . Remember that this is one of the main requirements for completing the perfect square of any quadratic equation.

$$\frac{y}{a} = \frac{ax^2 + bx + c}{a}$$

Next, we simplify.

$$\frac{y}{a} = \frac{ax^2}{a} + \frac{bx}{a} + \frac{c}{a}$$

$$\frac{y}{a} = x^2 + \frac{b}{a}x + \frac{c}{a}$$

Since we are assuming that the quadratic is not a perfect square, we want to force its independent term to be equal to the square of half of the linear term's coefficient (in the equation above, the square of half of $\frac{b}{a}$).

$$\frac{y}{a} = x^2 + \frac{b}{a}x + \frac{c}{a}$$

We want this to be equal to:

$$(\frac{\frac{b}{a}}{2})^2 = (\frac{\frac{b}{a}}{\frac{2}{1}})^2 = (\frac{b}{2a})^2 = \frac{b^2}{4a^2}$$

The best way to guarantee that the equation's independent value is equal to $\frac{b^2}{4a^2}$ is to first subtract $\frac{c}{a}$ from both sides of the equation so that it is not standing in the way of the value we want to have in its place, the value of $\frac{b^2}{4a^2}$. We thus first subtract $\frac{c}{a}$ from both sides, and then we add $\frac{b^2}{4a^2}$ to both sides, guaranteeing that the right side is a perfect square.

$$\frac{y}{a} - \frac{c}{a} = x^2 + \frac{b}{a}x + \frac{c}{a} - \frac{c}{a}$$

$$\frac{y}{a} - \frac{c}{a} = x^2 + \frac{b}{a}x$$

$$\frac{y}{a} - \frac{c}{a} = x^2 + \frac{b}{a}x$$

Now, we add $\frac{b^2}{4a^2}$ to both sides...

$$\frac{y}{a} - \frac{c}{a} + \frac{b^2}{4a^2} = x^2 + \frac{b}{a}x + \frac{b^2}{4a^2}$$

At this point, we know for certain that the right side is a perfect square, so we can factor it as reviewed earlier (using half of the linear term's coefficient as the value to use inside the set of parentheses):

$$\frac{y}{a} - \frac{c}{a} + \frac{b^2}{4a^2} = (x + \frac{\frac{b}{a}}{2})^2$$

We can simplify the fraction on the right side, and combine the two right-most fractions that are on the left side of the equation, as follows. First, multiply the first fraction by $\frac{4a}{4a}$ so that both fractions have the same denominator, as we reviewed in earlier chapters of the book:

$$\frac{y}{a} - \frac{(4a)(c)}{(4a)(a)} + \frac{b^2}{4a^2} = (x + \frac{b}{2a})^2$$

Then simplify and assign the negative sign of the first fraction on the left side to its numerator to help ease the computation...

$$\frac{y}{a} + \frac{-4ac}{4a^2} + \frac{b^2}{4a^2} = (x + \frac{b}{2a})^2$$

Next, combine the fractions...

$$\frac{y}{a} + \frac{-4ac + b^2}{4a^2} = (x + \frac{b}{2a})^2$$

And then switch the terms of the numerator so that its leading term is positive...

$$\frac{y}{a} + \frac{b^2 - 4ac}{4a^2} = (x + \frac{b}{2a})^2$$

Now, we can solve for **y** so that the equation is expressed in vertex form. To do this, we need to subtract the right-most fraction on the left side of the equation from both sides of the equation. I have to write the equation using two lines given its length, so don't be thrown off by that:

$$\frac{y}{a} + \frac{b^2 - 4ac}{4a^2} - \frac{b^2 - 4ac}{4a^2} =$$
$$(x + \frac{b}{2a})^2 - \frac{b^2 - 4ac}{4a^2}$$

Next, we can cancel the fractions on the left side:

$$\frac{y}{a} = (x + \frac{b}{2a})^2 - \frac{b^2 - 4ac}{4a^2}$$

Finally, we can multiply by **a** both sides of the equation to solve for **y** :

$$(a)(\frac{y}{a}) = (a)((x + \frac{b}{2a})^2 - \frac{b^2 - 4ac}{4a^2})$$

$$y = a(x + \frac{b}{2a})^2 - a\frac{b^2 - 4ac}{4a^2}$$

$$y = a(x + \frac{b}{2a})^2 - \frac{b^2 - 4ac}{4a}$$

If you recall, the vertex form provides the vertex point of the graph. It is of the form:

$$y = a(x - h)^2 + k$$

...and the vertex it specifies is:

$$(h , k)$$

Therefore, based on the equation we obtained above, which is in vertex form, the vertex is given by:

$$\textbf{vertex @ } (-\frac{b}{2a} , -\frac{b^2 - 4ac}{4a})$$

The negative signs are included because the vertex from expects to see the **h** value subtracting (and in the equation we derived it is adding), and the **k** value adding (and in the equation we derived it is subtracting). It is conveniently written as follows (by incorporating the negative signs into the numerator of the fractions, and in the case of the output value's formula, rearranging the terms as well):

$$\text{vertex @ } (\frac{-b}{2a} , \frac{4ac - b^2}{4a})$$

Note that the axis of symmetry of the parabola will always pass through the vertex point, and earlier in the chapter I stated that the axis of symmetry is always found at

$$x = -\frac{b}{2a}$$

See the connection?

While it may be a bit of a stretch to memorize the output value of the coordinate of the vertex ($\frac{4ac - b^2}{4a}$), it is very easy to memorize its input instead: $\frac{-b}{2a}$. This means that if you have a quadratic equation written in standard form, you can always figure out its vertex point by finding its location on the **x-axis** using the formula, and then by plugging in the **x** value found into the equation itself to know the output value.

Let's try this using the last quadratic equation we worked with:

$$y = x^2 + 5x + 6$$

We know its zeros are **-2** and **-3**, which means that its axis of symmetry must pass through exactly in the middle of these two points (since the points correspond to a symmetry pair). We thus know that this equation's vertex is

located at (**-2.5** , **?**). The **-2.5** is derived by adding **-2** and **-3** and dividing by **2**, as we reviewed earlier in the chapter.

So let's put the vertex formula to the test.

We know that this equation's **a**, **b**, and **c** values are **1**, **5**, and **6** respectively. Plugging in these values into the formula we derived we obtain the following (remember that we will only compute the location of the vertex as far as the **x-axis** is concerned; its location along the **y-axis** will be computed using the quadratic equation itself):

$$\text{vertex @ } (\frac{-b}{2a} , \frac{4ac - b^2}{4a})$$

$$\frac{-(5)}{2(1)} = \frac{-5}{2} = -2.5$$

We thus know that this equation's vertex must be located at:

$$(-2.5 , ?)$$

To find the location of the vertex along the **y-axis**, we can simply plug in **-2.5** into the quadratic equation, so that we can know this input's corresponding output value.

$$y = x^2 + 5x + 6$$

$$x = -2.5$$

$$y = (-2.5)^2 + 5(-2.5) + 6$$

$$y = 6.25 - 12.5 + 6$$

$$y = -.25$$

Therefore, the equation's vertex is located at:

$$(-2.5 \, , -.25 \,)$$

So knowing how to compute the **x** value (in other words, the input value) of the coordinate of the vertex for any quadratic equation written in standard form using this equation is very practical; without this, you would have to complete the perfect square and express it in vertex form, a much more complicated process (typically).

⚓

The **VERTEX** of a quadratic equation that is written in standard form

$$y = ax^2 + bx + c$$

may be found using the following formula:

$$(\frac{-b}{2a} , \frac{-b^2 + 4ac}{4a})$$

The location of the vertex along the **x-axis** is given by

$$\frac{-b}{2a}$$

while the location of the vertex along the **y-axis** is given by

$$\frac{-b^2 + 4ac}{4a}$$

The following graph can help you clearly see what each of the vertex formula components specify regarding the location of the parabola's vertex:

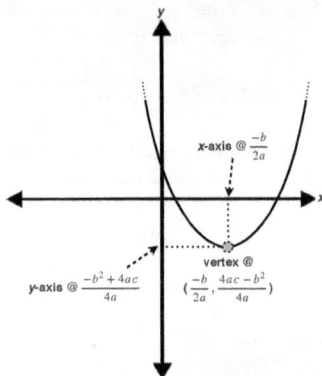

I want to stress the fact that it is more practical to find the **x** portion of the coordinate of the vertex, and then to plug that value in to the quadratic equation to find the **y** portion of the coordinate.

Let's work on another example. Try to solve it by yourself before looking at my solution process.

Find the vertex of the equation

$$y = -x^2 + 4x + 5$$

Ready to check your answer?

We know that this equation's **a**, **b**, and **c** values are **−1**, **4**, and **5** respectively. Plugging in these values into the formula we derived we obtain the following (remember that we will only compute the location of the vertex as far as the **x-axis** is concerned; its location along the **y-axis** will be computed using the quadratic equation itself):

$$\text{vertex @ } (\frac{-b}{2a} , \frac{4ac - b^2}{4a})$$

$$\frac{-(4)}{2(-1)} = \frac{-4}{-2} = 2$$

We thus know that this equation's vertex must be located at:

$$(2 , ?)$$

To find the location of the vertex along the **y-axis** (in other words, to find the **y** component of the coordinate of the vertex) we can simply plug in **2** into the quadratic equation, so that we can know this input's corresponding output value.

$$y = -x^2 + 4x + 5$$

$$x = 2$$

$$y = -(2)^2 + 4(2) + 5$$

$$y = -4 + 8 + 5$$

$$y = 9$$

Therefore, the equation's vertex is located at:

$$(2 , 9)$$

We also know that because this equation's **a** value is equal to **−1** , the parabola points down.

Let's find the zeros (or roots, or solutions, or **x-intercepts**) of the equation. Try to use the quadratic formula (or any one of the other methods we reviewed in this chapter) in order to find them. Note that if the zeros are **REAL**, then we would know that they correspond to the parabola's **x-intercepts**. If they are **COMPLEX**, then we know that the parabola does not cross the **x-axis** (although since we already know that this equation has its vertex at **(2 , 9)** and that the parabola points down, it follows unequivocally that this equation's parabola must cross the x-axis, and therefore, that its two zeros or roots or solutions must be **REAL** and distinct).

After you find its zeros, find two additional points that are part of this equation's parabola so that you can draw a precise graph of the equation.

Ready to check your work?

Let's begin with the quadratic formula. Our goal is to find this equation's zeros (or roots, or solutions), so that we may specify its **x-intercepts**.

Step 1. We must extract the equation's **a** , **b** , and **c** values: these will be used to replace the corresponding letters of the quadratic formula. To do this, we must make sure that the quadratic equation is solved for the variable **y** . Since the given equation is solved for **y** we may extract its values without having to do any extra work. They are:

$$y = -x^2 + 4x + 5$$

$$a = -1 \qquad b = 4 \qquad c = 5$$

Step 2. Replace the a, b, and c values of the quadratic formula with the extracted values from **Step 1**, being careful to use sets of parentheses for each of the values being replaced:

$$x = \frac{-b \pm \sqrt{b^2 - 4ac}}{2a}$$

$$a = -1 \qquad b = 4 \qquad c = 5$$

$$x = \frac{-(4) \pm \sqrt{(4)^2 - 4(-1)(5)}}{2(-1)}$$

Step 3. Solve the quadratic equation. Remember that it is very important to follow the correct order of operations, and to apply all the rules and principles that we have reviewed so far in the book.

$$x = \frac{-(4) \pm \sqrt{(4)^2 - 4(-1)(5)}}{2(-1)}$$

$$x = \frac{-4 \pm \sqrt{16 - (-4)(5)}}{-2}$$

$$x = \frac{-4 \pm \sqrt{16 - (-20)}}{-2}$$

$$x = \frac{-4 \pm \sqrt{36}}{-2}$$

$$x = \frac{-4 \pm 6}{-2}$$

We can simplify the fraction above, but we must be careful to branch off the two answers that the add/subtract symbol yields: one branch will contain an addition symbol, while the other a subtraction symbol, as follows:

$$x = \frac{-4 \pm 6}{-2}$$

$$x_1 = \frac{-4 + 6}{-2} \qquad x_2 = \frac{-4 - 6}{-2}$$

$$x_1 = \frac{2}{-2} \qquad x_2 = \frac{-10}{-2}$$

$$x_1 = -1 \qquad x_2 = 5$$

The **x-intercepts**, as we have stated previously, correspond to a location on the **x-y plane**, and should thus be defined by a coordinate whose output is **0** (in other words, whose **y** component is **0**). Therefore, the roots/zeros/solutions found correspond to the **x-intercepts** of:

$$(-1, 0) \quad \text{and} \quad (5, 0)$$

Observe that based on these two **x-intercepts** found, we would expect this equation's axis of symmetry to be located exactly between them: this leads to the computation

$$\textbf{axis of symmetry @ } x = \frac{-1 + 5}{2} = \frac{4}{2} = 2$$

...which coincides with the **x** portion of the coordinate of the vertex we found earlier for this equation, as should be, since in all parabolas its axis of symmetry will always pass through the vertex.

The last thing we need to do in order to draw a precise sketch of this equation is find two additional coordinates. I will use the input values of **0** and its corresponding symmetry pair **x** value of **4** .

$$x = 0$$
$$y = -(0)^2 + 4(0) + 5$$
$$y = 0 + 0 + 5$$
$$y = 5$$

Coordinate ---> (0 , 5)
which also implies ---> (4 , 5)

Did you notice something about the input value I chose to use? It corresponds to the **y-intercept**, and the standard form of a quadratic equation explicitly provides it:

Quadratic equation in
Standard Form

$$y = ax^2 + bx + c$$

y-intercept @
(0 , c)

Recall that the standard form of a quadratic equation only provides the **a** value (which states whether the parabola points up or down and the arm-aperture value) and the **y-intercept** point.

Therefore, when I stated earlier that I would use **0** and **4** as the input values for finding two additional coordinates, I did not have to set **x** to **0** and plug in to the equation. I could have just extracted the **y-intercept** value from the equation, and used that instead.

Using the vertex, the x-intercepts, and the two additional points found above, we can now sketch a precise graph of the equation:

$$y = -x^2 + 4x + 5$$

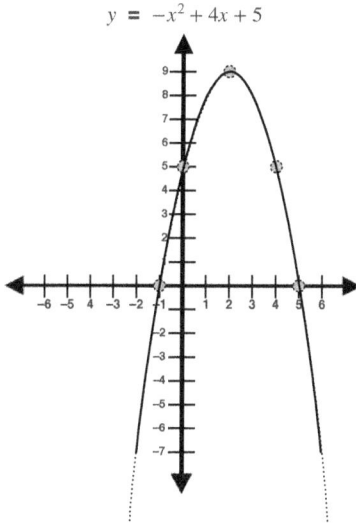

How does your graph compare to the one above? Were you able to use the quadratic formula to find the **x-intercepts**? You may have used a different **x** value in order to find the two additional points that you needed for the graph, but other than that, your graph and the one above should be very similar.

In this chapter we reviewed practically everything that there is to know about quadratic equations, how to solve them, and how to sketch their graphs. As you can readily see it is the longest chapter of the book (by far!) simply because quadratic equations can come in so many different forms, and because their exponential nature forces us to incorporate elements such as **COMPLEX** numbers during their solving process.

Before moving on to the last chapter of the book I want to summarize all that we reviewed in this chapter, and then to ask you to solve quadratic equation problems by yourself (I will, as usual, solve them as well so that you can compare your answers).

Quadratic equations summary

Quadratic equations are of the form

$$y = ax^2 + bx + c$$

...where the input variable must be raised to the second power, and its coefficient (a) must be equal to any **REAL** number other than 0. The other coefficients (b and c) may be equal to any **REAL** number. Note that **COMPLEX** numbers are not allowed for the values of a, b, or c.

All quadratic equations may be rewritten in any of the following forms, regardless of how they are initially presented. The three forms we reviewed are:

Quadratic Equation in Standard Form

$$y = ax^2 + bx + c$$
where $a \neq 0$
y-intercept @ (0 , c)
axis of symmetry @ $x = \dfrac{-b}{2a}$
vertex @ ($\dfrac{-b}{2a}$, $\dfrac{4ac - b^2}{4a}$)

➡️

Quadratic Equation in Vertex Form

$$y = a(x - h)^2 + k$$
where $a \neq 0$
vertex @ (h , k)
axis of symmetry @ $x = h$

Quadratic Equation in Intercept Form

$$y = a(x - m)(x - n)$$
where $a \neq 0$
x-intercepts @ (m , 0) and (n , 0)
axis of symmetry @ $x = \dfrac{m + n}{2}$

For all forms, regarding the equation's graph (parabola):

if a is + : **points up** ; if a is − : **points down**

| a | = **arm aperture**

A quadratic equation's graph is always a parabola. It may point up or down (see examples below). All parabolas have a vertex point, which the axis of symmetry always passes through:

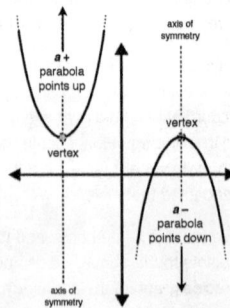

690

All parabolas will have a **y-intercept** point: in other words, we know for a fact that all quadratic equations have a valid **(0 , ?)** coordinate. This stems from the fact that the **domain** of all quadratic equations is **all** REAL numbers, which means that we may use any value we wish to use between negative infinity and positive infinity for the variable **x**. On the other hand, not all parabolas have **x-intercepts** (the parabola on the left on the diagram above is such an example). This stems from the fact that the range of a quadratic equation will always be either between negative infinity and a specific **REAL** number, or between a specific **REAL** number and positive infinity. Thus, the output **0** may not be part of a given quadratic's equation range.

Note that if we want to know the location of the **axis of symmetry** for a given quadratic equation, we can simply solve it for **y** so that it is expressed in **Standard Form**, and then use the following formula:

$$\textbf{axis of symmetry} \ @ \ x = \frac{-b}{2a}$$

It is important to note that it is very easy to express any quadratic equation in **Standard Form** (simply solve for **y**) and that it is possible to express any quadratic equation in **Vertex Form** without the aid of a calculator or a computer (simply by completing the perfect square). As far as expressing a quadratic in **Intercept Form** is concerned, we reviewed a method that involves starting out with the **Standard Form** of the equation, and then looking for two numbers that multiplied equal its **c** value and that added equal its **b** value; those two numbers correspond to **m** and **n** in the formula

$$y = a(x - m)(x - n)$$

Note that it is not always easy to find the two numbers that work for a given quadratic equation, and furthermore, that the numbers sought may be either two **REAL** numbers or two **COMPLEX** numbers (if the latter, they would need to be complex conjugates of each other).

The good news is that since **m** and **n** correspond to the zeros (or roots, or solutions, or **x-intercepts**) of the quadratic equation, another method that may be used (which is typically easier to apply for any given quadratic equation than the previous method) is to simply find the zeros of the equation using the quadratic formula: these correspond to the **m** and **n** values mentioned earlier.

The quadratic formula is:

$$x = \frac{-b \pm \sqrt{b^2 - 4ac}}{2a}$$

...where **a**, **b**, and **c** correspond to the same values of the **Standard Form**, which means that in order to use the quadratic formula, the quadratic equation must be written in **Standard Form** (by solving for **y**).

A very important skill regarding quadratic equations involves the ability to solve them. Solving a quadratic equation means that we are interested in finding the value or values for **x** that when plugged in to the equation, yield a specific **y** value (note that it is always possible to set the equation to **0** in order to solve it... For example, if we needed to solve $10 = x^2 + 8x + 26$ we could simply subtract **10** from both sides of the equation, and we would thus have the quadratic equation set to zero: $0 = x^2 + 8x + 16$, see next paragraph below).

In order to solve a quadratic equation, it is necessary to first set it to zero (as mentioned above). Once this is done, any of the methods we reviewed may be used to find the values of x that produce such an output value (that of 0, since that's what the variable y has essentially been set to during the solving process). The most commonly used methods are:

Method 1. Fully factoring the quadratic equation, and then use the ***zero product property*** to solve each of the sets of parentheses. This means setting up and solving two linear equations. Recall one of the examples we worked on earlier:

$$0 = x^2 - 6x + 8$$

$$0 = (x - 2)(x - 4)$$

$(x - 2) = 0$	$(x - 4) = 0$
$x - 2 = 0$	$x - 4 = 0$
$x - 2 + 2 = 0 + 2$	$x - 4 + 4 = 0 + 4$
$x = 2$	$x = 4$

In this example, we needed to find the values of x that when plugged in to $x^2 - 6x + 8$ produce an output of 0; we thus factored the equation and then used the zero product property to find such values: $x_1 = 2$ and $x_2 = 4$.

Method 2. Express the equation in vertex form, and then solve for x. The same example above could thus be solved as follows:

$$0 = x^2 - 6x + 8$$

$$1 = x^2 - 6x + 9$$

$$1 = (x - 3)^2$$

Remember that when the exponent affecting the variable x is canceled by taking the square root of both sides of the equation, it is necessary to introduce the ***plus/minus*** sign on the side opposite to where the variable x is located. This is the source of the two solutions that are typical of a quadratic equation (though for some quadratics, both solutions are actually equal to each other, in which case the quadratic equation is said to have only one **REAL** solution). Continuing with the example above, we thus proceed as follows:

$$1 = (x - 3)^2$$

$$\sqrt{1} = \sqrt{(x - 3)^2}$$

$$\pm\sqrt{1} = (x - 3)$$

$1 = x - 3$	$-1 = x - 3$
$x_1 = 4$	$x_2 = 2$

Notice that we find the same solutions as those found using the previous method, as should be, since it is the exact same equation.

Once again, these solutions are values for x that if plugged back into the equation, produce an output value of 0. Hence the fact that the solutions of a quadratic equation are also called the zeros, or roots of the quadratic equation. Furthermore, since the solutions' values correspond to the input value (x), and both produce an output value (y) of 0, the solutions correspond to the **x-intercept**. Thus the following are all equivalent terms when used in the context of quadratic equations:

Solutions
Zeros
Roots
x-intercepts

We know, of course, that the *x*-intercepts, strictly speaking, correspond to points on the graph, and therefore are of coordinate form (using the letters *m* and *n* of the Intercept Form):

x-intercepts:

$(m , 0)$ and $(n , 0)$

If the solutions are **REAL** numbers, then the *x*-intercepts exist (the equation's graph crosses the *x*-axis). If they are **COMPLEX**, then they must be conjugates of each other, and the equation's parabola never crosses the *x*-axis.

Let's end the chapter with a set of quadratic equation problems. Try to work them out by yourself, and only after you are finished solving each one read my answers so that you may check your work. As always, you can look back to the appropriate sections of the chapter (or the book) as needed.

Quadratic equation problems

--> Find the roots of the following quadratic equations.

a) $y = x^2 + 9x + 20$
b) $y = -3x^2 + 4x - 6$
c) $y = -5(-x + 2)(2x + 3)$
d) $y = 4(x - 3)^2 - 16$
e) $2y - 4x = 2x^2 - 10x - 20$

--> Sketch a precise graph the following quadratic equations by finding their vertex, *x*-intercepts, and two additional points:

f) $y = x^2 + 4x - 5$

--> Solve the following prompt:

g) An engineer needs to solve the following quadratic equation for $y = -3$ so that she may determine the length, in centimeters, of a component that will form part of a robot's arm:

$$y = x^2 - 4x - 8$$

Find the length of the component.

--> Use the vertex formula to find the vertex of the following equations:

h) $y = -6x^2 + 24x + 1$
i) $-3y = 6x^2 + 3x - 12$

Ready to check your answers? Compare your work with my answers.

--> Find the roots of the following quadratic equations.

a) $y = x^2 + 9x + 20$

Because this quadratic can be easily factored by using the "pair of numbers" method, I will find its roots by first factoring the equation using said method, and then by solving each parentheses set using the zero product property. We need two numbers that multiplied equal 20 and that added equal **9** ; the pair of numbers that works is **5** and **4** . These correspond to the **m** and **n** values of the Intercept Form, and since **a = 1** we can rewrite the equation as

$$y = (x + 5)(x + 4)$$

Next, we set the equation to **0** (make **y** be equal to **0**), and then using the zero product property, solve each parentheses set for **x** having set each to **0** as well:

$$0 = (x + 5)(x + 4)$$

$$(x + 5) = 0 \qquad (x + 4) = 0$$
$$x + 5 = 0 \qquad x + 4 = 0$$
$$x + 5 - 5 = 0 - 5 \qquad x + 4 - 4 = 0 - 4$$
$$x = -5 \qquad x = -4$$

We have found the equation's two roots (or zeros, or solutions, or **x-intercepts**):

$$x_1 = -5 \qquad x_2 = -4$$

b) $y = -3x^2 + 4x - 6$

To find this equation's roots, I will use the quadratic formula (do note that we could complete the perfect square as well, so it's simply a matter of preference).

Step 1. We must extract the equation's **a** , **b** , and **c** values: these will be used to replace the corresponding letters of the quadratic formula. To do this, we must make sure that the quadratic equation is solved for the variable **y** . Since the given equation is solved for y we may extract its values without having to do any extra work. They are:

$$y = -3x^2 + 4x - 6$$

$$a = -3 \qquad b = 4 \qquad c = -6$$

Step 2. Replace the **a** , **b** , and **c** values of the quadratic formula with the extracted values from **Step 1** , being careful to use sets of parentheses for each of the values being replaced:

$$x = \frac{-b \pm \sqrt{b^2 - 4ac}}{2a}$$

$$a = -3 \qquad b = 4 \qquad c = -6$$

$$x = \frac{-(4) \pm \sqrt{(4)^2 - 4(-3)(-6)}}{2(-3)}$$

Step 3. Solve the quadratic equation. Remember that it is very important to follow the correct order of operations, and to apply all the rules and principles that we have reviewed so far in the book.

$$x = \frac{-(4) \pm \sqrt{(4)^2 - 4(-3)(-6)}}{2(-3)}$$

$$x = \frac{-4 \pm \sqrt{16 - (-12)(6)}}{-6}$$

$$x = \frac{-4 \pm \sqrt{16 - (-72)}}{-6}$$

$$x = \frac{-4 \pm \sqrt{88}}{-6}$$

$$x = \frac{-4 \pm \sqrt{(4)(22)}}{-6}$$

$$x = \frac{-4 \pm 2\sqrt{22}}{-6}$$

Note that I simplified $\sqrt{88}$ as $2\sqrt{22}$ as we reviewed in **Chapter 10**. We can now simplify the fraction above, but we must be careful to branch off the two answers that the add/subtract symbol yields: one branch will contain an addition symbol, while the other a subtraction symbol, as follows:

$$x = \frac{-4 \pm 2\sqrt{22}}{-6}$$

$$x_1 = \frac{-4 + 2\sqrt{22}}{-6} \qquad x_2 = \frac{-4 - 2\sqrt{22}}{-6}$$

First, break up each fraction into the two fractions that they represent (based on the fact that the numerators contain two terms each)

$$x_1 = \frac{-4}{-6} + \frac{2\sqrt{22}}{-6} \qquad x_2 = \frac{-4}{-6} - \frac{2\sqrt{22}}{-6}$$

Next, simplify each fraction. For example, the left branch contains the fraction $\frac{-4}{-6}$ which simplifies to $\frac{2}{3}$ (recall that two negative numbers dividing yield a positive number, and that 4 and 6 are both divisible by 2). The fraction that contains the root symbol can also be simplified using a similar process, as we reviewed in earlier

chapters of the book. The branch on the right can also be simplified using the same techniques.

$$x_1 = \frac{2}{3} + \frac{-\sqrt{22}}{3} \qquad x_2 = \frac{2}{3} - \frac{-\sqrt{22}}{3}$$

Next, simplify the expression by using the Sign Table:

$$x_1 = \frac{2}{3} - \frac{\sqrt{22}}{3} \qquad x_2 = \frac{2}{3} + \frac{\sqrt{22}}{3}$$

And we have found the two roots of the equation. If we needed to express the roots as numbers with decimals (which would be approximations since these roots correspond to **IRRATIONAL** numbers, a calculator could be used to do so:

$$x_1 = -.8968... \qquad x_2 = 2.2301...$$

c) $y = -5(-x + 2)(2x + 3)$

This quadratic equation is set up perfectly for determining its roots. Because it is in **Intercept Form**, we can use the zero product property to find them quickly.

First, wet the equation to **0** by setting $y = 0$:

$$0 = -5(-x + 2)(2x + 3)$$

Next, take each parentheses set and set it to zero, and solve for x as we reviewed earlier in the chapter. We may ignore the value of a (in other words, the number **–5** that is a factor of the equation's right side) because as long as one of the "factors" of the right side is zero, then the product of the entire right side will equal zero, which is what the equation "requests". So we focus exclusively on each set of parentheses.

Observe.

$$0 = -5(-x + 2)(2x + 3)$$

$(-x + 2) = 0$	$(2x + 3) = 0$
$-x + 2 = 0$	$2x + 3 = 0$
$-x + 2 - 2 = 0 - 2$	$2x + 3 - 3 = 0 - 3$
$-x = -2$	$2x = -3$
$(-1)(-x) = (-2)(-1)$	$\dfrac{2x}{2} = \dfrac{-3}{2}$
$x = 2$	$x = \dfrac{-3}{2}$

$$x_1 = 2 \qquad x_2 = \frac{-3}{2}$$

And we have found the two roots of the equation.

d) $y = 4(x - 3)^2 - 16$

This quadratic equation is in Vertex Form. To find its roots, we can simply set it to zero (set $y = 0$) and then solve for x as we reviewed earlier in the chapter. Observe.

$$0 = 4(x - 3)^2 - 16$$

$$0 + 16 = 4(x - 3)^2 - 16 + 16$$

$$16 = 4(x - 3)^2$$

Next, we need to take the square root of both sides to get rid of the exponent that stands in the way of being able to solve for the variable x:

$$16 = 4(x - 3)^2$$

$$\sqrt{16} = \sqrt{4(x - 3)^2}$$

It is critical to take note of the fact that on the right side, we are now faced with the square root of the following: the product of **4** and $(x - 3)^2$; therefore, on the next step, when we simplify the right side, we must express it as the product of the square root of **4** times the square root of $4(x - 3)^2$, as we reviewed in **Chapter 10**. The exponent attached to the set of parentheses has no bearing at all on the factor **4** which is why this must be done. Thus the next step looks as follows:

$$\sqrt{16} = \sqrt{4}\sqrt{(x - 3)^2}$$

$$\sqrt{16} = 2\sqrt{(x - 3)^2}$$

Now, recall that when we cancel the exponent on the right side of the equation with the square root (its inverse operation), we must incorporate the plus/minus symbol on the other side because the variable was being squared, as we reviewed earlier in the chapter (note the dashed line indicating the operations that cancel each other out):

$$\sqrt{16} = 2\sqrt{(x - 3)^2}$$

$$\pm\sqrt{16} = 2(x - 3)$$

Now, we can simplify the left side, and continue solving for x:

$$\pm 4 = 2(x - 3)$$

$4 = 2(x - 3)$	$-4 = 2(x - 3)$
$\dfrac{4}{2} = \dfrac{2(x - 3)}{2}$	$\dfrac{-4}{2} = \dfrac{2(x - 3)}{2}$

$$2 = (x - 3) \qquad -2 = 2(x - 3)$$

$$2 = x - 3 \qquad -2 = 2x - 6$$

$$2 + 3 = x - 3 + 3 \qquad -2 + 6 = 2x - 6 + 6$$

$$5 = x \qquad 4 = 2x$$

$$\frac{4}{2} = \frac{2x}{2}$$

$$2 = x$$

And so we find that the two roots (or zeros, or solutions, or **x-intercepts**) of this equation are:

$$x_1 = 5 \qquad x_2 = 2$$

e) $2y - 4x = 2x^2 - 10x - 20$

To find the roots of this quadratic equation, first, we need to solve the equation for **y**.

$$2y - 4x + 4x = 2x^2 - 10x - 20 + 4x$$

$$2y = 2x^2 - 6x - 20$$

$$\frac{2y}{2} = \frac{2x^2 - 6x - 20}{2}$$

$$y = \frac{2x^2}{2} - \frac{6x}{2} - \frac{20}{2}$$

$$y = x^2 - 3x - 10$$

Next, we set it to zero by setting $y = 0$:

$$0 = x^2 - 3x - 10$$

Now, we have a choice to make. We can find the roots of the equation by using the quadratic formula, by completing the perfect square, or by trying to factor the equation in order to express it

in **Intercept Form**. A quick analysis tells us that the latter method is probably the fastest to use in this case, since the numbers **–5** and **2** multiply to **–10** and add to **–3**. These numbers can now be used to express this equation's right side in **Intercept Form** as follows:

$$0 = (x - 5)(x + 2)$$

Now we can use the zero product property, as we did earlier.

$$0 = (x - 5)(x + 2)$$

$$(x - 5) = 0 \qquad\qquad (x + 2) = 0$$
$$x - 5 = 0 \qquad\qquad x + 2 = 0$$
$$x - 5 + 5 = 0 + 5 \qquad x + 2 - 2 = 0 - 2$$
$$x = 5 \qquad\qquad x = -2$$

We could have simply extracted the roots (or zeros, or solutions, or **x-intercepts**) form each set of parentheses, as we reviewed earlier in the chapter, as follows:

$$0 = (x - 5)(x + 2)$$

$$x = 5 \qquad\qquad x = -2$$

This requires us to remember that the **general Intercept Form** is as follows:

$$0 = a(x - m)(x - n)$$

...and therefore, it expects to have the **m** and the **n** values subtracting. In this case, since the **m** value is subtracting (the number **5**), we know the root it embodies is **positive five**; however, the n value is adding (the number **2**), which means that the root it embodies is **negative two**, as the following diagram illustrates:

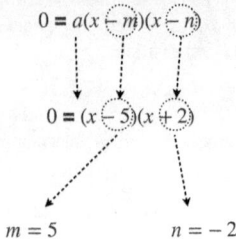

$$0 = a(x - m)(x - n)$$

$$0 = (x - 5)(x + 2)$$

$$m = 5 \qquad n = -2$$

The value of **a** (not explicitly shown above) is, of course, equal to **1**.

Regardless of the method you use, the correct roots of the equation are:

$$\mathbf{x_1} = 5 \qquad \mathbf{x_2} = -2$$

--> Sketch a precise graph the following quadratic equations by finding their vertex, **x-intercepts**, and two additional points:

f) $y = x^2 + 4x - 5$

The most practical approach is to factor this quadratic equation:

$$y = (x + 5)(x - 1)$$

Now we know that its two **x-intercepts** are:

$$(-5, 0) \text{ and } (1, 0)$$

Because these two points form a symmetry pair (they have the same output or **y** value), the equation's axis of symmetry must pass exactly in the middle of these two points:

$$\textbf{axis of symmetry @ } \mathbf{x} = \frac{-5 + 1}{2} = \frac{-4}{2} = -2$$

Next, we can use this **x** value and plug it in to the equation to find its output: we will then know the equation's vertex.

$$\textbf{if } x = -2 :$$

$$y = (-2)^2 + 4(-2) - 5$$

$$y = 4 + (-8) - 5$$

$$y = 4 - 8 - 5$$

$$y = -4 - 5$$

$$y = -9$$

And so the vertex point is:

$$(-2, -9)$$

Finally, we need two additional coordinates in order to sketch a precise graph of this quadratic equation. Clearly, the best course of action is to use the **y-intercept** point that the Standard Form explicitly provides:

$$y = x^2 + 4x - 5$$

$$(0, -5)$$

...which implies (using the concept of symmetry pair points):

$$(-4, -5)$$

We know this last coordinate is part of this equation's parabola because it is located equally far apart from the axis of symmetry (but on the opposite side of the **y-intercept** point), and will thus have the same output as the **y-intercept** point. We can now sketch a precise graph using all these elements we were able to define.

Graph of
$y = x^2 + 4x - 5$

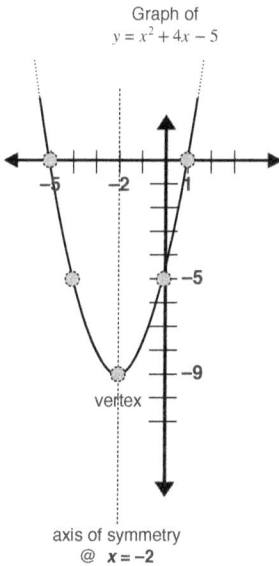

vertex

axis of symmetry
@ $x = -2$

--> Solve the following prompt:

g) An engineer needs to solve the following quadratic equation for $y = -3$ so that she may determine the length, in centimeters, of a component that will form part of a robot's arm:

$$y = x^2 - 4x - 8$$

Find the length of the component.

To solve this, we need to understand that the engineer is trying to determine the input value (x) that will yield an output value of -3 . In other words, if $y = -3$, which x value (or values) produce such an output? We would expect her to proceed as follows.

First, set the equation to the desired output value.

$$-3 = x^2 - 4x - 8$$

Next, set to zero and solve. Remember that we set the equation to zero because it would allow us to use the zero product property, a powerful method of solving quadratic equations. Or, if we wanted to use the quadratic formula instead, we need to have it set to zero as well. The only technique that doesn't require setting the equation to zero is the complete the perfect square method; however, I do not typically use that method to solve quadratic equations, unless they are already expressed in such form.

So let's set it zero... Simply add **3** to both sides of the equation, and the goal will have been achieved.

$$-3 + 3 = x^2 - 4x - 8 + 3$$

$$0 = x^2 - 4x - 5$$

For the next step, a quick analysis tells us that this equation is easily factored: the two numbers **-5** and **1** multiply to the independent term's value of **-5** and add to the middle term's coefficient of **-4** . Therefore, we know that **m** and **n** are equal to **-5** and **1** respectively.

We can thus rewrite the equation as follows:

$$0 = x^2 - 4x - 5$$

$$0 = (x - 5)(x + 1)$$

Next, we can extract the equation's roots (or zeros, or solutions, or x-intercepts) directly; just remember to consider the fact whether the values are adding or subtracting inside each set of parentheses, as we reviewed throughout the chapter.

$$0 = (x - 5)(x + 1)$$

$$m = 5 \qquad n = -1$$

$$x_1 = 5 \qquad x_2 = -1$$

So how would the engineer interpret these two answers? Clearly, since she is determining the length of a part that is going to be used to build a robot's arm, the solution she needs must be positive. Therefore, she would disregard the **−1** answer, and determine that the length must be equal to **5 centimeters**. To prove that this length is correct, we can revisit the original equation that we set out to solve to begin with, plug in this solution found, and check for consistency. Observe.

Original equation:

$$-3 = x^2 - 4x - 8$$

Check if **x = 5** solves the equation:

$$-3 \overset{?}{=} (5)^2 - 4(5) - 8$$

$$-3 \overset{?}{=} 25 - 20 - 8$$

$$-3 \overset{?}{=} 5 - 8$$

$$-3 \overset{\checkmark}{=} -3$$

As you can see, the solution works perfectly well. You can check if you like that the other root, the value of **−1**, also works, although in the context of this problem, the root has no significance.

--> Use the vertex formula to find the vertex of the following equations:

j) $y = -6x^2 + 24x + 1$

In order to use the vertex formula

$$\textbf{vertex} \ @ \ (\ \frac{-b}{2a} \ , \ \frac{4ac - b^2}{4a} \ )$$

...we need to make sure that the equation is solved for **y** because we need to be able to specify the equation's **a**, **b**, and **c** values. Because this equation is solved for **y**, we can extract the needed values:

$$a = -6 \ ; \ b = 24 \ ; \ c = 1$$

Next, we find the **x** component of the vertex coordinate:

$$x = \frac{-b}{2a} = \frac{-(24)}{2(-6)} = \frac{-24}{-12} = 2$$

Next, as I mentioned earlier, it is usually easier to simply use the equation itself to find the **y** component of the vertex coordinate. In this example, I will use both methods to prove to you that both work and are equivalent.

Method 1. Plug in the **x** value found in the previous step into the original equation to determine the **y** component of the vertex. This works because we are in essence finding the input/output pair that corresponds to the equation's vertex.

If $x = 2$:

$$y = -6(2)^2 + 24(2) + 1$$

$$y = -6(4) + 48 + 1$$

$$y = -24 + 48 + 1$$

$$y = 24 + 1$$

$$y = 25$$

And we have the exact location of the vertex:

Vertex @ $(2 , 25)$

Let's use the other method to prove that they are both equivalent to each other.

Method 2. Use the vertex formula to find the **y** component of the vertex:

$$y = \frac{4ac - b^2}{4a} = \frac{4(-6)(1) - (24)^2}{4(-6)}$$

$$= \frac{-24 - 576}{-24} = \frac{-600}{-24} = 25$$

As you can see, the formula for computing the **y** component of the vertex is perfectly valid; however, you can see that the previous method is a bit easier to use.

k) $-3y = 6x^2 + 3x - 12$

This equation is not solved for **y** ; therefore, before being able to use the vertex formula we must solve it for **y** as follows:

$$-3y = 6x^2 + 3x - 12$$

$$\frac{-3y}{-3} = \frac{6x^2 + 3x - 12}{-3}$$

$$y = \frac{6x^2}{-3} + \frac{3x}{-3} - \frac{12}{-3}$$

$$y = -2x^2 + (-x) - (-4)$$

$$y = -2x^2 - x + 4$$

Now, we can extract the necessary values to be used in the **x** component of the vertex formula (**a** and **b** only; the value of **c** is not needed for this component of the vertex coordinate):

$$a = -2 \quad ; \quad b = -1$$

...and plug these values into the vertex formula for computing the **x** component:

$$x = \frac{-b}{2a} = \frac{-(-1)}{2(-2)} = \frac{1}{-4} = -.25$$

Now, with the **x** component of the vertex coordinate established, we can find the **y** component by plugging in **x** into the equation (using this method, as opposed to using the vertex formula for this **y** component):

$$y = -2x^2 - x + 4$$

$$y = -2(\frac{1}{-4})^2 - (\frac{1}{-4}) + 4$$

$$y = -2(\frac{1}{16}) + \frac{1}{4} + 4$$

$$y = \frac{-2}{16} + \frac{1}{4} + \frac{(4)(4)}{(1)(4)}$$

$$y = \frac{-1}{8} + \frac{1}{4} + \frac{16}{4}$$

$$y = \frac{-1}{8} + \frac{17}{4}$$

$$y = \frac{-1}{8} + \frac{(17)(2)}{(4)(2)}$$

$$y = \frac{-1}{8} + \frac{34}{8}$$

$$y = \frac{33}{8}$$

$$y = 4.125$$

In the case of this particular equation, finding its vertex was a bit more challenging because we had to deal with fractions. However, you should be able to follow the steps above and find the answer yourself based on all the rules and principles that we have reviewed throughout the book.

Once the **y** component is determined, we may specify the vertex coordinate either by using the fraction version of the **x** and **y** values found, or by using their decimal equivalents, as follows:

<u>In fraction format</u>

Vertex @ $(\frac{-1}{4}, \frac{33}{8})$

<u>In decimal format</u>

Vertex @ $(-.25, 4.125)$

The prompt is thus solved.

Believe it or not, we have reached the end of the chapter (yes... finally!). You should give yourself a pat in the back and even treat yourself to something nice: you certainly deserve it. Although most find quadratic equations quite challenging and confusing, I am certain that if you read this chapter carefully and tried to work out by yourself all–or at least most–of the problems presented, that you are on track to master it.

The last chapter of the book now follows. Similar to the program we followed with linear equations, we will next be reviewing **quadratic functions** as opposed to **quadratic equations**. You will quickly realize how simple it is to work with quadratic functions after having worked with both linear functions and quadratic equations.

See you on the final chapter of this book!

21

QUADRATIC FUNCTIONS

Welcome to the last chapter of the book! It's been quite a journey exploring and reviewing a *very small part* of the immense field of this truly amazing body of knowledge that is mathematics. Rest assured that all that we covered in the book amounts to an important part of the core principles and key concepts that mathematics is built upon.

The last concept we need to cover before formally saying goodbye is that of quadratic functions. You should be able to foresee what this chapter is all about because we explored the same transition with linear equations (to linear functions).

Recall that a linear equation in point slope form has the following format:

$$y = mx + b$$

...and when we defined a linear function, we simply replaced the **y** (output variable) with the function notation $f(x)$ as follows:

$$f(x) = mx + b$$

Recall that this function notation simply provides a way to explicitly provide the input/output relationship a name (in this case, the letter "**f**" although it could've been called "**g**" or "**h**", etc...), specify explicitly the input variable that defines the equation (in this case the variable "**x**" although once again, any other variable letter may be used), and to provide a way to explicitly state the input value being used to evaluate the function (such as when we write $f(2)$ or $f(-1)$ which implies using an input value of **2** or **−1** respectively).

Other than this, a linear function behaves just like its close relative, the linear equation. In fact, while a linear equation yields coordinates, for example, of the from

$$(x, y)$$

...if it is defined in terms of the independent variable **x** and the dependent variable **y**, the equivalent linear function would yield coordinate pairs of the form

$$(x, f(x))$$

You can thus see how a function replaces the output variable–typically **y**–with $f(x)$. The difference is that the notation $f(x)$ is dynamic: the **x** inside the set of parentheses can be replaced with a number that the function will be evaluated for, as stated earlier.

Recall that a quadratic equation of two variables was defined as

$$y = ax^2 + bx + c$$

where **a** ≠ **0** ; **b** and **c** any **REAL** number

We can now define a **quadratic function** as follows:

⚓

A Quadratic Function of two variables is defined as

$$f(x) = ax^2 + bx + c$$

a ≠ 0
b and **c** any **REAL** number

Its graph is a **parabola:**

a ---> if + : up ; if − : down
| **a** | = *arm aperture*

y-intercept @ (**0** , **c**)
axis of symmetry @ $x = \dfrac{-b}{2a}$
vertex @ ($\dfrac{-b}{2a}$, $\dfrac{4ac - b^2}{4a}$)

After reading the previous chapter, I am sure you will agree with the fact that **quadratic equations** and **quadratic functions** are practically the same (barring the more formal mathematical aspects contained in their definition of what an equation is as opposed to a function).

Remember the first quadratic equation we worked with on the previous chapter? It was the following equation (called the *parent* quadratic equation):

$$y = x^2$$

This equation's defining values are

$$a = 1 \quad b = 0 \quad c = 0$$

...and because *a* is equal to **1** its graph is a parabola that points up, with an arm aperture of |1| = **1** .

So how is this equation written in function format? Simple: replace the **y** with $f(x)$ as follows:

$$f(x) = x^2$$

This function behaves exactly the same as its equation format equivalent. It forms a quadratic relationship between its input and output values, its graph is a parabola that points up, and its vertex is at the origin (coordinate **(0 , 0)**). Everything we reviewed in the previous chapter regarding quadratic equations applies to quadratic functions as well. This means that quadratic functions can be expressed in Standard Form (like this one above), or in Vertex Form, or in Intercept Form (fully factored form). Furthermore, setting a quadratic function to zero means doing the same thing as setting a quadratic equation to zero: simply set $f(x)$ equal to **0** .

Let's analyze the following quadratic function so you can see how making the switch from

equation format to function format is not complicated at all.

$$f(x) = x^2 + 2x - 8$$

First, we should always define its **a** value. Recall that in the process of factoring the equation or setting it to zero to find its roots, the original **a** value may be lost, so it is a good idea to define it as soon as the function is solved for $f(x)$ (yes, the $f(x)$ notation also permits multiplying elements, dividing elements, exponents, roots, adding terms, subtracting terms, etc... as we will see shortly). Thus, we state that

$$a = 1$$

Before doing anything else, let's find a few input/output value pairs that are part of this function's solution set. Let say we want to know the output given the following input values: **–10** , **0** , $\frac{3}{5}$, π , and **8.5** . Using function notation, we are trying to find the following:

$$f(-10) = ?$$

$$f(0) = ?$$

$$f(\frac{3}{5}) = ?$$

$$f(\pi) = ?$$

$$f(8.5) = ?$$

...and you should know what to do to solve them. So try to find the output values for the given values specified above, and express the input/output pairs in coordinate form. Then check your work to see if you were able to do this correctly.

Ready? Let's find the output values.

$$f(-10) = ?$$

$$f(-10) = (-10)^2 + 2(-10) - 8$$

$$f(-10) = 100 + (-20) - 8$$

$$f(-10) = 100 - 20 - 8$$

$$f(-10) = 80 - 8$$

$$f(-10) = 72$$

Coordinate: **(−10 , 72)**

Next input value (note that this particular input value corresponds to the **y-intercept** point, which the **Standard Form** explicitly gives away: it is equal to the independent term or **c** value, which is **−8** ; therefore, we don't need to actually replace **x** with **0** , but I will do it nonetheless):

$$f(0) = ?$$

$$f(0) = (0)^2 + 2(0) - 8$$

$$f(0) = 0 + 0 - 8$$

$$f(-10) = -8$$

Coordinate: **(0 , −8)**

Next input value:

$$f(\frac{3}{5}) = ?$$

$$f(\frac{3}{5}) = (\frac{3}{5})^2 + 2(\frac{3}{5}) - 8$$

$$f(\frac{3}{5}) = \frac{9}{25} + \frac{6}{5} - 8$$

$$f(\frac{3}{5}) = \frac{9}{25} + \frac{30}{25} - 8$$

$$f(\frac{3}{5}) = \frac{39}{25} - 8$$

$$f(\frac{3}{5}) = \frac{39}{25} - \frac{200}{25}$$

$$f(\frac{3}{5}) = \frac{-161}{25}$$

Coordinate: **($\frac{3}{5}$, $\frac{-161}{25}$)**

In decimal form: **(.6 , −6.44)**

Next input value:

$$f(\pi) = ?$$

$$f(\pi) = (\pi)^2 + 2(\pi) - 8$$

$$f(\pi) = \pi^2 + 2\pi - 8$$

$$f(\pi) = \pi(\pi + 2) - 8$$

Coordinate: **(π , $\pi(\pi + 2) - 8$)**

In decimal form: **(3.14159... , 8.15278...)**

Next input value:

$$f(8.5) = ?$$

$$f(8.5) = (8.5)^2 + 2(8.5) - 8$$

$$f(8.5) = 72.25 + 17 - 8$$

$$f(8.5) = 81.25$$

Coordinate: **(8.5 , 81.25)**

706

Recall that we can create an input/output table for this function. Using the coordinates we found above, the table would look as follows:

$$f(x) = x^2 + 2x - 8$$

Input x	Output $f(x)$
-10	72
0	-8
$\frac{3}{5}$	$\frac{-161}{25}$
π	$\pi(\pi+2)$ - 8
8.5	81.25

Next, let's find the function's roots (or zeros, or solutions, or **x-intercepts**). We can start by setting it to zero:

$$f(x) = 0$$

This literally translates to "we need the function "*ef* of **x** " to be equal to zero". Of course, since we also know that $f(x)$ is equal to $x^2 + 2x - 8$ we can thus express the following:

$$f(x) = 0 = x^2 + 2x - 8$$

From this moment on, we can simply state

$$0 = x^2 + 2x - 8$$

...and try to solve it using any of the methods we reviewed on the previous chapter. In this case, since it is easily factored, I will use the method that requires finding two numbers whose product equals the independent term's value and whose sum equals the middle term's (linear term's) coefficient. I suggest you try this by yourself before reading on.

Ready to check your answers? Let's see... To factor it fully (express the right side of this equation in **Intercept Form**), we need a couple of numbers that multiplied should equal **–8** and that added should equal **2** . The only pair of **REAL** numbers that work are **4** and **–2** . As we reviewed in the previous chapter, these can now be used in the **Intercept Form** as follows:

$$0 = (x + 4)(x - 2)$$

By the way, if you are wondering how we transitioned from the function $f(x) = x^2 + 2x - 8$ to the equation $0 = x^2 + 2x - 8$, it stems from the fact that we set the function's output value to **0** , so we created an equation that we are now attempting to solve.

Now that we have the fully factored form of the equation's right side, and because it is set to zero, we can use the **zero product property** and find the solutions we are looking for. Alternatively, we can simply extract these values from the **Intercept Form** as reviewed in the previous chapter. I will use both methods.

Method 1. Use the **zero product property**.

$$0 = (x + 4)(x - 2)$$

$(x + 4) = 0$	$(x - 2) = 0$
$x + 4 = 0$	$x - 2 = 0$
$x + 4 - 4 = 0 - 4$	$x - 2 + 2 = 0 + 2$
$x = -4$	$x = 2$

We now have the roots of the function:

$$x_1 = -4 \qquad x_2 = 2$$

Recall that if we know the function's roots (in the previous chapter, we would speak of "the quadratic equation's roots") we know its **x-intercepts**. Therefore, we can unequivocally state that the following two coordinates are part of this function's graph (which is another way of saying that this function's graph, a parabola, passes through the following two coordinates):

x-intercepts @
$(-4,0)$ and $(2,0)$

Let's review the other method that we could have used to define the function's roots once the quadratic was switched to fully-factored form (**Intercept Form**).

Method 2. Extract the zeros directly from the fully factored form.

We know that in general, the Intercept Form has the following set-up:

$$f(x) = (x - m)(x - n)$$

...with roots that are equal to **m** and **n** . Therefore, comparing our equation with this template, we find that:

$$0 = (x + 4)(x - 2)$$

$$m = -4 \qquad n = 2$$

Observe that in this case, **m** is negative because the general form expects to see the **m** value subtracting, so if it is adding, it means that **m** must be negative. On the other hand, in this case **n** is positive because it is being subtracted, as the general form expects. Moving

on, this analysis leads us to conclude that the roots of this function are:

$$x_1 = -4 \qquad x_2 = 2$$

Note that as expected, we found the same answers as the previous method.

Next, we want to find the function's vertex. Note that since we have already found the two roots of the function, we know for a fact that the parabola's axis of symmetry must pass through a point that lies exactly halfway between them. This is very helpful because the coordinate of the vertex has the same **x** component as the **x** value of the axis of symmetry. We can thus establish the axis of symmetry as follows:

axis of symmetry @ $\dfrac{-4+2}{2} = \dfrac{-2}{2} = -1$

In equation form @ $x = -1$

We now know that the vertex must be a point whose position along the **x-axis** is equal to **-1** ; knowing this, we can state that it is of the form $(-1, ?)$. Therefore, all we need to establish is the output of the function when the input value is equal to **-1** (which means, using function notation, that we need to find $f(-1)$):

$$f(-1) = ?$$
$$f(-1) = (-1)^2 + 2(-1) - 8$$
$$f(-1) = 1 + (-2) - 8$$
$$f(-1) = 1 - 2 - 8$$
$$f(-1) = -1 - 8$$
$$f(-1) = -9$$

Vertex @ $(-1, -9)$

We had the option of using the vertex formula instead (specifically the formula that allows us to find the x component of the vertex). Why would we choose this option? Because if we do not know the roots of the function, or if we don't know any two symmetry points, this would be the best course of action.

Recall that the vertex formula is:

$$\textbf{vertex @ } (\frac{-b}{2a} , \frac{4ac - b^2}{4a})$$

Since we are only going to use the x component formula, we only need to establish the function's a and b values: **1** and **2** respectively. Plugging these values into the corresponding formula we obtain:

$$\textbf{Vertex @ } x = \frac{-b}{2a}$$

$$x = \frac{-(2)}{2(1)} = \frac{-2}{2} = -1$$

We obtain, of course, the same x value that we computed using the previous method. From this moment on, the steps are the same as before: find the function's output given this input value, and the input/output pair corresponds to the vertex coordinate.

The last item on the agenda is to graph the function. The process, of course, is the same as when graphing quadratic equations. The only relevant difference is that the **x-y plane** becomes the **x-f(x) plane** , as we discussed in **Chapter 17**.

Try to graph the function by yourself.

Ready to check your sketch? Observe.

Graph of
$f(x) = x^2 + 2x - 8$

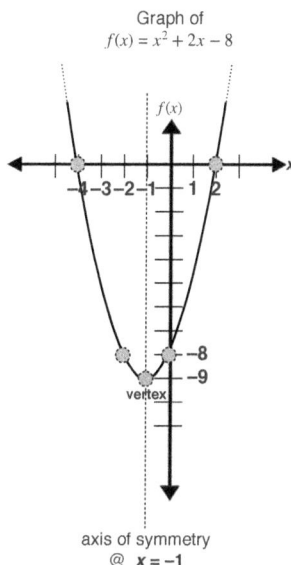

axis of symmetry
@ $x = -1$

The only observation I'd like to make is that I used the coordinate $(0, -8)$ and its symmetry pair point—the coordinate $(-2, -8)$—as the two extra points that I recommend using apart from the vertex and **x-intercepts** when sketching a precise parabola. Recall that if we know there is a point at $(0, -8)$ that is part of the parabola, then there must be a symmetry point equally far apart from the axis of symmetry—but on its opposite side—with the exact same output value; thus, since $(0, -8)$ is one unit to the **right** of the axis of symmetry, then it follows that its symmetry pair point must be located one unit to the **left** of the axis of symmetry, at $(-2, ?)$; we then assign it the same y component as its

symmetry pair point (in this case, **−8**) , and the values of the point are fully established:
(**−2** , **−8**) .

See how simple it is to work with a quadratic function once you've mastered quadratic equations?

Let's move on to another important aspect of working with quadratic functions.

I mentioned briefly earlier in the chapter that function notation does allow for the $f(x)$ to be multiplied, divided, raised to a power, having terms adding to it or subtracting to it, etc... It seems a bit odd at first especially since I told you that you should treat the "$f(x)$" as a label, not as "*ef* **multiplying** parentheses *ex*", for this would be incorrect. But this does not mean that we should not be allowed to, say, multiply a function times two, for example. Observe.

$$f(x) = x^2 + 2x - 8$$

This is the same function that we graphed earlier. Let's say we want to obtain double the output of this function for all of its domain. We could do this:

$$g(x) = 2(x^2 + 2x - 8)$$

Notice how I changed the function's name? Why do you think I did this? If you answered along the lines of "...because the original function's name is ef, and if we change the function, it would be relevant to change its name so that we do not make the mistake of thinking it is the same function..." then you would be absolutely correct!

But this means that we could do the following if we wanted to keep the same function's name:

$$2(f(x)) = 2(x^2 + 2x - 8)$$

This essentially means that we can treat the function-notation label "$f(x)$" as a mathematical entity just as if it were a variable, or a number, or a grouping symbol, etc...

Just remember to keep the "*ef* of *ex*" always intact: never think of the "parentheses *ex*" portion of the label as an entity by itself, as if the "*ef*" portion of the label was a (multiplying) factor: that is not correct at all.

Thus, say you have the following expression:

$$-3(g(x)) = 9x^2 + 6x - 30$$

We could solve for $g(x)$ in order to know what this particular function stood for exactly, because it certainly isn't equal to $9x^2 + 6x - 30$ (can you see why? It's because the function $g(x)$ has a multiplying element of **−3** that prevents us from knowing explicitly what the function is actually equal to). Let's solve the function for $g(x)$:

$$-3(g(x)) = 9x^2 + 6x - 30$$

$$\frac{-3(g(x))}{-3} = \frac{9x^2 + 6x - 30}{-3}$$

$$g(x) = \frac{9x^2}{-3} + \frac{6x}{-3} - \frac{30}{-3}$$

$$g(x) = -3x^2 + (-2x) - (-10)$$

$$g(x) = -3x^2 - 2x + 10$$

And there it is: the function $g(x)$ clearly defined. If you compare this right side with the original right side that we started out with, you should be able to clearly see the difference, supporting the idea that the original right side expression of

$9x^2 + 6x - 30$ was not what $g(x)$ actually stood for in this particular set-up.

Try to specify by yourself what the following functions are explicitly equal to.

a) $2(f(x)) = -8x^2 - 20x + 1$

b) $\dfrac{-3(g(x))}{2} = 6x^2 - 3x - 9$

c) $-(h(x)) = x^2 - \pi x + \sqrt{3}$

d) $f(x) - 3x^2 = 5x^2 - x - 2$

e) $\dfrac{g(x)}{3} + x^2 - 1 = 2x^2 - 4x - 10$

Remember to treat the function notation label "*ef of ex*" or "*gee* of *ex*" etc... as what it is: an entity that cannot be separated. Ready to check your work? Observe.

a) $2(f(x)) = -8x^2 - 20x + 1$

$$2(f(x)) = -8x^2 - 20x + 1$$

$$\frac{2(f(x))}{2} = \frac{-8x^2 - 20x + 1}{2}$$

$$f(x) = -4x^2 - 10x + \frac{1}{2}$$

b) $\dfrac{-3(g(x))}{2} = 6x^2 - 3x - 9$

$$\frac{-3(g(x))}{2} = 6x^2 - 3x - 9$$

$$(2)(\frac{-3(g(x))}{2}) = (2)(6x^2 - 3x - 9)$$

$$-3(g(x)) = 12x^2 - 6x - 18$$

$$\frac{-3(g(x))}{-3} = \frac{12x^2 - 6x - 18}{-3}$$

$$g(x) = -4x^2 - (-2x) - (-6)$$

$$g(x) = -4x^2 + 2x + 6$$

c) $-(h(x)) = x^2 - \pi x + \sqrt{3}$

$$-(h(x)) = x^2 - \pi x + \sqrt{3}$$

$$(-1)(-(h(x))) = (-1)(x^2 - \pi x + \sqrt{3})$$

$$h(x) = -x^2 + \pi x - \sqrt{3}$$

d) $f(x) - 3x^2 = 5x^2 - x - 2$

$$f(x) - 3x^2 = 5x^2 - x - 2$$

$$f(x) - 3x^2 + 3x^2 = 5x^2 - x - 2 + 3x^2$$

$$f(x) = 8x^2 - x - 2$$

e) $\dfrac{g(x)}{3} + x^2 - 1 = 2x^2 - 4x - 10$

$$\frac{g(x)}{3} + x^2 - 1 = 2x^2 - 4x - 10$$

$$\frac{g(x)}{3} + x^2 - 1 - x^2 + 1 = 2x^2 - 4x - 10 - x^2 + 1$$

$$\frac{g(x)}{3} = x^2 - 4x - 9$$

$$(3)(\frac{g(x)}{3}) = (3)(x^2 - 4x - 9)$$

$$g(x) = 3x^2 - 12x - 27$$

By now you should be able to isolate a function in order to know what it is explicitly equal to. I avoided situations where the function is being raised to a power (exponent), including set ups where a root of the function is involved. The reason I avoided these is that in oder to solve functions where those operations are involved we would need to do a more in-depth analysis of functions, beyond the scope of the book. However, I am sure you can visualize how a set up such as

$$(f(x))^2 = 9x^2$$

...would involve thinking about the concept of the plus/minus sign, and the role it would play on defining the function's domain and range, because in order to isolate the function "*ef* of *ex*" we would need to take the square root of both sides of the function, and that, as we reviewed in several chapters of the book, would involve thinking about the plus/minus sign since there is an even power (exponent) impacting a variable that would get canceled out in the process of solving for the label "*ef* of *ex*". If you need to review this very specific part of function behavior, there are multiple online reputable resources that you may now use to study this on your own: armed with the knowledge we have reviewed in this book, you should not have any major issues following the formal treatment of this area of mathematics.

The last concept we will review regarding quadratic functions is the idea that they may be expressed in different formats (just like quadratic equations): they can be expressed in Standard Form, Intercept Form, or Vertex Form. Thus, if we start out with the following function:

$$f(x) = x^2 + 2x - 8$$

...which is explicitly written in Standard Form, we may rewrite it using the other forms as we reviewed on the previous chapter. Therefore, the following are all equivalent:

Standard Form

$$f(x) = x^2 + 2x - 8$$

Vertex Form

$$f(x) = (x + 1)^2 - 9$$

Intercept Form

$$f(x) = (x + 4)(x - 2)$$

We can thus present the following diagram regarding quadratic functions.

⚓

Quadratic Function in Standard Form

$$f(x) = ax^2 + bx + c$$
where $a \neq 0$

$f(x)$-intercept @ $(0 , c)$

axis of symmetry @ $x = \dfrac{-b}{2a}$

vertex @ $(\dfrac{-b}{2a} , \dfrac{4ac - b^2}{4a})$

Quadratic Function in Vertex Form

$$f(x) = a(x - h)^2 + k$$
where $a \neq 0$
vertex @ (h , k)
axis of symmetry @ $x = h$

Quadratic Function in Intercept Form

$$f(x) = a(x - m)(x - n)$$
where $a \neq 0$
x-intercepts @ $(m , 0)$ and $(n , 0)$
axis of symmetry @ $x = \dfrac{m + n}{2}$

For all forms:
a ---> if + : up ; if – : down
$| a | = $ arm aperture

Believe it or not, we have reached the end of the book. It has been quite a journey, but one that I truly hope allowed you to better understand and master some of the basic principles and rules of math.

Do know that mathematics is extremely vast. There is a tremendous body of work that has been and continues to be published year after year. There isn't a single mathematics book that can cover it all. But rather than feel discouraged by this, or overwhelmed, rest assured that if you were able to follow most (if not all) of the rules and principles covered in this book, then you are ready to take it to the next level if that is part of your academic (and professional) goals.

Remember that I warned you at the beginning of the book that I would not offer a formal treatment of mathematics, even of its basic principles and rules that serve as its foundation, to the probable despair of most, if not all, mathematicians out there roaming the world. But I defend the approach I decided to use for this book on the grounds that if a mathematics student is able to first understand that mathematics must be seen as a language, and that it is grounded in a "particulate" principle in the sense that it treats individual entities or elements as such, and that whenever you see a mathematical expression you must be able to see and identify its elementary components (numbers in any of the defined formats: **INTEGERS** or decimal form or fractions or **IRRATIONALS**, variables, grouping symbols, signs (positive/negative), operations such as addition and subtraction and multiplication and division and exponents and roots and absolute value, etc..., then you will be more likely to succeed in a course that does take the more formal approach. Of course, we reviewed important concepts such as the **order of operations** (**PEMDAS**) and how bypassing this leads to incorrect computations, the idea of **equations** and what they represent (including those that use inequality symbols or the not-equal-to relational symbol), of **functions** as well, and an introduction to the concept of points on a **graph** and how to **sketch** equations and functions, including a closer study of linear equations and functions and of quadratic equations and functions.

Best of luck on your future mathematical endeavors, and may you find it a more logical field than how you thought of it before reading this book, and a lot more easier to use and work with and study.

Goodbye my friends!

Ciao amici!

Aloha 'oe!

Dowidzenia, moi przyjaciele!

Sayonara, watashi no tomodachi!

Adios amigos!

Au revoir, mes amis!

Auf wiedersehen, mein freunde!

Zài jiàn!

Adeus, meus amigos!

Poka, druz'ya!

Adjö mina vänner!

Moikka ystäväni!

Alavidā!

Ahn nyeong hee ga se yo, chin-gu-duhl!

...

FINAL WORDS

I. Vis-a-vis the content...

In writing this book, I had to choose which areas of basic math should be included and which should be left out; otherwise, it would have quickly become a never-ending project. I am certain, however, that the material that the book does contain is what a reader needs to truly understand and master the foundations of mathematics. With this foundation in hand, if a reader needed to review, say, how to graph systems of inequalities, he or she could easily study it and master it using any of the many online (or in print) resources that are commonly available, because this book provides the core skills that will allow you to be an independent thinker and a self-taught student of math.

II. To err is human!

Great efforts were taken in an attempt to revise the text, diagrams, formulas, calculations, etc... that are contained in these pages, and if asked, I'd have to say that I am extremely confident that the book is at least 99.999% correct. However, if any mistakes did make it to the published version of the book, I offer you my most sincere apologies, and at the same time, release myself from any liability imaginable that anyone whatsoever may think of using against me in any court of law or its equivalent, on any part of the Milky Way Galaxy and beyond.

If you would like to help with future editions of this book, you may use the "Contact me" section of the website

www.quasarsphere.com

to kindly send me any errors that you may bump into as you work your way towards the last page of this *brick*. And I do mean "kindly": please do not be smug about it... I do have feelings, you know. Your help, though, would be massively welcome!

INDEX

COPYRIGHT

Copyright © 2015 by Henry Lindell.

Photographs: Henry Lindell.

All rights reserved. No part of this publication may be reproduced, distributed, or transmitted in any form or by any means, including photocopying, recording, or other electronic or mechanical methods, without the prior written permission of the publisher, except in the case of brief quotations embodied in critical reviews and certain other noncommercial uses permitted by copyright law. For permission requests, write to the publisher, addressed "Attention: Permissions Coordinator," at the address below.

Quasar Sphere

www.quasarsphere.com

Ordering Information:

Quantity sales: for details, contact the publisher at the address above.